Checklist of Beetles of the British Isles

2nd Edition

Edited by A.G. Duff

with an Introduction by D.A. Lott
and a chapter on Fossil Beetles by P.I. Buckland & P.C. Buckland

Pemberley Books (Publishing)
Iver
2012

Checklist of Beetles of the British Isles

First edition 2008 published by A.G. Duff
Revised edition 2012

ISBN: 978-0-9573357-0-7

Published by Pemberley Books (Publishing), United Kingdom.

Pemberley Books
18 Bathurst Walk, Iver SL0 9AZ United Kingdom

Tel.: +44(0)1753 631114
Fax: +44(0)1753 631115

E-mail: orders@pemberleybooks.com

WWW: www.pemberleybooks.com

Cover: *Platyrhinus resinosus* (Scopoli, 1763) (Anthribidae) © John Walters

Summary of changes in the 2012 edition

This section details the main changes in the current edition of the checklist, compared to the previous edition (Duff, 2008). For further details of these changes, please refer to the endnotes.

Family-group names:

- Nomenclature of family-group names follows Bouchard *et al.* (2011)
- Family PAELOBIIDAE is renamed to HYGROBIIDAE
- Family SCYDMAENIDAE becomes a subfamily of STAPHYLINIDAE
- Family BOLBOCERATIDAE becomes a subfamily of GEOTRUPIDAE
- Order of families in SCARABAEOIDEA changed
- Family ANOBIIDAE is renamed to PTINIDAE
- Family LANGURIIDAE deleted and included species moved to EROTYLIDAE

Species added to the main list (47):

- *Suphrodytes figuratus*
- *Somotrichus unifasciatus*
- *Hololepta plana*
- *Ochthebius nilssoni*
- *Catops borealis*
- *Brachygluta klimschi*
- *Mycetoporus ambiguus*
- *Mycetoporus reichei*
- *Acrotona convergens*
- *Gyrophaena transversalis*
- *Anotylus hammondi*
- *Bledius lohsei*
- *Carpelimus erichsoni*
- *Carpelimus zealandicus*
- *Carpelimus alutaceus*
- *Edaphus beszedesi*
- *Paederidus ruficollis*
- *Philonthus alpinus*
- *Quedius lyszkowskii*
- *Quedius lucidulus*
- *Gyrohypnus wagneri*
- *Agrilus cuprescens*
- *Agrilus cyanescens*
- *Trachys subglaber*
- *Isorhipis melasoides*
- *Gastrallus laevigatus*
- *Mesocoelopus collaris*
- *Mirosternomorphus heali*
- *Colotes punctatus*
- *Meligethes matronalis*
- *Meligethes symphyti*
- *Typhaea haagi*
- *Berginus tamarisci*
- *Semanotus russicus*
- *Poecilium lividum*
- *Leiopus linnei*
- *Bruchus brachialis*
- *Aphthona pallida*
- *Longitarsus symphyti*
- *Pseudoperapion brevirostre*
- *Bradybatus fallax*
- *Cotaster uncipes*
- *Allopentarthrum elumbe*
- *Conarthrus praeustus*
- *Otiorhynchus cribricollis*
- *Sitona gressorius*
- *Polygraphus grandiclava*

Species deleted from the main list (9):

- *Hydaticus continentalis*
- *Abax parallelus*
- *Stenolophus comma*
- *Aglyptinus agathidioides*
- *Alevonota aurantiaca*
- *Bledius praetermissus*
- *Heterocerus hispidulus* (duplication error)
- *Protapion ryei*
- *Cathormiocerus maritimus*

Acknowledgements

I would like to thank everyone who has helped to create and maintain the *Checklist of Beetles of the British Isles*. During the early part of this project I was greatly assisted by draft checklist manuscripts written by Roger Booth and Eric Philp. For keeping track of new and deleted species, published updates to the British list were invaluable (Lott, 1995, 1998; Mann, 2000b, 2002; Owen, 1993a, 1994c). Recent published checklists were also very helpful covering the Irish fauna (Anderson, Nash & O'Connor, 1997; Nash, Anderson & O'Connor, 1997), water beetles (Foster, 2004, 2005; Foster & Friday, 2011), Carabidae (Luff, 2007), Leiodidae (Cooter, 1996b), Staphylinidae part (Lott, 2009a; Lott & Anderson, 2011), Dermestidae (Peacock, 1993), Nitiduloidea (Cooter, 1995), Kateretidae and Nitidulidae (Kirk-Spriggs, 1996), Ciidae (Orledge & Booth, 2006; Orledge, 2009), Anthicidae (Telnov, 2010), Scraptiidae (Levey, 2009), seed and leaf beetles (Cox, 2007) and weevils (Morris, 2003, 2011). The late Derek Lott deserves a special mention, not only as the original section author for the very large family Staphylinidae, but also for writing his stimulating introduction to the project, and for his unstinting encouragement. With his encyclopaedic knowledge of the British fauna, and nomenclatural issues in particular, a great debt is owed to Roger Booth for undertaking to author many individual families and for answering numerous questions. All the family authors are to be warmly thanked: Keith Alexander (Cantharidae), Max Barclay (Tetratomidae and Tenebrionidae), Roger Booth (numerous families), Jon Cooter (Leiodidae), Mike Cox (Megalopodidae, Orsodacnidae and Chrysomelidae), Garth Foster (water beetles), Colin Johnson (Ptiliidae, Clambidae, Monotomidae, Cryptophagidae, Corylophidae and Latridiidae), Brian Levey (Buprestidae, Mordellidae and Scraptiidae), the late Derek Lott (Staphylinidae), Martin Luff (Carabidae), Darren Mann (Scarabaeoidea), Howard Mendel (Elateroidea), Mike Morris (Curculionoidea), Glenda Orledge (Ciidae) and Martin Rejzek (Cerambycidae). Many thanks also to Phil and Paul Buckland for their chapter on beetles known only as fossils from Quaternary sediments.

Several coleopterists have reviewed all or part of the checklist. Tony (A.J.) Allen helped with literature, reviewed numerous drafts of the complete checklist and suggested several formatting improvements. Max Barclay reviewed the arrangement of *Otiorhynchus*. Martin Collier spotted several errors and suggested some format changes. Scotty Dodd spotted an important error in *Altica*. Peter Hammond is thanked for correcting several errors, for some formatting suggestions and for thoroughly reviewing the entire checklist. Mark Hill corrected some errors during the course of extracting data for the Biological Records Centre species dictionary. Peter Hodge assisted with literature and drew my attention to a change in *Globicornis*. Eric Philp spotted a duplication in Heteroceridae. Mark Telfer is to be thanked for his comments on various drafts of the checklist, and for maintaining the MapMate® Coleoptera species dictionary. Finally, the Board of Governors of *The Coleopterist* journal are to be thanked for allowing the checklist to be hosted on the journal's website at www.coleopterist.org.uk.

Andrew Duff

Contents

Introduction

D.A. Lott

Since publication of the Coleoptera volume of the second edition of Kloet & Hincks' checklist of British insects (Pope, 1977), a large number of additions and deletions to the list have been published, as well as numerous nomenclatural changes. In addition, recent advances in our understanding of the phylogeny of Coleoptera have led to major modifications of higher classification in checklists and catalogues that have been published more recently in Europe and North America. In the absence of any widely accepted, modern checklist of the British fauna, British workers have been using binomial names taken from a variety of often conflicting sources.

It is intended that this checklist should have a wider currency and be suitable for use in future conservation reviews, survey reports and taxonomic dictionaries for biological recording. It is, therefore, vital that the nomenclature used should be widely accepted by coleopterists working on the British fauna. It is also important that the checklist should be as useful as possible to those working at larger scales ranging from European through Palaearctic to the world fauna. It is envisaged that the checklist will be updated regularly to keep abreast of published changes and suggested amendments from correspondents. To this end, comments on the checklist are welcome and should be addressed to the Editor by e-mail to andrew.duff@virgin.net.

Scope

The checklist aims to include all species that have been reliably recorded from the British Isles as possible residents. Exotic species which are only known from casual importation, and have never formed established populations, are listed in an appendix. The British Isles is here taken to include Great Britain and Ireland, including the Isle of Man but not the Channel Islands.

The checklist aims to be comprehensive in its listing of subfamilies, genera and species. Tribes and subtribes are only listed when supported by a recent, authoritative review. Contemporary checklists vary in the extent to which they include subgenera and many subgeneric classifications are old and in need of review. Consequently, subgenera have only been listed when they enjoy current and widespread usage.

Rationale

It is considered that there are two main uses for a checklist:

- to provide current standard names for species, genera, families and any other levels that are in common usage
- to enable other commonly used names to be interpreted as current standard names.

It is also recognised that checklists have several subsidiary functions including:

- an inventory of biodiversity in a particular region
- the ordering of names into a standard sequence.

Three areas are considered to be outside this checklist's remit. Firstly, the exhaustive listing of old synonyms is more properly the job of a catalogue rather than a simple checklist. Secondly, new names and new approaches to classification should appear in comprehensive taxonomic reviews rather than checklists. Thirdly, the correction of errors is not the primary function of a checklist, although correctness will be an important factor in evaluating any recently proposed changes to classification or nomenclature; misspelt names which have become widely used, for example, are listed here in order to draw attention to the correct usage.

As far as possible, therefore, this checklist uses a classification that already exists in the current literature, although the opportunity has been taken, through the cooperation of several correspondents, to include some changes in advance of publication, in order to ensure future currency. There is, of course, no universally agreed classification because of differences in approach between regions and between individual workers. Inevitably, choices between conflicting practices have had to be made. The main objective in making these choices has been to construct a list of binomial names, whose content and arrangement will be broadly acceptable, or at least comprehensible, for the next ten to twenty years. In order to meet this aspiration, it has been necessary to address four factors that work against the stability of checklists fossilised at one moment in time. They are:

Additions and deletions of species

These changes are essentially unpredictable and will be communicated in future editions as they arise.

Nomenclatural changes

Replacement names for British taxa are frequently published in various notes, taxonomic reviews and foreign checklists. These changes arise from new synonymies, discoveries of senior homonyms and corrections of misidentifications. Corrections of misidentifications can result in the transfer of a misapplied name from one species to another. For example, Fauvel (1862) misidentified *Aleochara punctatella* Motschulsky as *A. obscurella* Gravenhorst and renamed the true *A.obscurella* Gravenhorst as *A. algarum* Fauvel. This error persisted for over 100 years until its discovery (Lohse, 1985b) led to the transfer of the name *A. obscurella* Gravenhorst back to the original species. Consequently the name *A. obscurella* Gravenhorst has been applied to two different species at different times. It is very easy to misinterpret such names in old publications.

The older the published list, the more difficult becomes the task of interpreting names. This checklist aims to provide the means for interpreting names used since around 1950 when Joy (1932) was the standard British identification guide. It must be recognised that the interpretation of names in older publications and manuscripts requires specialist knowledge that cannot be substituted by a simple checklist. Indeed the same can be said for any work dealing with taxonomically confused groups or using outdated reference sources.

Sometimes nomenclatural changes are contentious and lead to inconsistencies between different checklists. Occasionally, they prove to be wrong and are followed by reversions to previous usage in subsequent editions of the checklist. Unfortunately, the adoption of inappropriate type species for genera, the unearthing of ancient misidentifications and homonymies, and the reinstitution of old, long forgotten names, have too often led to the overturning of long-established nomenclature. Such changes are a regrettable cause of instability in taxonomic nomenclature and they are best approached in the context of a comprehensive taxonomic review. It is the duty of a checklist compiler to be cautious in this regard. It is better to err on the side of established usage, rather than indulge in taxonomic sorties that cause unwarranted instability.

It is to be hoped that recent changes in the ICZN code (ICZN, 1999) will prevent some of the worst excesses of past practice, by conserving widely used names under threat from disused 18th- and 19th-Century names. However, some well-established names of British taxa are known to be incorrect, or at least in need of investigation. Some of these names may well be the subject of future applications to the ICZN for their conservation. It is not intended here to initiate any radical changes that have not been instituted elsewhere.

Changes in the boundaries of higher taxa

The large volume of work on higher classification in recent decades has led to major advances in phylogeny and revolutionised our understanding of the relationships of taxa from family level down to species level. For example, the family Staphylinidae has been expanded to include several new subfamilies that were formerly families in their own right, whereas the Curculionidae has been broken up into several smaller families.

Some of these changes have arisen following cladistic analyses, which suggest that taxa are paraphyletic. Others appear to be based on more informal arguments. In either case, the ranking of a higher taxon, such as a genus, is largely based on opinion, because there is no universally accepted definition of a genus. Usually, the scale of morphological differences dictates the separation of genera, but numbers of species have also been used as an argument (see e.g. Seevers, 1978). It is therefore no surprise that the rank of a particular name as genus, subgenus or synonym has sometimes proved to be contentious and this has led to regional differences and to unstable binomial names. Inevitably, some judgment has to be made about which of two conflicting approaches is going to gain more currency, but, in general, this checklist follows widely accepted world and regional catalogues and subsequent authoritative taxonomic revisions.

Changes in the sequence of taxa in the list

Taxa are often ordered within checklists so that they are placed next to their closest relatives and so phylogenetic relationships provide the basis for a standard sequencing. Traditionally, basal groups precede more specialised taxa. Several existing checklists arrange taxa alphabetically rather than suggest phylogenetic relationships that are not backed up by a proper investigation. Pope (1977) ordered species alphabetically and ordered higher taxa so as to reflect relationships, or at least traditional ideas of relationships, and this general principle has been continued in this list except that junior synonyms are listed in ascending date order rather than alphabetically. It is expected that phylogenetic relationships will be of minor consequence to many checklist users. Nevertheless, advances in knowledge in this area have the capacity to disrupt the stability of any adopted standard arrangement of taxa. It should be noted that binomial names remain unaffected, whatever the arrangement of taxa.

The sequence of taxa in this checklist generally follows arrangements in other checklists that will be familiar to a wide range of users.

Format

The checklist is divided into three sections in order to list separately:

- extant or recently extinct (i.e. within historical times) native and naturalised species, including natural immigrants/vagrants and established (either temporarily or permanently) introductions
- species that are known only from the Quaternary fossil record
- non-established introductions.

The checklists are followed by a bibliography, and then an index to genus-group names (names of genera and subgenera, including their synonyms) and family-group names (for all categories higher than genus).

Within the main checklist, conventional Latin phrases are used to qualify certain names. Although written in a largely forgotten language, these phrases have the considerable merit of being very concise:

ad	until
ante	before
auctt.	authors (i.e. in entomological literature generally)
auctt. Brit.	British authors
auctt. Europ.	European authors
nec	nor of
non	but not of
partes maj.	for the most part
partim	partly
sensu	in the sense of
sensu lato	in the broad sense
sensu stricto	in the narrow sense

Question marks are used to indicate uncertainty. A question mark before a name indicates unsure identification or placement as a synonym. A question mark before a citation indicates doubt over whether the author used the name in the sense shown.

Endnotes are generally given for significant changes to the British list since Pope (1977), unless these have already been detailed in the family introduction. In addition, endnotes may draw attention to particular nomenclatural or taxonomic points. Relevant Opinions of the International Commission for Zoological Nomenclature are cited as "ICZN Op." in endnotes (see www.iczn.org/Opinions.htm).

Citation

Citation should follow this example format:

MANN, D.J. 2012. Scarabaeidae. In: Duff, A.G. (ed.) *Checklist of Beetles of the British Isles*. 2nd edition. Pemberley Books.

Extant or recently extinct native and naturalised species

Order COLEOPTERA Linnaeus, 1758

Suborder MYXOPHAGA Crowson, 1955

Superfamily SPHAERIUSOIDEA Erichson, 1845

1. Family SPHAERIUSIDAE Erichson, 1845 [1]
MICROSPORIDAE Crotch, 1873

Family author: G.N. Foster

SPHAERIUS Waltl, 1838 [2]
MICROSPORUS Kolenati, 1846
acaroides Waltl, 1838
   obsidianus (Kolenati, 1846)

Suborder ADEPHAGA Clairville, 1806

2. Family GYRINIDAE Latreille, 1810

Family author: G.N. Foster

Subfamily GYRININAE Latreille, 1810 [3]

Tribe GYRININI Latreille, 1810

GYRINUS Geoffroy, 1762 [4]

Subgenus GYRINULUS Zaitzev, 1908
minutus Fabricius, 1798

Subgenus GYRINUS Geoffroy, 1762
aeratus Stephens, 1835
   opacus sensu auctt. Brit. ad 1914 non Sahlberg, C.R., 1819
   thomsoni Zaitzev, 1907
   edwardsi Sharp, 1914
caspius Ménétries, 1832
   elongatus Aubé, 1838 non Marsham, 1802
distinctus Aubé, 1838
   colymbus sensu auctt. Brit. non Erichson, 1837
marinus Gyllenhal, 1808
natator (Linnaeus, 1758)
   mergus Ahrens, 1812
opacus Sahlberg, C.R., 1819
   lecontei Omer-Cooper, 1930 non Fall, 1921
   blairi Omer-Cooper, 1931
paykulli Ochs, 1927
   bicolor sensu Paykull, 1798 non Fabricius, 1787
substriatus Stephens, 1828
   natator sensu auctt. Brit. partes maj. non (Linnaeus, 1758)
   oblitus Sharp, 1914
   fowleri Omer-Cooper, 1930
suffriani Scriba, 1855
urinator Illiger, 1807

Tribe ORECTOCHILINI Régimbart, 1882

ORECTOCHILUS Dejean, 1833
villosus (Müller, O.F., 1776)

3. Family HALIPLIDAE Aubé, 1836

Family author: G.N. Foster

BRYCHIUS Thomson, C.G., 1859
elevatus (Panzer, 1793)

PELTODYTES Régimbart, 1879
CNEMIDOTUS sensu Erichson, 1832 non Illiger, 1802
caesus (Duftschmid, 1805)
   impressus sensu (Panzer, 1794) non (Fabricius, 1787)

HALIPLUS Latreille, 1802 [5]

Subgenus HALIPLIDIUS Guignot, 1928
confinis Stephens, 1828
   pallens Fowler, 1887
   halberti Bullock, 1928
obliquus (Fabricius, 1787)
varius Nicolai, 1822 [6]

Subgenus NEOHALIPLUS Netolitzky, 1911
lineatocollis (Marsham, 1802)

Subgenus HALIPLUS Latreille, 1802
apicalis Thomson, C.G., 1868
   striatus Sharp, 1869
fluviatilis Aubé, 1836
furcatus Seidlitz, 1887
heydeni Wehncke, 1875
   ruficollis sensu auctt. Brit. partim non (De Geer, 1774)
immaculatus Gerhardt, 1877
lineolatus Mannerheim, 1844
   nomax Balfour-Browne, F., 1911
   browneanus Sharp, 1913
ruficollis (De Geer, 1774)
   fulvicollis sensu Edwards, 1911 non Erichson, 1837
sibiricus Motschulsky, 1860
   borealis Gerhardt, 1877 non LeConte, 1850
   wehnckei Gerhardt, 1877 [7]

Subgenus LIAPHLUS Guignot, 1928
flavicollis Sturm, 1834
fulvus (Fabricius, 1801)
laminatus (Schaller, 1783)
   cinereus Aubé, 1836
mucronatus Stephens, 1828
variegatus Sturm, 1834

4. Family NOTERIDAE Thomson, C.G., 1860 [8]

Family author: G.N. Foster

NOTERUS Clairville, 1806
clavicornis (De Geer, 1774)
   capricornis sensu auctt. Brit. non (Herbst, 1784)
   sparsus (Marsham, 1802)
crassicornis (Müller, O.F., 1776)
   capricornis (Herbst, 1784)
   clavicornis sensu auctt. Brit. partim non (De Geer, 1774)

5. Family HYGROBIIDAE Régimbart, 1878 (1837) [9]

PAELOBIIDAE Erichson, 1837

Family author: G.N. Foster

HYGROBIA Latreille, 1804
HYGRIOBIA Latreille, 1804 [10]
PAELOBIUS Schönherr, 1808
PELOBIUS Erichson, 1832 (misspelling)
hermanni (Fabricius, 1775)
herrmanni (Fabricius, 1775) [10]
tarda (Herbst, 1779)

6. Family DYTISCIDAE Leach, 1815 [11]

Family author: G.N. Foster

Subfamily AGABINAE Thomson, C.G., 1867

AGABUS Leach, 1817
Subgenus ACATODES Thomson, C.G., 1859
arcticus (Paykull, 1798)
congener (Thunberg, 1794)
sturmii (Gyllenhal in Schönherr, 1808)
sturmi auctt. (misspelling)

Subgenus AGABUS Leach, 1817
labiatus (Brahm, 1790)
femoralis (Paykull, 1798)
uliginosus (Linnaeus, 1761)
dispar Bold, 1849
undulatus (Schrank, 1776)
abbreviatus (Fabricius, 1787)

Subgenus GAURODYTES Thomson, C.G., 1859
affinis (Paykull, 1798)
biguttatus (Olivier, 1795) [12]
bipustulatus (Linnaeus, 1767)
solieri Aubé, 1837
brunneus (Fabricius, 1798)
conspersus (Marsham, 1802)
didymus (Olivier, 1795)
guttatus (Paykull, 1798)
melanarius Aubé, 1837
nebulosus (Forster, 1771)
paludosus (Fabricius, 1801)
striolatus (Gyllenhal, 1808)
unguicularis (Thomson, C.G., 1867)

ILYBIUS Erichson, 1832
aenescens Thomson, C.G., 1870
angustior sensu auctt. non (Gyllenhal, 1808)
ater (De Geer, 1774) [13]
chalconatus (Panzer, 1796) [14]
fenestratus (Fabricius, 1781)
fuliginosus (Fabricius, 1792)
guttiger (Gyllenhal, 1808)
montanus (Stephens, 1828) [14]
melanocornis (Zimmermann, 1915)
quadriguttatus (Lacordaire, 1835)
obscurus (Marsham, 1802) non (Panzer, 1795)
subaeneus Erichson, 1837
wasastjernae (Sahlberg, C.R., 1824) [14] [15]

PLATAMBUS Thomson, C.G., 1859
maculatus (Linnaeus, 1758)
pulchellus (Heer, 1839)
immaculatus Donisthorpe, 1899

Subfamily COLYMBETINAE Erichson, 1837

COLYMBETES Clairville, 1806
fuscus (Linnaeus, 1758)

RHANTUS Dejean, 1833
RANTUS Dejean, 1833 [16]

Subgenus NARTUS Zaitzev, 1907
grapii (Gyllenhal, 1808)

Subgenus RHANTUS Dejean, 1833
bistriatus (Bergsträsser, 1778) [17]
adspersus (Fabricius, 1801) non (Panzer, 1796)
aberratus Gemminger & Harold, 1868
exsoletus (Forster, 1771)
frontalis (Marsham, 1802)
notatus sensu (Fabricius, 1781) non (Bergsträsser, 1778)
suturalis (MacLeay, 1825)
punctatus (Geoffroy, 1785) non (Scopoli, 1763)
pulverosus (Stephens, 1828)
suturellus (Harris, 1828)
bistriatus sensu auctt. Brit. non (Bergsträsser, 1778)

Subfamily COPELATINAE Branden, 1885

LIOPTERUS Dejean, 1833 [18]
COPELATUS sensu auctt. non Erichson, 1832
haemorrhoidalis (Fabricius, 1787)
ruficollis (Schaller, 1783) non (De Geer, 1774)
schalleri (Gmelin in Linnaeus, 1790)
agilis (Fabricius, 1792)
oblongus (Illiger, 1801) non (Herbst, 1784)

Subfamily DYTISCINAE Leach, 1815

Tribe ACILIINI Thomson, C.G., 1867

ACILIUS Leach, 1817
canaliculatus (Nicolai, 1822)
fasciatus sensu Fowler, 1887 non (De Geer, 1774)
sulcatus (Linnaeus, 1758)
scoticus Stephens, 1828

GRAPHODERUS Dejean, 1833
bilineatus (De Geer, 1774)
cinereus (Linnaeus, 1758)
zonatus (Hoppe, 1795)

Tribe CYBISTRINI Sharp, 1880 [19]

CYBISTER Curtis, 1827
lateralimarginalis (De Geer, 1774)

Tribe DYTISCINI Leach, 1815

DYTISCUS Linnaeus, 1758
circumcinctus Ahrens, 1811
circumflexus Fabricius, 1801
dimidiatus Bergsträsser, 1778
lapponicus Gyllenhal, 1808
marginalis Linnaeus, 1758
semisulcatus Müller, O.F., 1776
  punctulatus Fabricius, 1777

Tribe HYDATICINI Sharp, 1880

HYDATICUS Leach, 1817 [20]
seminiger (De Geer, 1774)
transversalis (Pontoppidan, 1763)

Subfamily HYDROPORINAE Aubé, 1836

Tribe BIDESSINI Sharp, 1880

BIDESSUS Sharp, 1882
minutissimus (Germar, 1824)
unistriatus (Goeze, 1777)

HYDROGLYPHUS Motschulsky, 1853
  GUIGNOTUS Houlbert, 1934
geminus (Fabricius, 1792)
  pusillus (Fabricius, 1781) non (Müller, O.F., 1776)

Tribe HYDROPORINI Aubé, 1836

DERONECTES Sharp, 1882
latus (Stephens, 1829)

GRAPTODYTES Seidlitz, 1887
bilineatus (Sturm, 1835)
  hopffgarteni (Schilsky, 1892)
flavipes (Olivier, 1795)
  concinnus (Stephens, 1835)
granularis (Linnaeus, 1767)
pictus (Fabricius, 1787)

HYDROPORUS Clairville, 1806 [21]
  STERNOPORUS Falkenström, 1930
  HYDROPORIDIUS Guignot, 1949
angustatus Sturm, 1835
discretus Fairmaire & Brisout de Barneville, 1859 [22]
  neuter Fairmaire & Laboulbène, 1854
elongatulus Sturm, 1835
erythrocephalus (Linnaeus, 1758)
ferrugineus Stephens, 1829
glabriusculus Aubé, 1838
gyllenhalii Schiødte, 1841
  gyllenhali auctt. (misspelling)
  piceus sensu auctt. Europ. non Stephens, 1828
incognitus Sharp, 1869
longicornis Sharp, 1871
  parallelus Sharp, 1869 non Aubé, 1838
longulus Mulsant & Rey, 1861
  celatus Clark, 1862
marginatus (Duftschmid, 1805)
melanarius Sturm, 1835

  monticola Sharp, 1869
memnonius Nicolai, 1822
morio Aubé, 1838
  melanocephalus sensu (Gyllenhal, 1808) non
    (Marsham, 1802)
necopinatus Fery, 1999 [23]
  ssp. *roni* Fery, 1999
  cantabricus sensu auctt. Brit. non Sharp, 1882
neglectus Schaum, 1845
nigrita (Fabricius, 1792)
obscurus Sturm, 1835
obsoletus Aubé, 1838
palustris (Linnaeus, 1761)
  tinctus Clark, 1862
planus (Fabricius, 1782)
  ater (Forster, 1771) [24]
pubescens (Gyllenhal, 1808)
rufifrons (Müller, O.F., 1776)
  rufifrons (Duftschmid, 1805) non (Müller, O.F.,
    1776)
  piceus Stephens, 1828
  duftschmidti Rye, 1872
scalesianus Stephens, 1828
striola (Gyllenhal, 1826)
  vittula Erichson, 1837
tessellatus (Drapiez, 1819)
  lituratus sensu Fowler, 1887 non (Fabricius, 1781)
tristis (Paykull, 1798)
umbrosus (Gyllenhal, 1808)

NEBRIOPORUS Régimbart, 1906 [25]
  POTAMODYTES Zimmermann, 1919 non
    Grouvelle, 1896
  POTAMONECTES Zimmermann, 1921

Subgenus NEBRIOPORUS Régimbart, 1906
assimilis (Paykull, 1798)
depressus (Fabricius, 1775)
elegans (Panzer, 1794)

Subgenus ZIMMERMANNIUS Guignot, 1941
canaliculatus (Lacordaire, 1835) [26]

OREODYTES Seidlitz, 1887
alpinus (Paykull, 1798) [27]
davisii (Curtis, 1831)
  davisi auctt. (misspelling)
  borealis sensu auctt. non (Gyllenhal, 1826)
sanmarkii (Sahlberg, C.R., 1826)
  sanmarki auctt. (misspelling)
  rivalis (Gyllenhal, 1827)
septentrionalis (Gyllenhal, 1826)

PORHYDRUS Guignot, 1945
lineatus (Fabricius, 1775)

SCARODYTES des Gozis, 1914
halensis (Fabricius, 1787)

STICTONECTES Brinck, 1943
  STICTONOTUS Zimmermann, 1930 non Foerster,
    1856
lepidus (Olivier, 1795)

STICTOTARSUS Zimmermann, 1919
duodecimpustulatus (Fabricius, 1792)

BOREONECTES Angus, 2010
multilineatus (Falkenström, 1922)
    griseostriatus sensu auctt. non (De Geer, 1774)

SUPHRODYTES des Gozis, 1914
dorsalis (Fabricius, 1787)
figuratus (Gyllenhal, 1826) [28]

Tribe HYDROVATINI Sharp, 1880

HYDROVATUS Motschulsky, 1853
    OXYNOPTILUS Schaum in Schaum &
    Kiesenwetter, 1868
clypealis Sharp, 1876
cuspidatus (Kunze, 1818) [29]

Tribe HYGROTINI Portevin, 1929

HYGROTUS Stephens, 1828

Subgenus COELAMBUS Thomson, C.G., 1860
confluens (Fabricius, 1787)
impressopunctatus (Schaller, 1783)
nigrolineatus (Steven, 1808) [30]
    lautus (Schaum, 1843)
novemlineatus (Stephens, 1829)
parallellogrammus (Ahrens, 1812)
    parallelogrammus auctt. (misspelling)

Subgenus HYGROTUS Stephens, 1828
decoratus (Gyllenhal, 1810)
inaequalis (Fabricius, 1777)
quinquelineatus (Zetterstedt, 1828)
versicolor (Schaller, 1783)

Tribe HYPHYDRINI Gistel, 1848

HYPHYDRUS Illiger, 1802
ovatus (Linnaeus, 1761)

Tribe LACCORNINI Wolfe & Roughley, 1990

LACCORNIS des Gozis, 1914
    AGAPORUS Zimmermann, 1919
oblongus (Stephens, 1835)

Subfamily LACCOPHILINAE Gistel, 1848

LACCOPHILUS Leach, 1815
hyalinus (De Geer, 1774)
    interruptus (Panzer, 1795)
minutus (Linnaeus, 1758)
    variolosus (Herbst, 1784)
    obscurus (Panzer, 1795)
poecilus Klug, 1834
    variegatus (Germar & Kaulfuss, 1816) non
        (Fourcroy, 1785)
    obsoletus sensu auctt. non Westhoff, 1881
    ponticus Sharp, 1882

7. Family CARABIDAE Latreille, 1802

Family author: M.L. Luff

The currently used nomenclature on the British Carabidae is based on that in Lindroth (1974) and Pope (1977); see also Luff (2007). More recent lists that have updated the nomenclature of some or all of the British fauna include Lindroth (1985, Fennoscandia), Lohse & Lucht (1989, central Europe), Turin (1990, 2000, Netherlands), Kryzhanovskij et al. (1995, Russia), Hansen (1996, Denmark), Hůrka (1996, Czech & Slovak republics) and Anderson et al. (1997, 2000, Ireland). These have now all been superseded by the Palaearctic list edited by Löbl & Smetana (2003). There have also been specialised lists for the genus Carabus, sensu lato (Turin et al., 1993; Deuve, 1994; Casale & Březina, 2003). The most recent list of central European Carabidae (Müller-Motzfeld, 2004) does not include synonyms. The following list is based on that of Löbl & Smetana (2003), with only a few more recent amendments provided by Dr Roger Booth of the Natural History Museum, London.

The limits of the family Carabidae have remained constant in regard to the British fauna, since the Cicindelidae were included (as subfamily Cicindelinae) by Lindroth (1974). On the larger scale of the European or world fauna, however, there have been changes that do not affect the more limited British list. In particular, Erwin & Sims (1984) and Erwin (1985) recognised the separate family Trachypachidae, but included the former families Paussidae and Rhysodidae within the Carabidae. Lawrence & Newton (1995) followed the first two of these changes, but retained the Rhysodidae as a separate family.

There is also no general agreement as to the arrangement and status of supra-generic taxa within the Carabidae. In an introduction to the classification of Caraboidea, Ball et al. (1998) compare many of the various arrangements that have been advocated. They conclude that points of conflict far outnumber any generally held areas of agreement. The present list is based on the relatively simple system used by Müller-Motzfeld (2004) which divided the family into only four subfamilies, with the great majority of British species in the Carabinae. This is in contrast to the classification of Lawrence & Newton (1985), who had many subfamilies, and consider the broscines to be a tribe within the subfamily Trechinae. However all other authors either treat them as a separate higher taxon, e.g. family Broscidae alongside family Trechidae (Deuve, 1993), or as a subfamily separate from the Trechines e.g. subfamily Broscinae, whereas Trechitae are considered a supertribe within the subfamily Psydrinae (Erwin, 1985) or a separate subfamily Trechinae. Detailed studies of the broscines have since supported their separate status (Roig-

Juňent, 1998). The present list treats the broscines as a tribe within Carabinae.

A smaller confusion surrounds the placement of the Platynini (often including Sphodrini). Traditional British and European classifications have included these in the Pterostichini, e.g. Hansen (1996) for Denmark, Turin (2000) for the Netherlands. However Lawrence & Newton (1985) not only agreed with Erwin (1985) in considering the Platynini as a separate tribe, but placed them near the Perigonini, Zuphiini and other 'truncatipennes' groups. The latter placement can be supported on larval characters (Arndt, 1998). However although retaining the Sphodrini and Platynini, I prefer to place them immediately after the Pterostichini, as in the traditional arrangement. The tribe Licinini is here considered to include Badistriini, following Kryzhanovskij et al. (1995). Subtribe names are not included in the list, although the Harpalini and Platynini, in particular, are often subdivided at this level.

The treatment of the major genera, Carabus, Bembidion, Pterostichus, Amara, Harpalus and Agonum in the present list has tried to follow the divisions currently generally accepted by most European workers and as used in Löbl & Smetana (2003). This inevitably has resulted in some inconsistencies between these genera. The limits of Carabus are generally agreed, at least for the British fauna, so no 'new' generic names are used. The names and arrangement of subgenera of Carabus follow Casale & Březina (2003).

Bembidion, although the largest genus of European carabids, is seldom split into separate genera. Only Ocys (two British species) and

Cillenus (a single species) are separated by Kryzhanovskij et al. (1995), Hůrka (1996) and Hansen (1996). I have used these genera, as well as treating Bracteon (two species) as a distinct genus, based at least partly on larval features (Luff, 1993). Most European taxonomic works (excepting Hansen, 1996) now consider Poecilus to be a separate genus from Pterostichus, and this is followed here. The remaining taxa within Pterostichus are considered as subgenera, although generic status for some or all may well be justified. Similarly, only Curtonotus is separated from Amara as a distinct genus, following Kryzhanovskij et al. (1995). I follow Telfer (2001b) in considering Ophonus as a distinct genus, but have not used the genus Pseudoophonus, which remains within Harpalus (see Kryzhanovskij et al., 1995). Agonum is the only one of these traditionally large genera which is now generally split into a number of genera. However Europhilus remains as a subgenus of Agonum.

Other than changes in synonymies given in the list, 12 species have been added since Pope (1977): the references for these are given in Table 1, as well as in the endnotes in the main list. Two species have been deleted from Pope's list as they are now considered to be non-established introductions: Abax parallelus and Stenolophus comma (= plagiatus), while some other names have been lost due to synonymy or homonymy. The list also includes species now considered extinct. Introduced species which have never formed established populations here are listed in the appendix "Non-established introductions" on page 131.

Table 1: Carabidae species added to the British list since Pope (1977).

Species	Reason for addition	Literature source
Asaphidion curtum	taxonomic split	Speight, Martinez & Luff (1986)
Asaphidion stierlini	taxonomic split	Speight, Martinez & Luff (1986)
Bembidion caeruleum	immigrant	Telfer (2001a)
Bembidion inustum	? rare & overlooked	Levey & Pavett (1999)
Calathus cinctus	taxonomic split	Anderson & Luff (1994)
Agonum lugens	? rare & overlooked	Anderson (1985)
Harpalus griseus	immigrant	Owen (1996)
Ophonus subsinuatus	overlooked museum specimens	Telfer (2001b)
Acupalpus maculatus	immigrant	Telfer (2003)
Somotrichus unifasciatus	temporarily established introduction	Denton (2007)
Cymindis macularis	? immigrant	Hammond (1982a)
Microlestes minutulus	immigrant	Eversham & Collier (1997)

Subfamily CICINDELINAE Latreille, 1802

CICINDELA Linnaeus, 1758
campestris Linnaeus, 1758
hybrida Linnaeus, 1758
maritima Latreille & Dejean, 1822
sylvatica Linnaeus, 1758
  silvatica auctt. (misspelling)

CYLINDERA Westwood, 1831
  CICINDELA sensu auctt. partim non Linnaeus, 1758
germanica (Linnaeus, 1758)

Subfamily BRACHININAE Bonelli, 1810

BRACHINUS Weber, 1801

Subgenus BRACHINUS Weber, 1801
crepitans (Linnaeus, 1758)

Subgenus BRACHYNIDIUS Reitter, 1919
sclopeta (Fabricius, 1792)

Subfamily OMOPHRONINAE Bonelli, 1810

OMOPHRON Latreille, 1802
  SCOLYTUS Fabricius, 1790 non Müller, O.F., 1764
limbatum (Fabricius, 1777)

Subfamily CARABINAE Latreille, 1802

Tribe CARABINI Latreille, 1802

CALOSOMA Weber, 1801
inquisitor (Linnaeus, 1758)
sycophanta (Linnaeus, 1758)

CARABUS Linnaeus, 1758 [31]

Subgenus LIMNOCARABUS Géhin, 1876
clatratus Linnaeus, 1761
  clathratus auctt. (misspelling)

Subgenus EUCARABUS Géhin, 1885
arvensis Herbst, 1784
ssp. *sylvaticus* Dejean, 1826
  silvaticus auctt. (misspelling)
  anglicanus Motschulsky, 1865/6

Subgenus CARABUS Linnaeus, 1758
granulatus Linnaeus, 1758
ssp. *granulatus* Linnaeus, 1758
ssp. *hibernicus* Lindroth, 1956

Subgenus MORPHOCARABUS Géhin, 1885
monilis Fabricius, 1792
  consitus sensu auctt. non Panzer, 1809
  insularis Born, 1908

Subgenus ARCHICARABUS Seidlitz, 1887
nemoralis Müller, O.F., 1764

Subgenus TACHYPUS Weber, 1801
auratus Linnaeus, 1761

Subgenus HEMICARABUS Géhin, 1885
nitens Linnaeus, 1758

Subgenus OREOCARABUS Géhin, 1885
glabratus Paykull, 1790
ssp. *lapponicus* Eorn, 1909

Subgenus MESOCARABUS Thomson, C.G., 1875
problematicus Herbst, 1786
ssp. *feroensis* Lapouge, 1910
ssp. *harcyniae* Sturm, 1815
  catenulatus sensu auctt. non Scopoli, 1763
  gallicus Géhin 1885

Subgenus CHAETOCARABUS Thomson, C.G., 1875
intricatus Linnaeus, 1761

Subgenus MEGODONTUS Solier, 1848
violaceus Linnaeus, 1758
ssp. *purpurascens* Fabricius, 1787
  exasperatus sensu auctt. non Duftschmid, 1812
ssp. *sollicitans* Hartert, 1907
  britannicus Born, 1908
  browni Deuve, 1999

Tribe CYCHRINI Perty, 1830

CYCHRUS Fabricius, 1794
caraboides (Linnaeus, 1758)
ssp. *rostratus* (Linnaeus, 1761)

Tribe NEBRIINI Laporte, 1834

LEISTUS Frölich, 1799

Subgenus POGONOPHORUS Latreille, 1802
montanus Stephens, 1827
ssp. *rhaeticus* Heer, 1837
rufomarginatus (Duftschmid, 1812)
spinibarbis (Fabricius, 1775)

Subgenus LEISTOPHORUS Reitter, 1905
fulvibarbis Dejean 1826

Subgenus LEISTUS Frölich, 1799
ferrugineus (Linnaeus, 1758)
  rufescens sensu auctt. partim non (Fabricius, 1775)
terminatus (Hellwig in Panzer, 1793)
  rufescens (Fabricius, 1775) non (Strøm, 1768)
  praeustus (Fabricius, 1792) [32]

NEBRIA Latreille, 1802

Subgenus PARANEBRIA Jeannel, 1937
livida (Linnaeus, 1758)
  lateralis (Fabricius, 1787)
  sabulosa (Fabricius, 1787)

Subgenus NEBRIA Latreille, 1802
brevicollis (Fabricius, 1792)
salina Fairmaire & Laboulbène, 1854
  degenerata Schaufuss, 1862
  iberica Oliveira, 1876
  klinckowstroem Mjöberg, 1915

Subgenus BOREONEBRIA Jeannel, 1937
nivalis (Paykull, 1790)
rufescens (Strøm, 1768)
  gyllenhali (Schönherr, 1806)
  balbii Bonelli, 1810

EURYNEBRIA Ganglbauer, 1891
complanata (Linnaeus, 1767)

PELOPHILA Dejean, 1821
borealis (Paykull, 1790)

Tribe NOTIOPHILINI Motschulsky, 1850

NOTIOPHILUS Duméril, 1806
aesthuans Motschulsky, 1864
    aestuans auctt. (misspelling)
    pusillus Waterhouse, G.R., 1833 non (von Schreber,
      1759)
aquaticus (Linnaeus, 1758)
    pusillus (von Schreber, 1759)
    strigifrons (Baudi, 1864)
    blacki Edwards, J., 1913
biguttatus (Fabricius, 1779)
germinyi Fauvel, 1863
    hypocrita sensu auctt. non Putzeys, 1866
palustris (Duftschmid, 1812)
quadripunctatus Dejean, 1826
    quadriguttatus Fowler, 1886 (error)
rufipes Curtis, 1829
substriatus Waterhouse, G.R., 1833

Tribe ELAPHRINI Latreille, 1802

BLETHISA Bonelli, 1810
    HELOBIUM Leach, 1815
multipunctata (Linnaeus, 1758)

ELAPHRUS Fabricius, 1775

Subgenus ELAPHRUS Fabricius, 1775
cupreus Duftschmid, 1812
lapponicus Gyllenhal, 1810
uliginosus Fabricius, 1792

Subgenus TRICHELAPHRUS Semenov, 1926
riparius (Linnaeus, 1758)

Tribe LORICERINI Bonelli, 1810

LORICERA Latreille, 1802
    LOROCERA auctt. (misspelling)
pilicornis (Fabricius, 1775)
    coerulescens sensu auctt. non (Linnaeus, 1758)

Tribe SCARITINI Bonelli, 1810

CLIVINA Latreille, 1802
collaris (Herbst, 1784)
    contracta (Geoffroy in Fourcroy, 1785)
fossor (Linnaeus, 1758)

DYSCHIRIUS Bonelli, 1810

Subgenus DYSCHIRIUS Bonelli, 1810
angustatus (Ahrens, 1830)
obscurus (Gyllenhal, 1827)
thoracicus (Rossi, 1790)
    arenosus Stephens, 1827

Subgenus DYSCHIRIODES Jeannel, 1941
aeneus (Dejean, 1825)
extensus Putzeys, 1846
    elongatulus Dawson, 1856
globosus (Herbst, 1784)

    gibbus (Fabricius, 1792)
impunctipennis Dawson, 1854
luedersi Wagner, 1915
    aeneus sensu auctt. non (Dejean, 1825)
    ?*tristis* Stephens, 1827
    unicolor sensu auctt. non Motschulsky, 1844
nitidus (Dejean, 1825)
politus (Dejean, 1825)
salinus Schaum, 1843

Tribe BROSCINI Hope, 1838

BROSCUS Panzer, 1813
cephalotes (Linnaeus, 1758)

MISCODERA Eschscholtz, 1830
arctica (Paykull, 1798)

Tribe TRECHINI Bonelli, 1810

PERILEPTUS Schaum, 1860
    BLEMUS sensu Laporte, 1840 non Dejean, 1821
areolatus (Creutzer, 1799)

AEPUS Leach, 1819
    AEPOPSIS Jeannel, 1922
marinus (Strøm, 1783)
    fulvescens Samouelle, 1819
robinii (Laboulbène, 1849)
    robini auctt. (misspelling)

TRECHUS Clairville, 1806
    BLEMUS sensu Stephens, 1827 non Dejean, 1821

Subgenus EPAPHIUS Leach, 1819
rivularis (Gyllenhal, 1810)
secalis (Paykull, 1790)

Subgenus TRECHUS Clairville, 1806
fulvus Dejean, 1831
    lapidosus (Dawson, 1849)
obtusus Erichson, 1837
quadristriatus (Schrank, 1781)
    minutus (Fabricius, 1781)
    rubens sensu Clairville, 1806 non (Fabricius, 1792)
rubens (Fabricius, 1792)
    paludosus (Gyllenhal, 1810)
subnotatus Dejean, 1831

THALASSOPHILUS Wollaston, 1854
    TRECHUS sensu Fowler, 1886 partim non Clairville,
      1806
longicornis (Sturm, 1825)

BLEMUS Dejean, 1821
    LASIOTRECHUS Ganglbauer, 1892
discus (Fabricius, 1792)

TRECHOBLEMUS Ganglbauer, 1891
micros (Herbst, 1784)

Tribe BEMBIDIINI Stephens, 1827

TACHYS Dejean, 1821

Subgenus PARATACHYS Casey, 1918
EOTACHYS Jeannel, 1941
bistriatus (Duftschmid, 1812)
?*pallidulus* (Antoine, 1943) non (Ménétries, 1846) [33]
micros (Fischer von Waldheim, 1828)
gregarius Chaudoir, 1846
obtusiusculus (Jeannel, 1941)
piceus Edmonds, 1934 non Dalla Torre, 1877
edmondsi Moore, 1956 [33]

Subgenus TACHYS Dejean, 1821
scutellaris Stephens, 1828

ELAPHROPUS Motschulsky, 1839

TACHYS sensu auctt. partim non Dejean, 1821
parvulus (Dejean, 1831)
walkerianus (Sharp, 1913)

ASAPHIDION des Gozis, 1886

TACHYPUS sensu auctt. non Weber, 1801
curtum (Heyden, 1870) [34]
flavipes sensu auctt. Brit. partim non (Linnaeus, 1761)
flavipes (Linnaeus, 1761)
pallipes (Duftschmid, 1812)
stierlini (Heyden, 1880) [35]
flavipes sensu auctt. Brit. partim non (Linnaeus, 1761)

OCYS Stephens, 1828

BEMBIDION sensu auctt. partim non Latreille, 1802
harpaloides (Audinet-Serville, 1821)
rufescens (Guérin-Méneville, 1823)
quinquestriatus (Gyllenhal, 1810)

CILLENUS Leach, 1819

BEMBIDION sensu auctt. partim non Latreille, 1802
lateralis Samouelle, 1819

BRACTEON Bedel, 1879

BEMBIDION sensu auctt. partim non Latreille, 1802
CHRYSOBRACTEON Netolitzky, 1914
argenteolum (Ahrens, 1812)
litorale (Olivier, 1790)
littorale (Olivier, 1791)
paludosum (Panzer, 1794)

BEMBIDION Latreille, 1802

BEMBIDIUM auctt. (misspelling)

Subgenus NEJA Motschulsky, 1864
nigricorne Gyllenhal, 1827

Subgenus METALLINA Motschulsky, 1850
lampros (Herbst, 1784)
celere (Fabricius, 1792)
properans (Stephens, 1828)
velox Erichson, 1837 non (Linnaeus, 1761)
coeruleotinctum Reitter, 1908
cyaneotinctum Sharp, 1913
caeruleipenne Saunders, 1936

Subgenus PRINCIDIUM Motschulsky, 1864
punctulatum Drapiez, 1821

Subgenus ACTEDIUM Motschulsky, 1864
pallidipenne (Illiger, 1802)
ruficolle (Illiger, 1801) non (Panzer, 1796)

Subgenus TESTEDIUM Motschulsky, 1864
bipunctatum (Linnaeus, 1761)

Subgenus EUPETEDROMUS Netolitzky, 1911
dentellum (Thunberg, 1787)
flammulatum (Clairville, 1806)

Subgenus NOTAPHUS Dejean, 1821
obliquum Sturm, 1825
semipunctatum (Donovan, 1806)
adustum Schaum, 1860
varium (Olivier, 1795) [36]
ustulatum sensu Sturm, 1825 non (Linnaeus, 1758)
nebulosum (Stephens, 1828)

Subgenus NOTAPHEMPHANES Netolitzky, 1920
NOTHAPHEMPHANES auctt. (misspelling)
ephippium (Marsham, 1802)

Subgenus PLATAPHUS Motschulsky, 1864
prasinum (Duftschmid, 1812)

Subgenus TRICHOPLATAPHUS Netolitzky, 1914
BLEPHAROPLATAPHUS Netolitzky, 1920
virens Gyllenhal, 1827

Subgenus BEMBIDIONETOLITZKYA Strand, E., 1929
DANIELA Netolitzky, 1910 non Koch, 1891
atrocaeruleum (Stephens, 1828)
atrocoeruleum auctt. (misspelling)
caeruleum Audinet-Serville, 1826 [37]
coeruleum auctt. (misspelling)
geniculatum Heer 1837/8
redtenbacheri Daniel, K., 1902
tibiale (Duftschmid, 1812)

Subgenus OCYDROMUS Clairville, 1806
PERYPHUS Dejean, 1821
bruxellense Wesmael, 1835
rupestre sensu auctt. non (Linnaeus, 1767)
bualei Jacquelin du Val, 1852
andreae sensu auctt. non (Fabricius, 1787)
cruciatum sensu auctt. non Dejean, 1831
ssp. *anglicanum* Sharp, 1869
ssp. *polonicum* Müller, J., 1930
decorum (Zenker in Panzer, 1800)
deletum Audinet-Serville, 1821
nitidulum (Marsham, 1802) non (Schrank, 1781)
dalmatinum var. *latinum* sensu MacKechnie-Jarvis, 1932 non Netolitzky, 1911
femoratum Sturm, 1825
fluviatile Dejean, 1831
lunatum (Duftschmid, 1812)
maritimum (Stephens, 1835)
concinnum sensu auctt. non (Stephens, 1828)
dorsuarium Bedel, 1879
monticola Sturm, 1825
saxatile Gyllenhal, 1827
vectense Fowler, 1886
stephensii Crotch, 1866
stephensi auctt (misspelling)
affine (Stephens, 1835) non Say, 1825
testaceum (Duftschmid, 1812)
tetracolum Say, 1825
ustulatum sensu auctt. non (Linnaeus, 1758)

littorale sensu auctt. non (Olivier, 1791)

Subgenus NEPHA Motschulsky, 1864
illigeri Netolitzky, 1914
    quadriguttatum sensu auctt. Brit. non (Illiger, 1798)
    nec (Fabricius, 1775)
    tetragrammum sensu auctt. non Chaudoir, 1846
    genei sensu auctt. Brit. non Küster, 1847

Subgenus SINECHOSTICTUS Motschulsky, 1864
    SYNECHOSTICTUS auctt. (misspelling)
stomoides Dejean, 1831
    atroviolaceum sensu auctt. non Dufour, 1820

Subgenus PSEUDOLIMNAEUM Kraatz, 1888
inustum Jacquelin du Val, 1857 [38]

Subgenus LYMNAEUM Stephens, 1828
nigropiceum (Marsham, 1802)

Subgenus SEMICAMPA Netolitzky, 1910
gilvipes Sturm, 1825
schuppelii Dejean, 1831
    schueppeli auctt. (misspelling)

Subgenus DIPLOCAMPA Bedel, 1896
assimile Gyllenhal, 1810
clarkii (Dawson, 1849)
    clarki auctt. (misspelling)
fumigatum (Duftschmid, 1812)

Subgenus EMPHANES Motschulsky, 1850
minimum (Fabricius, 1792) [39]
normannum Dejean, 1831

Subgenus BEMBIDION Latreille, 1802
    LOPHA Dejean, 1821 non Bolten, 1798
humerale Sturm, 1825
quadrimaculatum (Linnaeus, 1761)
    quadriguttatum (Fabricius, 1775)
quadripustulatum Audinet-Serville, 1821
    quadriguttatum sensu (Olivier, 1795) non (Fabricius, 1775)
    antiquorum Crotch, 1871

Subgenus TREPANEDORIS Netolitzky, 1918
doris (Panzer, 1796)

Subgenus TREPANES Motschulsky, 1864
    LEJA Dejean, 1821 [32]
articulatum (Panzer, 1795)
octomaculatum (Goeze, 1777)
    sturmii (Panzer, 1804)
    sturmi auctt. (misspelling)

Subgenus PHYLA Motschulsky, 1844
    PHILA Motschulsky, 1846 [40]
obtusum Audinet-Serville, 1821

Subgenus PHILOCHTHUS Stephens, 1828
    PHILOCTHUS auctt. (misspelling)
aeneum Germar, 1824
biguttatum (Fabricius, 1779)
guttula (Fabricius, 1792)
iricolor Bedel, 1879
    riparium sensu Fowler, 1886 partim non (Olivier, 1795)
lunulatum (Geoffroy in Fourcroy, 1785)
    riparium (Olivier, 1795)
mannerheimii Sahlberg, C.R., 1827
    mannerheimi auctt. (misspelling)
    haemorrhoum sensu auctt. non (Stephens, 1828)
    unicolor Chaudoir, 1850

Tribe POGONINI Laporte, 1834

POGONUS Dejean, 1821
chalceus (Marsham, 1802)
littoralis (Duftschmid, 1812) [41]
    litoralis auctt. (misspelling)
luridipennis (Germar, 1822)

Tribe PATROBINI Kirby, 1837

PATROBUS Dejean, 1821
assimilis Chaudoir, 1844
    clavipes Thomson, C.G., 1857
atrorufus (Strøm, 1768)
    excavatus (Paykull, 1790)
    rufipes sensu (Duftschmid, 1812) non (Fabricius, 1792)
septentrionis Dejean, 1828

Tribe PTEROSTICHINI Bonelli, 1810

STOMIS Clairville, 1806
pumicatus (Panzer, 1795)

POECILUS Bonelli, 1810
    PTEROSTICHUS sensu auctt. partim non Bonelli, 1810
    FERONIA Latreille, 1817
cupreus (Linnaeus, 1758)
    affinis (Sturm, 1824)
    erythropus (Dejean, 1828)
    dinniki (Lutshnik, 1912)
    caesica Donisthorpe, 1931
kugelanni (Panzer, 1797)
    dimidiatus (Olivier, 1795) non (Rossi, 1790)
lepidus (Leske, 1785)
versicolor (Sturm, 1824)
    caerulescens sensu auctt. non (Linnaeus, 1758)

PTEROSTICHUS Bonelli, 1810

Subgenus PTEROSTICHUS Bonelli, 1810
cristatus (Dufour, 1820)
ssp. ***parumpunctatus*** Germar, 1824

Subgenus STEROPUS Stephens, 1828
aethiops (Panzer, 1796) [42]
madidus (Fabricius, 1775)
    concinnus (Sturm, 1818)

Subgenus PEDIUS Motschulsky, 1850
longicollis (Duftschmid, 1812) [43]
    inaequalis (Marsham, 1802) non (Panzer, 1795)
    ochraceus (Sturm, 1824)

Subgenus LYPEROSOMUS Motschulsky, 1850
aterrimus (Herbst, 1784)

Subgenus ADELOSIA Stephens, 1835
macer (Marsham, 1802)
    picimanus (Duftschmid, 1812)

Subgenus PLATYSMA Bonelli, 1810
niger (Schaller, 1783) [44]
    scotus (Jeannel, 1942)

Subgenus BOTHRIOPTERUS Chaudoir, 1838
adstrictus Eschscholtz, 1823

orinomum (Stephens, 1828)
   vitreus (Dejean, 1828)
oblongopunctatus (Fabricius, 1787)
quadrifoveolatus Letzner, 1852
   angustatus (Duftschmid, 1812) non (Fabricius, 1787)

Subgenus OMASEUS Stephens, 1828
melanarius (Illiger, 1798)
   vulgaris sensu auctt. non (Linnaeus, 1758)

Subgenus PSEUDOMASEUS Chaudoir, 1838
anthracinus (Panzer, 1795)
gracilis (Dejean, 1828)
minor (Gyllenhal, 1827)
nigrita (Paykull, 1790)
rhaeticus Heer, 1837/8 [45]
   nigrita sensu auctt. partim non (Paykull, 1790)

Subgenus LAGARUS Chaudoir, 1838
vernalis (Panzer, 1795) [46]
   crenatus (Duftschmid, 1812) non (Gmelin in
      Linnaeus, 1790)

Subgenus ARGUTOR Stephens, 1828
diligens (Sturm, 1824)
   strenuus sensu Dawson, 1854 non (Panzer, 1796)
strenuus (Panzer, 1796)
   erythropus (Marsham, 1802)

ABAX Bonelli, 1810 [47]

parallelepipedus (Piller & Mitterpacher, 1783)
   parallelopipedus auctt. (misspelling)
   ater (Villers, 1790)
   striola (Fabricius, 1792)

Tribe SPHODRINI Laporte, 1834

PLATYDERUS Stephens, 1828
   PLATYDERES Stephens, 1827 [32]
depressus (Audinet-Serville, 1821)
   ruficollis (Marsham, 1802) non (Fabricius, 1787)

SYNUCHUS Gyllenhal, 1810
   TAPHRIA Latreille, 1819
   ODONTONYX Stephens, 1827
vivalis (Illiger, 1798)
   nivalis (Panzer, 1797) non (Paykull, 1790)

CALATHUS Bonelli, 1810

Subgenus AMPHIGYNUS Haliday, 1841
rotundicollis Dejean, 1828
   piceus sensu (Marsham, 1802) non (Linnaeus,
      1758)

Subgenus CALATHUS Bonelli, 1810
ambiguus (Paykull, 1790)
   fuscus (Fabricius, 1792)
cinctus Motschulsky, 1850 [48]
   melanocephalus sensu auctt. partim non (Linnaeus,
      1758)
   mollis sensu auctt. partim non Marsham, 1802
   erythroderus Gemminger & Harold, 1868
erratus (Sahlberg, C.R., 1827)
   fulvipes sensu (Gyllenhal, 1810) non (Fabricius,
      1792)
   flavipes sensu (Duftschmid, 1812) non (Geoffroy in
      Fourcroy, 1785)
fuscipes (Goeze, 1777)

cisteloides (Panzer, 1793)
melanocephalus (Linnaeus, 1758)
   ochropterus (Duftschmid, 1812)
   nubigena Haliday, 1838
micropterus (Duftschmid, 1812)
mollis (Marsham, 1802) [49]
   ochropterus sensu auctt. non (Duftschmid, 1812)

SPHODRUS Clairville, 1806
leucophthalmus (Linnaeus, 1758)
   leucopthalmus Fowler, 1886 (misspelling)
   planus (Fabricius, 1792)

LAEMOSTENUS Bonelli, 1810
   LAEMOSTHENES Schaufuss, 1865

Subgenus LAEMOSTENUS Bonelli, 1810
complanatus (Dejean, 1828)

Subgenus PRISTONYCHUS Dejean, 1828
terricola (Herbst, 1784)
   subcyaneus (Illiger, 1801)

Tribe PLATYNINI Bonelli, 1810

OLISTHOPUS Dejean, 1828
   ODONTONYX sensu auctt. non Stephens, 1827
rotundatus (Paykull, 1790) [50]
   rotundicollis (Marsham, 1802)

OXYPSELAPHUS Chaudoir, 1843
   AGONUM sensu auctt. partim non Bonelli, 1810
   ANCHUS LeConte, 1854
obscurus (Herbst, 1784)
   ?*obscurus* (Müller, O.F., 1776)
   oblongus (Fabricius, 1792)

PARANCHUS Lindroth, 1974
   AGONUM sensu auctt. partim non Bonelli, 1810
albipes (Fabricius, 1796)
   ruficornis (Goeze, 1777) non (De Geer, 1774)
   oblongus (Fabricius, 1792) non (Fabricius, 1792)
   pallipes (Fabricius, 1801) non (Fabricius, 1787)

ANCHOMENUS Bonelli, 1810
   AGONUM sensu auctt. partim non Bonelli, 1810
dorsalis (Pontoppidan, 1763)
   prasinus (Thunberg, 1784)

PLATYNUS Bonelli, 1810
   AGONUM sensu auctt. partim non Bonelli, 1810
assimilis (Paykull, 1790)
   ?*junceus* (Scopoli, 1763)
   angusticollis (Fabricius, 1801)

BATENUS Motschulsky, 1864
   AGONUM sensu auctt. partim non Bonelli, 1810
livens (Gyllenhal, 1810)

SERICODA Kirby, 1837
   AGONUM sensu auctt. partim non Bonelli, 1810
quadripunctata (De Geer, 1774)

AGONUM Bonelli, 1810

Subgenus EUROPHILUS Chaudoir, 1859
fuliginosum (Panzer, 1809)
gracile Sturm, 1824
micans Nicolai, 1822
piceum (Linnaeus, 1758)
scitulum Dejean, 1828
thoreyi Dejean, 1828
    pelidnum sensu (Paykull, 1792) non (Herbst, 1784)
    puellum Dejean, 1828

Subgenus AGONUM Bonelli, 1810
chalconotum Ménétries, 1832
    sahlbergii (Chaudoir, 1850)
    sahlbergi auctt. (misspelling)
    archangelicum Sahlberg, J., 1874
emarginatum (Gyllenhal, 1827)
    afrum (Duftschmid, 1812) non (Thunberg, 1784)
    moestum sensu auctt. non (Duftschmid, 1812) nec
        (Gmelin in Linnaeus, 1790)
ericeti (Panzer, 1809)
gracilipes (Duftschmid, 1812)
    elongatum Dejean, 1828
lugens (Duftschmid, 1812) [51]
marginatum (Linnaeus, 1758)
muelleri (Herbst, 1784)
    parumpunctatum (Fabricius, 1792)
    chalybeum Sturm, 1824
nigrum Dejean, 1828
    atratum sensu auctt. non (Duftschmid, 1812)
    dahli (Preudhomme de Borre, 1879)
sexpunctatum (Linnaeus, 1758)
versutum Sturm, 1824
viduum (Panzer, 1796)
    moestum sensu Fowler & Donisthorpe, 1913 non
        (Duftschmid, 1812)

Tribe ZABRINI Bonelli, 1810

ZABRUS Clairville, 1806

tenebrioides (Goeze, 1777)
    gibbus (Fabricius, 1794)

AMARA Bonelli, 1810

Subgenus ZEZEA Csiki, 1929
    TRIAENA LeConte, 1847 non Hübner, 1818
plebeja (Gyllenhal, 1810)
    plebeia auctt. (misspelling)
strenua Zimmermann, 1832
    vectensis Dawson, 1849

Subgenus AMARA Bonelli, 1810
aenea (De Geer, 1774)
    trivialis (Gyllenhal, 1810)
anthobia Villa & Villa, 1833
communis (Panzer, 1797)
    vulgaris sensu Dawson, 1854 non (Linnaeus, 1758)
convexior Stephens, 1828
    continua Thomson, C.G., 1873
curta Dejean, 1828
eurynota (Panzer, 1796)
    eyrinota auctt. (misspelling)
    acuminata (Paykull, 1798)
famelica Zimmermann, 1832
familiaris (Duftschmid, 1812)
lucida (Duftschmid, 1812)
lunicollis Schiødte, 1837

    vulgaris sensu auctt. non (Linnaeus, 1758)
montivaga Sturm, 1825
nitida Sturm, 1825
ovata (Fabricius, 1792) [52]
    obsoleta sensu Dejean, 1828 non (Duftschmid,
        1812)
    adamantina Kolenati, 1845
similata (Gyllenhal, 1810)
    obsoleta (Duftschmid, 1812)
spreta Dejean, 1831
tibialis (Paykull, 1798)

Subgenus CELIA Zimmermann, 1832
bifrons (Gyllenhal, 1810)
    livida sensu Schiødte, 1841 non (Fabricius, 1792)
cursitans (Zimmermann, 1832)
    fuscicornis (Zimmermann, 1832)
fusca Dejean, 1828
    complanata Dejean, 1828
infima (Duftschmid, 1812) [53]
praetermissa (Sahlberg, C.R., 1827)
    rufocincta (Sahlberg, C.R., 1827)

Subgenus PARACELIA Bedel, 1899
quenseli (Schönherr, 1806)
    quenselii auctt. (misspelling)

Subgenus BRADYTUS Stephens, 1827
apricaria (Paykull, 1790)
consularis (Duftschmid, 1812)
fulva (Müller, O.F., 1776) [54]

Subgenus PERCOSIA Zimmermann, 1832
equestris (Duftschmid, 1812)
    patricia (Duftschmid, 1812)

CURTONOTUS Stephens, 1827

    CYRTONOTUS auctt. (misspelling)
    AMARA sensu auctt. partim non Bonelli, 1810
alpinus (Paykull, 1790)
aulicus (Panzer, 1796)
    spinipes sensu Schiødte, 1841 non (Linnaeus,
        1758)
convexiusculus (Marsham, 1802)

Tribe HARPALINI Bonelli, 1810

HARPALUS Latreille, 1802

Subgenus PSEUDOOPHONUS Motschulsky, 1844
    PSEUDOPHONUS auctt. (misspelling)
    PARDILEUS des Gozis, 1882
calceatus (Duftschmid, 1812)
griseus (Panzer, 1797) [55]
rufipes (De Geer, 1774)
    ruficornis (Fabricius, 1775)
    pubescens (Müller, O.F., 1776)

Subgenus HARPALUS Latreille, 1802
    HAPLOHARPALUS Schauberger, 1926
affinis (Schrank, 1781)
    aeneus (Fabricius, 1775) non (De Geer, 1774)
anxius (Duftschmid, 1812)
attenuatus Stephens, 1828
    consentaneus Dejean, 1829
cupreus Dejean, 1829
dimidiatus (Rossi, 1790)
    caspius sensu auctt. Brit. non (von Steven, 1806)
froelichii Sturm, 1818
    froelichi auctt. (misspelling)

honestus (Duftschmid, 1812)
　ignavus (Duftschmid, 1812)
laevipes Zetterstedt, 1828
　seriepunctatus sensu Gyllenhal, 1827 non Sturm,
　　1818
　quadripunctatus Dejean, 1829
　montivagus Reitter, 1900
latus (Linnaeus, 1758)
　erythrocephalus (Fabricius, 1787)
　metallescens Rye, 1874
neglectus Audinet-Serville, 1821
pumilus Sturm, 1818
　vernalis sensu (Duftschmid, 1812) non (Panzer,
　　1796)
　picipennis sensu auctt. Brit. non Duftschmid, 1812
　funestus Audinet-Serville, 1821
rubripes (Duftschmid, 1812)
　sobrinus Dejean, 1829
rufipalpis Sturm, 1818
　rufitarsis (Duftschmid, 1812) non (Illiger, 1801)
　ignavus sensu auctt. Brit. non (Duftschmid, 1812)
serripes (Quensel in Schönherr, 1806)
servus (Duftschmid, 1812)
smaragdinus (Duftschmid, 1812)
　discoideus Erichson, 1837
tardus (Panzer, 1796)
　rufimanus (Marsham, 1802)
　luteicornis sensu auctt. Brit. non (Duftschmid, 1812)

Subgenus CRYPTOPHONUS Brandmayr & Zetto
　Brandmayr, 1982
melancholicus Dejean, 1829
tenebrosus Dejean, 1829
　ssp. *centralis* Schauberger, 1929

OPHONUS Dejean, 1821
　HARPALUS sensu auctt. partim non Latreille, 1802

Subgenus OPHONUS Dejean, 1821
ardosiacus (Lutshnik, 1922)
　obscurus sensu (Sturm, 1818) non (Fabricius, 1792)
　diffinis sensu Joy, 1932 non (Dejean, 1829)
　rotundicollis sensu auctt. Brit. non Kolenati, 1845
azureus (Fabricius, 1775)
　subquadratus sensu auctt. Brit. non (Dejean, 1829)
sabulicola (Panzer, 1796)
stictus Stephens, 1828
　obscurus (Fabricius, 1792) non (Müller, O.F., 1776)
　monticola (Dejean, 1829)

Subgenus METOPHONUS Bedel, 1895
cordatus (Duftschmid, 1812) [56]
laticollis Mannerheim, 1825
　punctatulus (Duftschmid, 1812) non (Fabricius,
　　1792)
　nitidulus sensu Stephens, 1828 non (Schrank,
　　1781)
melletii (Heer, 1837/8)
　melleti auctt. (misspelling)
　rectangulus Thomson, C.G., 1870
　championi Sharp, 1912
　rupicoloides Sharp, 1912
　brevicollis sensu Jeannel, 1942 ? (Audinet-Serville,
　　1821)
parallelus (Dejean, 1829)
　melleti sensu Jeannel, 1942 non (Heer, 1837/8)
　zigzag sensu auctt. Brit. non Costa, 1882
puncticeps Stephens, 1828

rectangulus sensu Sharp, 1912 non Thomson, C.G.,
　1870
　angusticollis Müller, J., 1921
puncticollis (Paykull, 1798)
rufibarbis (Fabricius, 1792)
　brevicollis sensu auctt. partim non (Audinet-Serville,
　　1821)
　subpunctatus Stephens, 1828
　seladon (Schauberger, 1926)
rupicola (Sturm, 1818)
schaubergerianus (Puel, 1937)
　rufibarbis sensu auctt. non (Fabricius, 1792)
　brevicollis sensu auctt. partim non (Audinet-Serville,
　　1821)
subsinuatus Rey, 1886 [57]

ANISODACTYLUS Dejean, 1829
binotatus (Fabricius, 1787)
　spurcaticornis Dejean, 1829
nemorivagus (Duftschmid, 1812)
　atricornis (Stephens, 1835)
poeciloides (Stephens, 1828)
　pseudoaeneus sensu auctt. non Dejean, 1829

DIACHROMUS Erichson, 1837
germanus (Linnaeus, 1758)

SCYBALICUS Schaum, 1862
oblongiusculus (Dejean, 1829)

DICHEIROTRICHUS Jacquelin du Val, 1855
　DICHIROTRICHUS auctt. (misspelling)
gustavii Crotch, 1871
　gustavi auctt. (misspelling)
　pubescens (Paykull, 1790) non (Müller, O.F., 1776)
obsoletus (Dejean, 1829)

TRICHOCELLUS Ganglbauer, 1892
　BRADYCELLUS sensu Fowler, 1886 partim non
　　Erichson, 1837
cognatus (Gyllenhal, 1827)
placidus (Gyllenhal, 1827)

BRADYCELLUS Erichson, 1837
caucasicus (Chaudoir, 1846)
　collaris (Paykull, 1798) non (Herbst, 1784)
csikii Laczó, 1912
distinctus (Dejean, 1829)
harpalinus (Audinet-Serville, 1821)
ruficollis (Stephens, 1828)
　similis (Dejean 1829)
sharpi Joy, 1912
　distinctus sensu Fowler, 1886 non (Dejean, 1829)
verbasci (Duftschmid, 1812)

STENOLOPHUS Dejean, 1821 [58]
mixtus (Herbst, 1784)
　vespertinus (Panzer, 1796)
　ziegleri (Panzer, 1809)
skrimshiranus Stephens, 1828
teutonus (Schrank 1781)
　vaporariorum sensu (Fabricius, 1775) non
　　(Linnaeus, 1758)
　anglicus Schiødte, 1861-63

ACUPALPUS Latreille, 1829
brunnipes (Sturm, 1825)
   brunneipes auctt. (misspelling)
dubius Schilsky, 1888
   luridus sensu auctt. non Dejean, 1829
   luteatus sensu Joy, 1932 non (Duftschmid, 1812)
elegans (Dejean, 1829)
exiguus Dejean, 1829
flavicollis (Sturm, 1825)
   luridus Dejean, 1829
maculatus (Schaum, 1860) [59]
meridianus (Linnaeus, 1761)
parvulus (Sturm, 1825)
   dorsalis (Fabricius, 1787) non (Pontoppidan, 1763)
   derelictus (Dawson, 1854)

ANTHRACUS Motschulsky, 1850
   ACUPALPUS sensu auctt. partim non Latreille, 1829
consputus (Duftschmid, 1812)

Tribe CHLAENIINI Brullé, 1834

CHLAENIUS Bonelli, 1810
   CHLAENIELLUS Reitter, 1908
nigricornis (Fabricius, 1787)
   melanocornis Dejean, 1826
nitidulus (Schrank, 1781)
   schrankii (Duftschmid, 1812)
tristis (Schaller, 1783)
   holosericeus (Fabricius, 1787)
vestitus (Paykull, 1790)

CALLISTUS Bonelli, 1810
lunatus (Fabricius, 1775)

Tribe OODINI LaFerté-Sénectère, 1851

OODES Bonelli, 1810
helopioides (Fabricius, 1792)

Tribe LICININI Bonelli, 1810

LICINUS Latreille, 1802
depressus (Paykull, 1790)
punctatulus (Fabricius, 1792)
   punctulatus Kloet & Hincks, 1945 (error)
   silphoides sensu (Fabricius, 1792) non (Rossi, 1790)

BADISTER Clairville, 1806

Subgenus BADISTER Clairville, 1806
bullatus (Schrank, 1798)
   bipustulatus (Fabricius, 1792) non (Fabricius, 1775)
meridionalis Puel, 1925
   kineli Makolski, 1952
unipustulatus Bonelli, 1813

Subgenus TRIMORPHUS Stephens, 1828
sodalis (Duftschmid, 1812)
   humeralis Bonelli, 1813

Subgenus BAUDIA Ragusa, 1884
collaris Motschulsky, 1844
   anomalus (Perris, 1866)

   striatulus Hansen, 1944
dilatatus Chaudoir, 1837
peltatus (Panzer, 1797)

Tribe PANAGAEINI Bonelli, 1810

PANAGAEUS Latreille, 1802
bipustulatus (Fabricius, 1775)
   quadripustulatus Sturm, 1815
cruxmajor (Linnaeus, 1758)

Tribe PERIGONINI Horn, 1881

PERIGONA Laporte, 1835
nigriceps (Dejean, 1831)

Tribe MASOREINI Chaudoir, 1871

MASOREUS Dejean, 1821
wetterhallii (Gyllenhal, 1813)
   wetterhalli auctt. (misspelling)
   laticollis (Sturm, 1825)

Tribe LEBIINI Bonelli, 1810

LEBIA Latreille, 1802

Subgenus LAMPRIAS Bonelli, 1810
chlorocephala (Hoffmann, J., 1803)
   chrysocephala (Motschulsky, 1864)
cyanocephala (Linnaeus, 1758)

Subgenus LEBIA Latreille, 1802
cruxminor (Linnaeus, 1758)
marginata (Geoffroy in Fourcroy, 1785)
   haemorrhoidalis (Fabricius, 1787)
scapularis (Geoffroy in Fourcroy, 1785)
   turcica (Fabricius, 1787)

SOMOTRICHUS Seidlitz, 1887
unifasciatus (Dejean, 1831) [60]

DEMETRIAS Bonelli, 1810

Subgenus RISOPHILUS Leach, 1815
   AETOPHORUS Schmidt-Göbel, 1846
imperialis (Germar, 1824)

Subgenus DEMETRIAS Bonelli, 1810
atricapillus (Linnaeus, 1758)
monostigma Samouelle, 1819
   unipunctatus (Germar, 1824)

CYMINDIS Latreille, 1806

Subgenus CYMINDIS Latreille, 1806
axillaris (Fabricius, 1794)

Subgenus TARSOSTINUS Motschulsky, 1864
macularis Mannerheim in Fischer von Waldheim, 1824 [61]

Subgenus TARULUS Bedel, 1906
vaporariorum (Linnaeus, 1758)
   basalis Gyllenhal, 1810

PARADROMIUS Fowler, 1887
  DROMIUS sensu auctt. partim non Bonelli, 1810
linearis (Olivier, 1795)
longiceps (Dejean, 1826)

DROMIUS Bonelli, 1810
agilis (Fabricius, 1787) [62]
  bimaculatus Dejean, 1825
angustus Brullé, 1834
meridionalis Dejean, 1825
  discus Puel, 1919
quadrimaculatus (Linnaeus, 1758)

CALODROMIUS Reitter, 1905
  DROMIUS sensu auctt. partim non Bonelli, 1810
spilotus (Illiger, 1798)
  quadrinotatus (Zenker in Panzer, 1800) non
    (Fabricius, 1798)

PHILORHIZUS Hope, 1838
  DROMIUS sensu auctt. partim non Bonelli, 1810
  DROMIOLUS Reitter, 1905
melanocephalus (Dejean, 1825)
notatus (Stephens, 1827)
  nigriventris (Thomson, C.G., 1857)
quadrisignatus (Dejean, 1825)
sigma (Rossi, 1790)
vectensis (Rye, 1873)
  insignis sensu auctt. non (Lucas, 1846)

MICROLESTES Schmidt-Göbel, 1846
  BLECHRUS Motschulsky, 1847
maurus (Sturm, 1827)
  glabratus sensu auctt. non (Duftschmid, 1812)
minutulus (Goeze, 1777) [63]
  glabratus (Duftschmid, 1812)

LIONYCHUS Wissmann, 1846
quadrillum (Duftschmid, 1812)
  bipunctatus (Heer, 1838)
  unicolor Schilsky, 1888

SYNTOMUS Hope, 1838
  METABLETUS Schmidt-Göbel, 1846
foveatus (Geoffroy in Fourcroy, 1785)
  foveola (Gyllenhal, 1810)
obscuroguttatus (Duftschmid, 1812)
  atratus (Dejean, 1825)
truncatellus (Linnaeus, 1761)

Tribe ODACANTHINI Laporte, 1834

ODACANTHA Paykull, 1798
  COLLIURIS sensu auctt. non De Geer, 1774
melanura (Linnaeus, 1767)

Tribe DRYPTINI Bonelli, 1810

DRYPTA Latreille, 1796
dentata (Rossi, 1790)
  emarginata (Olivier, 1790)

Tribe ZUPHIINI Bedel, 1895

POLISTICHUS Bonelli, 1810
  POLYSTICHUS auctt. (misspelling)
connexus (Geoffroy in Fourcroy, 1785)
  fasciolatus sensu auctt. non (Rossi, 1790)
  vittatus Brullé, 1834

Suborder POLYPHAGA Emery, 1886

Superfamily HYDROPHILOIDEA Latreille, 1802

8. Family HELOPHORIDAE Leach, 1815 [64]

Family author: G.N. Foster

HELOPHORUS Fabricius, 1775
  ELOPHORUS Fabricius, 1775 [65]

Subgenus EMPLEURUS Hope, 1838
nubilus Fabricius, 1776
  nubilis auctt. (misspelling)
porculus Bedel, 1881
  rugosus sensu auctt. Brit. partim non Olivier, 1795
rufipes (Bosc d'Antic, 1791)
  rugosus Olivier, 1795

Subgenus CYPHELOPHORUS Kuwert, 1884
tuberculatus Gyllenhal, 1808
  scaber LeConte, 1850

Subgenus TRICHELOPHORUS Kuwert, 1886
alternans Gené, 1836
  intermedius Mulsant, 1844

Subgenus HELOPHORUS Fabricius, 1775
aequalis Thomson, C.G., 1868
  aquaticus sensu auctt. Brit. ad 1965 partim non
    (Linnaeus, 1758) [66]
grandis Illiger, 1798
  aquaticus sensu auctt. Brit. partim, Strand, A., 1962
    non (Linnaeus, 1758)

Subgenus ATRACTHELOPHORUS Kuwert, 1886
arvernicus Mulsant, 1846
brevipalpis Bedel, 1881 [67]
  creticus Kiesenwetter, 1858
  bulbipalpis Kuwert, 1890

Subgenus RHOPALHELOPHORUS Kuwert, 1886
dorsalis (Marsham, 1802)
  quadrisignatus Bach, 1851
flavipes Fabricius, 1792
  viridicollis Stephens, 1829
  aeneipennis sensu auctt. non Thomson, C.G., 1854
  sphagnicola Hardy, 1871
  shetlandicus Kuwert, 1890
  phalleterus Sharp, 1916
fulgidicollis Motschulsky, 1860
  mulsanti Cox, 1874 ex Rye, 1866
  flavipes var. *mulsanti* sensu auctt. Brit. partim
  exasperatus Sharp, 1916
granularis (Linnaeus, 1761)
  brevicollis Thomson, C.G., 1868
  ytenensis Sharp, 1916
griseus Herbst, 1793
  affinis (Marsham, 1802) non (Thunberg, 1794)

23

semifulgens sensu d'Orchymont, 1937 non Rey, 1885
laticollis Thomson, C.G., 1854
    pallidulus Thomson, C.G., 1868
longitarsis Wollaston, 1864
    erichsoni Bach, 1866
    semifulgens Rey, 1885
    diffinis Sharp, 1916
minutus Fabricius, 1775
nanus Sturm, 1836
obscurus Mulsant, 1844
    aeneipennis Thomson, C.G., 1854
    walkeri Sharp, 1916
strigifrons Thomson, C.G., 1868
    championi Sharp, 1915

9. Family GEORISSIDAE Laporte, 1840 [68]

Family author: G.N. Foster

GEORISSUS Latreille, 1809
    GEORYSSUS auctt. (misspelling)
    CATHAMMISTES Illiger, 1807 [69]
crenulatus (Rossi, 1794)
    pygmaeus (Fabricius, 1798)

10. Family HYDROCHIDAE Thomson, C.G., 1859 [70]

Family author: G.N. Foster

HYDROCHUS Leach, 1817
angustatus Germar, 1824
brevis (Herbst, 1793)
crenatus (Fabricius, 1792)
    carinatus Germar, 1824 [71]
elongatus (Schaller, 1783)
ignicollis Motschulsky, 1860
    elongatus sensu auctt. Brit. partim non (Schaller, 1783)
megaphallus van Berge Henegouwen, 1988 [72]
nitidicollis Mulsant, 1844
    interruptus sensu auctt. Brit. non von Heyden, 1870

11. Family SPERCHEIDAE Erichson, 1837 [73]

Family author: G.N. Foster

SPERCHEUS Kugelann in Illiger, 1798
emarginatus (Schaller, 1783)

12. Family HYDROPHILIDAE Latreille, 1802 [74]

Family author: G.N. Foster

Subfamily HYDROPHILINAE Latreille, 1802

ANACAENA Thomson, C.G., 1859
bipustulata (Marsham, 1802)
globulus (Paykull, 1798)
limbata (Fabricius, 1792)

    ovata (Reiche, 1861)
lutescens (Stephens, 1829) [75]
    limbata sensu auctt. partim non (Fabricius, 1792)
    nitida (Heer, 1841)

PARACYMUS Thomson, C.G., 1867
aeneus (Germar, 1824)
scutellaris (Rosenhauer, 1856)
    aeneus sensu (Stephens, 1829) non (Germar, 1824)
    nigroaeneus (Sahlberg, J., 1875)

BEROSUS Leach, 1817 [76]

Subgenus BEROSUS Leach, 1817
affinis Brullé, 1835
luridus (Linnaeus, 1761)
signaticollis (Charpentier, 1825)

Subgenus ENOPLURUS Hope, 1838
fulvus Kuwert, 1888
    spinosus sensu auctt. partim non (von Steven in Schönherr, 1808)

CHAETARTHRIA Stephens, 1833
seminulum (Herbst, 1797)
    coccinelloides (Stephens, 1829)
simillima Vorst & Cuppen, 2003 [77]
    seminulum sensu auctt. partim non (Herbst, 1797)

CYMBIODYTA Bedel, 1881
marginellus (Fabricius, 1792)
    marginella auctt. (misspelling)
    margipallens (Marsham, 1802)
    ovalis (Thomson, C.G., 1853)

ENOCHRUS Thomson, C.G., 1859
    PHILYDRUS Solier, 1834 non Duftschmid, 1805
    PHILHYDRUS Brullé, 1835 non Brookes, 1828
affinis (Thunberg, 1794)
    minutus sensu auctt.. ?sensu (Fabricius, 1801) non (Fabricius, 1775, 1792)
bicolor (Fabricius, 1792)
    maritimus (Thomson, C.G., 1853)
coarctatus (Gredler, 1863)
    suturalis (Sharp, 1872)
fuscipennis (Thomson, C.G., 1884) [78]
    melanocephalus sensu auctt. Brit. partim non (Olivier, 1792)
halophilus (Bedel, 1878) [79]
    melanocephalus sensu auctt. Brit. partim non (Olivier, 1792)
melanocephalus (Olivier, 1792)
    bicolor sensu (Fowler, 1887) non (Fabricius, 1792)
nigritus (Sharp, 1872) [80]
    affinis sensu auctt. partim non (Thunberg, 1794)
    isotae Hebauer, 1981 [81]
ochropterus (Marsham, 1802)
    frontalis (Erichson, 1837)
    nigricans (Zetterstedt, 1838)
quadripunctatus (Herbst, 1797)
    melanocephalus sensu auctt. Brit. partim non (Olivier, 1795)
    ytenensis (Sharp, 1915)
testaceus (Fabricius, 1801)

HELOCHARES Mulsant, 1844
lividus (Forster, 1771)
   griseus (Fabricius, 1787)
obscurus (Müller, O.F., 1776) [82]
punctatus Sharp, 1869
   obscurus sensu auctt. partim non (Müller, O.F., 1776)

HYDROBIUS Leach, 1815
fuscipes (Linnaeus, 1758)
   chalconotus Stephens, 1829
   subrotundus Stephens, 1829
   picicrus Thomson, C.G., 1884

LIMNOXENUS Motschulsky, 1853
   HYDROBIUS sensu Fowler, 1887 partim non Leach, 1815
niger (Gmelin in Linnaeus, 1790) [83]
   oblongus (Herbst, 1797)

HYDROCHARA Berthold, 1827
   HYDROPHILUS sensu Leach, 1817 non Geoffroy, 1762
   HYDROCHARIS Hope, 1838 [40]
caraboides (Linnaeus, 1758)

HYDROPHILUS Geoffroy, 1762
   HYDROUS Linnaeus, 1775
piceus (Linnaeus, 1758)

LACCOBIUS Erichson, 1837 [84]
atratus Rottenberg, 1874
   scutellaris sensu auctt. Brit. non Motschulsky, 1855
   regularis sensu auctt. Brit. non Rey, 1885
bipunctatus (Fabricius, 1775)
   alutaceus Thomson, C.G., 1868
colon (Stephens, 1829)
   biguttatus Gerhardt, 1877
   bipunctatus sensu Fowler, 1887 non (Fabricius, 1775)
minutus (Linnaeus, 1758)
simulatrix d'Orchymont, 1932 [85]
   simulator auctt. (misspelling)
   sculptus sensu Gentili & Chiesa, 1976 partim non d'Orchymont, 1936
sinuatus Motschulsky, 1849
   oblongus Gorham, 1907
striatulus (Fabricius, 1801)
   nigriceps Thomson, C.G., 1853
   sinuatus sensu auctt. Brit. ante 1907 non Motschulsky, 1849
   purpurascens Newbery, 1908
ytenensis Sharp, 1910
   atrocephalus sensu auctt. Brit. non Reitter, 1872 [86]

Subfamily SPHAERIDIINAE Latreille, 1802

COELOSTOMA Brullé, 1835
   CYCLONOTUM Erichson, 1837
orbiculare (Fabricius, 1775)

DACTYLOSTERNUM Wollaston, 1854
abdominale (Fabricius, 1792) [87]

CERCYON Leach, 1817 [88]
   CERYCON Rey, 1886
   ERCYCON Rey, 1886
   PARALIOCERCYON Ganglbauer, 1904

Subgenus DICYRTOCERCYON Ganglbauer, 1904
ustulatus (Preyssler, 1790)
   haemorrhous (Gyllenhal, 1808)

Subgenus PARACYCREON d'Orchymont, 1942
laminatus Sharp, 1873

Subgenus CERCYON Leach, 1817
alpinus Vogt, 1969 [89]
bifenestratus Küster, 1851
convexiusculus Stephens, 1829
   lugubris sensu (Paykull, 1798) non (Olivier, 1790)
   intermixtus Sharp, 1918
depressus Stephens, 1829
granarius Erichson, 1837
haemorrhoidalis (Fabricius, 1775)
   flavipes (Fabricius, 1792)
   aquatilis Donisthorpe, 1932 [90]
impressus (Sturm, 1807)
   atomarius (Fabricius, 1775) non (Linnaeus, 1767)
   haemorrhoidalis sensu (Herbst, 1792) non (Fabricius, 1775)
lateralis (Marsham, 1802)
littoralis (Gyllenhal, 1808)
   litoralis auctt. (misspelling)
marinus Thomson, C.G., 1853
   aquaticus Laporte, 1840 non Stephens, 1829
melanocephalus (Linnaeus, 1758)
nigriceps (Marsham, 1802)
   atricapillus (Marsham, 1802) [91]
obsoletus (Gyllenhal, 1808)
   lugubris sensu (Olivier, 1790) non (Geoffroy in Fourcroy, 1785)
pygmaeus (Illiger, 1801)
quisquilius (Linnaeus, 1761)
   flavus (Marsham, 1802)
sternalis (Sharp, 1918)
   subsulcatus sensu auctt. non Rey, 1885
   pumilio (Sharp, 1918)
terminatus (Marsham, 1802)
tristis (Illiger, 1801)
   boletophagus (Marsham, 1802)
   minutus sensu auctt. Brit. non (Gyllenhal, 1808)
unipunctatus (Linnaeus, 1758)

Subgenus PARACERCYON Seidlitz, 1888
analis (Paykull, 1798)
   flavipes sensu (Thunberg, 1794) non (Fabricius, 1792)

MEGASTERNUM Mulsant, 1844
concinnum (Marsham, 1802)
   obscurum (Marsham, 1802) non (Fabricius, 1792)
   boletophagum sensu auctt. non (Marsham, 1802)
   minutum (Gyllenhal, 1808)

CRYPTOPLEURUM Mulsant, 1844
crenatum (Kugelann, 1794)
minutum (Fabricius, 1775)
   atomarium sensu (Olivier, 1790) non (Linnaeus, 1767)
subtile Sharp, 1884

SPHAERIDIUM Fabricius, 1775
bipustulatum Fabricius, 1781
    ?quadrimaculatum (Marsham, 1802)
    ?semistriatum Laporte, 1840
lunatum Fabricius, 1792
    scarabaeoides sensu auctt. Brit. partim non
    (Linnaeus, 1758)
marginatum Fabricius, 1787 [92]
    bipustulatum sensu auctt. partim non Fabricius,
    1781
scarabaeoides (Linnaeus, 1758)
    quadrimaculatum sensu auctt. Brit. partim non
    (Marsham, 1802)

13. Family SPHAERITIDAE Shuckard, 1839

Family author: R.G. Booth

SPHAERITES Duftschmid, 1805
glabratus (Fabricius, 1792)

14. Family HISTERIDAE Gyllenhal, 1808 [93]

Family author: A.G. Duff

Subfamily ABRAEINAE MacLeay, 1819

Tribe ABRAEINI MacLeay, 1819

ABRAEUS Leach, 1817
granulum Erichson, 1839
perpusillus (Marsham, 1802)
    globosus (Hoffmann, J., 1803)

Tribe PLEGADERINI Portevin, 1929

PLEGADERUS Erichson, 1834
dissectus Erichson, 1839
vulneratus (Panzer, 1796)

Tribe ACRITINI Wenzel, 1944

ACRITUS LeConte, 1853
Subgenus ACRITUS LeConte, 1853
nigricornis (Hoffmann, J., 1803)
    minutus (Paykull, 1811) non (Herbst, 1792)
Subgenus PYCNACRITUS Casey, 1916
homoeopathicus Wollaston, 1857
    rhenanus Fuss, 1868

AELETES Horn, 1873
atomarius (Aubé, 1842)

HALACRITUS Schmidt, J., 1893
    ACRITUS sensu Fowler, 1889 partim non LeConte,
    1853
punctum (Aubé, 1843)

Tribe TERETRIINI Bickhardt, 1914

TERETRIUS Erichson, 1834
fabricii Mazur, 1972
    picipes sensu (Fabricius, 1792) non (Olivier, 1789)

Subfamily SAPRININAE Blanchard, 1845

SAPRINUS Erichson, 1834
aeneus (Fabricius, 1775)
immundus (Gyllenhal, 1827)
planiusculus Motschulsky, 1849
    cuspidatus Ihssen, 1949
semistriatus (Scriba, 1790)
    nitidulus (Fabricius, 1801)
subnitescens Bickhardt, 1909
virescens (Paykull, 1798)

HYPOCACCUS Thomson, C.G., 1867
    SAPRINUS sensu Fowler, 1889 partim non
    Erichson, 1834
Subgenus HYPOCACCUS Thomson, C.G., 1867
crassipes (Erichson, 1834) [94]
metallicus (Herbst, 1792)
rugiceps (Duftschmid, 1805)
    quadristriatus sensu (Hoffmann, J., 1803) non
    (Thunberg, 1794)
rugifrons (Paykull, 1798)
Subgenus BAECKMANNIOLUS Reichardt, 1926
    PACHYLOPUS sensu auctt. non Erichson, 1834
dimidiatus (Illiger, 1807)
ssp. *maritimus* (Stephens, 1830)

GNATHONCUS Jacquelin du Val, 1857
buyssoni Auzat, 1917
communis (Marseul, 1862)
    schmidti Reitter, 1894
    nidicola Joy, 1907
nannetensis (Marseul, 1862)
rotundatus (Kugelann, 1792)
    nanus (Scriba, 1790) non (Piller & Mitterpacher,
    1783)
    punctulatus Thomson, C.G., 1862

MYRMETES Marseul, 1862
paykulli Kanaar, 1979
    piceus (Paykull, 1809) non (Marsham, 1802)

Subfamily DENDROPHILINAE Reitter, 1909

Tribe DENDROPHILINI Reitter, 1909

DENDROPHILUS Leach, 1817
punctatus (Herbst, 1792)
pygmaeus (Linnaeus, 1758)
xavieri Marseul, 1873

KISSISTER Marseul, 1862
    CARCINOPS sensu Fowler, 1889 partim non
    Marseul, 1855
minimus (Laporte, 1840)

Tribe PAROMALINI Reitter, 1909

CARCINOPS Marseul, 1855
pumilio (Erichson, 1834)
    quatuordecimstriatus (Stephens, 1835)

PAROMALUS Erichson, 1834
    MICROLOMALUS Lewis, 1907
flavicornis (Herbst, 1792)
parallelepipedus (Herbst, 1792)
    parallelopipedus auctt. (misspelling)

Subfamily ONTHOPHILINAE MacLeay, 1819

ONTHOPHILUS Leach, 1817
punctatus (Müller, O.F., 1776)
    sulcatus (Geoffroy in Fourcroy, 1785)
    globulosus sensu Fowler, 1889 non (Olivier, 1789)
striatus (Forster, 1771)

Subfamily TRIBALINAE Bickhardt, 1914

EPIERUS Erichson, 1834
comptus Erichson, 1834 [95]

Subfamily HISTERINAE Gyllenhal, 1808

MARGARINOTUS Marseul, 1854
    HISTER sensu Fowler, 1889 partim non Linnaeus, 1758

Subgenus PTOMISTER Houlbert & Monnot, 1923
brunneus (Fabricius, 1775)
    impressus (Fabricius, 1798)
    cadaverinus (Hoffmann, J., 1803)
merdarius (Hoffmann, J., 1803)
striola (Sahlberg, C.R., 1819)
ssp. *succicola* (Thomson, C.G., 1862)

Subgenus STENISTER Reichardt, 1926
obscurus (Kugelann, 1792)
    stercorarius (Hoffmann, J., 1803)

Subgenus PARALISTER Bickhardt, 1917
neglectus (Germar, 1813)
purpurascens (Herbst, 1792)
ventralis (Marseul, 1854)
    carbonarius sensu auctt. Brit. non (Hoffmann, J., 1803)

Subgenus PROMETHISTER Kryzhanovskij, 1966
    GRAMMOSTETHUS sensu auctt. partim non Lewis, 1906
marginatus (Erichson, 1834)

HISTER Linnaeus, 1758
bissexstriatus Fabricius, 1801
illigeri Duftschmid, 1805
    sinuatus sensu Illiger, 1798 non Fabricius, 1792
quadrimaculatus Linnaeus, 1758
    gagates Illiger, 1807
    aethiops Heer, 1841
quadrinotatus Scriba, 1790
unicolor Linnaeus, 1758

ATHOLUS Thomson, C.G., 1859
    HISTER sensu Fowler, 1889 partim non Linnaeus, 1758
    PERANUS Lewis, 1906
bimaculatus (Linnaeus, 1758)
duodecimstriatus (Schrank, 1781)

HOLOLEPTA Paykull, 1811
plana (Sulzer, 1776) [96]

Subfamily HAETERIINAE Marseul, 1857

HAETERIUS Dejean, 1833
    HETAERIUS Erichson, 1834
ferrugineus (Olivier, 1789)

Superfamily STAPHYLINOIDEA Latreille, 1802

15. Family HYDRAENIDAE Mulsant, 1844

Family author: G.N. Foster

Subfamily HYDRAENINAE Mulsant, 1844

HYDRAENA Kugelann, 1794
britteni Joy, 1907
flavipes Sturm, 1836
    minutissima sensu Stephens, 1829 non (Weber & Mohr, 1804) [97]
    ?*atricapilla* (Waterhouse, G.R., 1833)
gracilis Germar, 1824
    elongata Curtis, 1830
nigrita Germar, 1824
palustris Erichson, 1837
pulchella Germar, 1824
pygmaea Waterhouse, G.R., 1833
riparia Kugelann, 1794
rufipes Curtis, 1830
    angustata sensu auctt. Brit. non Sturm, 1836
    longior Rey, 1885
testacea Curtis, 1830

LIMNEBIUS Leach, 1815
    LIMNOBIUS auctt. (misspelling)
aluta Bedel, 1881
    picinus sensu auctt. Brit. partim non (Marsham, 1802)
    minutissimus sensu Balfour-Browne, J., 1938 non (Germar, 1824)
crinifer Rey, 1885 [98]
nitidus (Marsham, 1802)
    picinus (Marsham, 1802)
papposus Mulsant, 1844
truncatellus (Thunberg, 1794)

Subfamily OCHTHEBIINAE Thomson, C.G., 1859

ENICOCERUS Stephens, 1829 [99]
    HENICOCERUS Agassiz, 1846
    OCHTHEBIUS sensu auctt. partim non Leach, 1815
exsculptus (Germar, 1824)
    viridiaeneus Stephens, 1829

gibsoni Curtis, 1830
tristis Curtis, 1830
sulcicollis Sturm, 1836
lividipes Fairmaire & Laboulbène, 1854

OCHTHEBIUS Leach, 1815

Subgenus ASIOBATES Thomson, C.G., 1859
auriculatus Rey, 1886
bicolon Germar, 1824
   rufimarginatus Stephens, 1829
dilatatus Stephens, 1829
   bicolon sensu Stephens, 1829 non Germar, 1824
   impressicollis Laporte, 1840
   kaninensis sensu Hellén, 1946 non Poppius, 1909

Subgenus HOMALOCHTHEBIUS Kuwert, 1887
aeneus Stephens, 1835
lejolisii Mulsant & Rey, 1861
   lejolisi auctt. (misspelling)
   subinteger sensu auctt. non Mulsant & Rey, 1861
minimus (Fabricius, 1792)
   pygmaeus sensu auctt. Brit. non (Fabricius, 1792)
   impressus (Marsham, 1802)

Subgenus HYMENODES Mulsant, 1844 [100]
nanus Stephens, 1829
   aeratus Stephens, 1829
nilssoni Hebauer, 1986 [101]
poweri Rye, 1869
   metallescens sensu Balfour-Browne, J., 1958 non
      Rosenhauer, 1847
punctatus Stephens, 1829

Subgenus OCHTHEBIUS Leach, 1815 [102]
lenensis Poppius, 1907
marinus (Paykull, 1798)
pusillus Stephens, 1835
   margipallens (Latreille, 1807) non (Marsham, 1802)
   viridis sensu Fowler & Donisthorpe, 1913 non
      Peyron, 1858
viridis Peyron, 1858
   margipallens sensu Fowler & Donisthorpe, 1913
      non Latreille, 1807
ssp. *viridis* Peyron, 1858
ssp. *fallaciosus* Ganglbauer, 1901

AULACOCHTHEBIUS Kuwert, 1887 [99]

   OCHTHEBIUS sensu auctt. partim non Leach, 1815
exaratus (Mulsant, 1844)

Family author: C. Johnson

Higher classification, nomenclature and synonymy
follow that in the Palaearctic catalogue (Johnson,
2004), except that the listing below follows what
seems to me to be a more natural system, instead
of an alphabetical one. When thorough
phylogenetic studies on the family have been
carried out, however, we may expect further
changes in higher classification. Regarding
synonymy, only important foreign names are
included below, but an attempt has been made to
include all names in the British literature,
especially misidentifications and misspellings.

Subfamily NOSSIDIINAE Sörensson, in prep.

NOSSIDIUM Erichson, 1845
pilosellum (Marsham, 1802)
   brunneum (Marsham, 1802)
   nitidulum (Marsham, 1802)

Subfamily PTILIINAE Erichson, 1845

Tribe PTENIDIINI Flach, 1889

PTENIDIUM Erichson, 1845 [103]
   ANISARTHRIA Stephens, 1830

Subgenus GRESSNERIUM Ganglbauer, 1899
gressneri Erichson, 1845

Subgenus MATTHEWSIUM Flach, 1889
laevigatum Erichson, 1845
turgidum Thomson, C.G., 1855

Subgenus WANKOWIZIUM Flach, 1889
brenskei Flach, 1887
intermedium Wankowicz, 1869
   vankoviezii Matthews, A., 1870
   wankowiezii auctt. (misspelling)

Subgenus PTENIDIUM Erichson, 1845
formicetorum Kraatz, 1851
   myrmecophilum (Motschulsky, 1845) non (Allibert,
      1844)
   kraatzii Matthews, A., 1872
   kraatzi Matthews, A., 1872 (misspelling)
   nigrifrons Britten, 1926 [104]
fuscicorne Erichson, 1845
   picipes Matthews, A., 1860
   rugosum Britten, 1910
longicorne Fuss, 1868
   brisoutii Matthews, A., 1872
   brisouti auctt. (misspelling)
punctatum (Gyllenhal, 1827)
pusillum (Gyllenhal, 1808) [103]
   melas (Marsham, 1802)
   perpusillum (Marsham, 1802)
   nitidum (Stephens, 1830)
   punctulum (Stephens, 1830)
   apicale Erichson, 1845
   terminale Haldeman, 1848
   atamaroides Motschulsky, 1869
   atomaroides Motschulsky, 1869 (misspelling)

evanescens sensu Matthews, A., 1872 ? (Marsham, 1802)

Subgenus GILLMEISTERIUM Flach, 1889
nitidum (Heer, 1841)
    minutissimum (Stephens, 1830)
    laevigatum sensu Matthews, A., 1872 non Erichson, 1845

Tribe PTILIINI Erichson, 1845

EURYPTILIUM Matthews, A., 1872
gillmeisteri Flach, 1889 [105]
saxonicum (Gillmeister, 1845)
    marginatum sensu auctt. non Aubé, 1850

PTILIOLA Haldeman, 1848
    NANOPTILIUM Flach, 1889 [106]
brevicollis (Matthews, A., 1860)
kunzei (Heer, 1841)
    nana (Stephens, 1830)
    trisulcata (Stephens, 1830)
    rugulosa (Allibert, 1844)

PTILIOLUM Flach, 1888

Subgenus PTILIOLUM Flach, 1888
fuscum (Erichson, 1845)
    spencei sensu auctt. Brit. partim non (Allibert, 1844)
    flachi (Reitter, 1909)
marginatum (Aubé, 1850)
    spencei sensu auctt. Brit. partim non (Allibert, 1844)
    lederi (Flach, 1888)
sahlbergi (Flach, 1888)
spencei (Allibert, 1844)
    angustatum (Erichson, 1845)
    oblongum (Gillmeister, 1845)
    foersteri (Matthews, A., 1872)

Subgenus EUPTILIUM Flach, 1889
caledonicum (Sharp, 1871)
    croaticum (Matthews, A., 1872)
schwarzi (Flach, 1887)
    asperum (Britten, 1917)

ACTIDIUM Matthews, A., 1868
aterrimum (Motschulsky, 1845)
    concolor (Sharp, 1867)
coarctatum (Haliday, 1855)

OLIGELLA Motschulsky, 1869
foveolata (Allibert, 1844)
    ?minima (Herbst, 1793)
    excavata (Erichson, 1845)
    clandestina (Haliday, 1855)
insignis (Matthews, A., 1861)
intermedia Besuchet, 1971

PTILIUM Gyllenhal, 1827 [107]
affine Erichson, 1845
    incognitum Matthews, A., 1889
caesum Erichson, 1845 [108]
exaratum (Allibert, 1844)
    canaliculatum Erichson, 1845
horioni Rosskothen, 1934
    exaratum sensu auctt. Brit. partim non Allibert, 1844

myrmecophilum (Allibert, 1844)
    inquilinum Erichson, 1845

MILLIDIUM Motschulsky, 1855 [109]
    PTILIUM sensu auctt. partim non Gyllenhal, 1827
minutissimum (Ljungh, 1804)
    trisulcatum (Aubé, 1833)

MICROPTILIUM Matthews, A., 1872 [110]
palustre Kuntzen, 1914 [111]
pulchellum (Allibert, 1844)

Tribe PTINELLINI Reitter, 1906 (1891)

MICRIDIUM Motschulsky, 1869 [112]
halidaii (Matthews, A., 1868)
    halidayi auctt. (misspelling)

PTINELLA Motschulsky, 1844
    PLITIUM Besuchet, 1971
aptera (Guérin-Méneville, 1839)
    tenella sensu Matthews, A., 1872 non Erichson, 1845
    angustula sensu Matthews, A., 1872 non Gillmeister, 1845
    tenuis Csiki, 1911
britannica Matthews, A., 1858
    brittannica auctt. (misspelling)
cavelli (Broun, 1893)
denticollis (Fairmaire, 1858)
    maria Matthews, A. 1862
    mariae auctt. (misspelling)
errabunda Johnson, 1975
limbata (Heer, 1841)
    testacea (Heer, 1841)
    proteus Matthews, A., 1862
simsoni (Matthews, A., 1878)
    subvariolosa (Britten, 1932) [113]
    flaviventris sensu (Donisthorpe, 1939) non (Motschulsky, 1869)
taylorae Johnson, 1977

PTERYX Matthews, A., 1858
suturalis (Heer, 1841)
    mutabilis Matthews, A., 1858

Subfamily ACROTRICHINAE Reitter, 1909 (1856)

Tribe NEPHANINI Portevin, 1929

NEPHANES Thomson, C.G., 1859
    TITAN Matthews, A., 1858 [114]
    ELACHYS Matthews, A., 1860
titan (Newman, 1834)
    abbreviatella (Heer, 1841)
    curta (Allibert, 1844)

SMICRUS Matthews, A., 1872
    MICRUS sensu Matthews, A., 1858 non Motschulsky, 1848
filicornis (Fairmaire & Laboulbène, 1855)

BAEOCRARA Thomson, C.G., 1859
BOEOCRARA auctt. (misspelling)
variolosa (Mulsant & Rey, 1861)
litoralis (Thomson, C.G., 1855) non (Motschulsky, 1845)
thomsoni (Sharp, 1866)

Tribe ACROTRICHINI Reitter, 1909 (1856)

ACTINOPTERYX Matthews, A., 1872
fucicola (Allibert, 1844)
mollis (Haliday, 1855)

ACROTRICHIS Motschulsky, 1848 [115]
ACRATRICHIS auctt. (misspelling)
TRICHOPTERYX Kirby, 1826 non Hübner, 1825

Subgenus ACROTRICHIS Motschulsky, 1848
arnoldi Rosskothen, 1935
atomaria (De Geer, 1774)
minima (Marsham, 1802) non (Herbst, 1793)
minutissima sensu (Marsham, 1802) ? (Linnaeus, 1767)
minuta (Stephens, 1830)
convexa (Matthews, A., 1858)
convexiuscula sensu (Matthews, A., 1872) non Motschulsky, 1851 [116]
thoracica sensu (Matthews, A., 1872) non (Waltl, 1838)
seminitens (Matthews, A., 1877)
brevicornis sensu (Fowler, 1884) non (Motschulsky, 1868)
brevipennis (Erichson, 1845)
clavipes (Gillmeister, 1845)
kirbii (Matthews, A., 1865) [116]
carbonaria (Matthews, A., 1873) [116]
cephalotes (Allibert, 1844) [117]
chevrolatii sensu auctt. non (Allibert, 1844)
chevrolati auctt. (misspelling)
pygmaea (Erichson, 1845)
parallelogramma (Gillmeister, 1845)
cognata (Matthews, A., 1877)
platonoffi Renkonen, 1945
danica Sundt, 1958
atomaria sensu Joy, 1932 non (De Geer, 1774)
dispar (Matthews, A., 1865)
bovina sensu auctt. Brit. non (Motschulsky, 1845)
fascicularis (Herbst, 1793)
cantiana (Matthews, A., 1871) [116]
laetitiae (Matthews, A., 1873) [116]
laetitia auctt. (misspelling)
henrici (Matthews, A., 1872)
fratercula (Matthews, A., 1878)
insularis (Mäklin, 1852)
intermedia (Gillmeister, 1845) [118]
suffocata (Haliday, 1855) [116]
lata sensu (Matthews, A., 1865) non (Motschulsky, 1845)
josephi (Matthews, A., 1872) [119]
subcognata Johnson, 1975
lucidula Rosskothen, 1935
montandonii (Allibert, 1844) [120]
montandoni auctt. (misspelling) [116]
picicornis (Mannerheim, 1843)
rivularis (Allibert, 1844)
longicornis (Mannerheim, 1844)

similis (Gillmeister, 1845)
jansoni (Matthews, A., 1866) [116]
angusta (Matthews, A., 1889) [116]
norvegica Strand, A., 1941
parva Rosskothen, 1935
silvatica sensu Sundt, 1958 partim non Rosskothen, 1935
pumila (Erichson, 1845)
kirbii sensu (Matthews, A., 1865) partim [116]
fuscula (Matthews, A., 1871) [116]
guerinii sensu (Matthews, A., 1872) partim ? (Allibert, 1844) [116]
guerini auctt. (misspelling)
longicornis sensu Matthews, A., 1872 non Mannerheim, 1844
brevipennis sensu Joy, 1932 partim non (Erichson, 1845)
rosskotheni Sundt, 1971
fraterna Johnson, 1975
rugulosa Rosskothen, 1935
volans sensu (Matthews, A., 1878) non Motschulsky, 1845
sjobergi sensu Johnson, 1967 non Sundt, 1958 [121]
sjoebergi auctt. (misspelling)
sericans (Heer, 1841)
picicornis sensu auctt. non (Mannerheim, 1843)
depressa (Gillmeister, 1845)
bovina (Motschulsky, 1845)
brevis (Motschulsky, 1845) [116]
obscoena Wollaston, 1857
obscaena auctt. (misspelling)
ambigua (Matthews, A., 1865) [116]
chevrierii sensu (Matthews, A., 1866) non ? (Allibert, 1844)
waterhousii (Matthews, A., 1866) [116]
edithia (Matthews, A., 1871) [116]
longula (Matthews, A., 1871) [116]
guerinii sensu (Matthews, A., 1872) partim ? (Allibert, 1844) [116]
guerini auctt. (misspelling)
poweri (Matthews, A., 1872)
chevrolatii sensu Johnson, 2003 ? (Allibert, 1844) [116]

silvatica Rosskothen, 1935
sitkaensis (Motschulsky, 1845)
fratercula sensu auctt. non (Matthews, A., 1878)
strandi Sundt, 1958
laetitiae sensu (Fowler, 1879) non (Matthews, A., 1873) [116]
thoracica (Waltl, 1838)
attenuata (Gillmeister, 1845) [116]
anthracina (Matthews, A., 1865)
brevicornis Motschulsky, 1868

Subgenus CTENOPTERYX Flach, 1889
grandicollis (Mannerheim, 1844)
sanctaehelenae Johnson, 1972 [122]

17. Family LEIODIDAE Fleming, 1821

Family author: J. Cooter

Subfamily LEIODINAE Fleming, 1821 [123]

Tribe SOGDINI Lopatin, 1961

SOGDA Lopatin, 1961
HYDNOBIUS sensu auctt. partim non Schmidt, W.L.E., 1841
TRICHOHYDNOBIUS Vogt, 1961
suturalis (Zetterstedt, 1828)
perrisii (Fairmaire, 1855)
perrisi auctt. (misspelling)

HYDNOBIUS Schmidt, W.L.E., 1841
latifrons (Curtis, 1840)
strigosus Schmidt, W.L.E., 1841
punctatus (Sturm, 1807)
punctatissimus (Stephens, 1829)
spinipes (Gyllenhal, 1813)

TRIARTHRON Märkel, 1840
maerkelii Märkel, 1840
maerkeli auctt. (misspelling)

Tribe LEIODINI Fleming, 1821

LEIODES Latreille, 1796
LIODES Gemminger & Harold, 1868 (misspelling)
OOSPHAERULA Ganglbauer, 1899
TRICHOSPHAERULA Fleischer, 1904
badia (Sturm, 1807)
similata (Rye, 1870)
calcarata (Erichson, 1845)
polita (Marsham, 1802) non (Fuessly, 1775)
?maritima Stephens, 1832
nigrescens Fleischer, 1906
ciliaris (Schmidt, W.L.E., 1841)
cinnamomea (Panzer, 1793)
lycoperdi Stephens, 1829
stephensi Stephens, 1832
ferruginea (Fabricius, 1787)
?nigricollis Stephens, 1832
ovalis (Schmidt, W.L.E., 1841)
scita (Erichson, 1845)
flavescens (Schmidt, W.L.E., 1841)
stenocoryphe Joy, 1911
furva (Erichson, 1845)
gallica (Reitter, 1884)
brunnea sensu auctt. Brit. non (Sturm, 1807)
gyllenhalii Stephens, 1829
brunnea (Gyllenhal, 1810) non (Sturm, 1807)
parvula (Sahlberg, C.R., 1833)
parvula sensu Fowler, 1889 partim
litura Stephens, 1832
punctulata sensu auctt. non Gyllenhal, 1810
ornata (Fairmaire, 1855)
longipes (Schmidt, W.L.E., 1841)
curta (Fairmaire & Laboulbène, 1855)
curta var. *donisthorpei* Fleischer, 1911
lucens (Fairmaire, 1855)
picea Illiger sensu Stephens, 1829 non (Panzer, 1797)
oblonga sensu auctt. non Erichson, 1845

lunicollis (Rye, 1872)
macropus (Rye, 1873)
curvipes sensu Reitter, 1884 non (Schmidt, W.L.E., 1841)
nigrita (Schmidt, W.L.E., 1841)
scita sensu Reitter, 1884 non Erichson, 1845
obesa (Schmidt, W.L.E., 1841)
dubia sensu Joy, 1911 partim non (Kugelann, 1794)
oblonga (Erichson, 1845)
grandis (Fairmaire & Laboulbène, 1855)
anglica (Rye, 1873)
picea (Panzer, 1797)
rufipennis (Paykull, 1798)
thoracica Stephens, 1829
maxillosa Stephens, 1829
?testacea Stephens, 1829
dyllwynii Stephens, 1832
?suturalis Stephens, 1832
clavicornis (Rye, 1875)
davidiana (Joy, 1911)
algirica sensu Donisthorpe, 1911 non (Rye, 1875)
dubia sensu auctt. partim non (Kugelann, 1794)
dubia var. *subglobosa* sensu Joy, 1911 non Reitter, 1884
dubia var. *bicolor* sensu Joy, 1911 non (Schmidt, W.L.E., 1841)
rugosa Stephens, 1829
arenaria Stephens, 1832
silesiaca (Kraatz, 1852)
strigipennis Daffner, 1983
flavicornis sensu auctt. Brit. non (Brisout de Barneville, 1863) [124]
parvula sensu Fowler, 1889 partim non (Sahlberg, C.R., 1833)
triepkii (Schmidt, W.L.E., 1841)
triepkei auctt. (misspelling)
pallens sensu auctt. Brit. non (Sturm, 1807) [125]
rubiginosa sensu auctt. Brit. non (Schmidt, W.L.E., 1841)
curvipes (Schmidt, W.L.E., 1841)
rotundata sensu Donisthorpe, 1933 non (Erichson, 1845)

LIOCYRTUSA Daffner, 1982
CYRTUSA sensu auctt. non Erichson, 1842
minuta (Ahrens, 1812)
vittata (Curtis, 1840)
pauxilla (Schmidt, W.L.E., 1841)

Tribe PSEUDOLIODINI Portevin, 1926

AGARICOPHAGUS Schmidt, W.L.E., 1841
cephalotes Schmidt, W.L.E., 1841
conformis Erichson, 1845

COLENIS Erichson, 1845
immunda (Sturm, 1807)
dentipes (Gyllenhal, 1810)
aciculata (Stephens, 1829)

Tribe AGATHIDIINI Westwood, 1838

ANISOTOMA Panzer, 1797
LIODES Erichson, 1845 non von Heyden, 1826
castanea (Herbst, 1792)
glabra (Fabricius, 1792)

axillaris sensu Stephens, 1829 non Gyllenhal, 1810
humeralis (Fabricius, 1792)
armata Stephens, 1829
orbicularis (Herbst, 1792)
seminulum (Kugelann, 1794)

AMPHICYLLIS Erichson, 1845

globus (Fabricius, 1792)
ruficollis (Olivier, 1792)
ferrugineus (Sturm, 1807)

AGATHIDIUM Panzer, 1797

Subgenus CYPHOCEBLE Thomson, C.G., 1859
arcticum Thomson, C.G., 1862
rhinoceros Sharp, 1866
nigrinum Sturm, 1807
staphylaeum (Gyllenhal, 1810)
ferrugineum sensu Stephens, 1829 non Sturm, 1807
seminulum sensu Stephens, 1829 partim non (Linnaeus, 1758)

Subgenus NEOCEBLE des Gozis, 1886
confusum Brisout de Barneville, 1863
piceum Thomson, C.G., 1862 non Erichson, 1845
clypeatum Sharp, 1866
convexum Sharp, 1866
piceum Erichson, 1845 non Melsheimer, 1844
globosum Mulsant & Rey, 1861 non (Kugelann, 1794)
marginatum Sturm, 1807
seminulum sensu Stephens, 1829 partim non (Linnaeus, 1758)
mandibulare sensu Stephens, 1829 partim non Sturm, 1807
pumilum Hardy, 1852
nigripenne (Fabricius, 1792)
mandibulare sensu Stephens, 1829 partim non Sturm, 1807
rotundatum (Gyllenhal, 1827)
lycogalae Hardy, 1852
sphaerulum Reitter, 1898
reitteri Ganglbauer, 1899
varians Beck, 1817
mandibulare sensu Stephens, 1829 partim non Sturm, 1807

Subgenus AGATHIDIUM Panzer, 1797
atrum (Paykull, 1798)
atrum sensu Stephens, 1829 partim
globus sensu Stephens, 1829 partim non (Paykull, 1798)
rufipes Stephens, 1832
laevigatum Erichson, 1845
atrum sensu Stephens, 1829 partim
orbiculare sensu Stephens, 1829 non (Herbst, 1792)
nigrinum sensu Stephens, 1829 non Sturm, 1807
affine sensu Stephens, 1829
pisanum Brisout de Barneville, 1872
badium sensu auctt. Brit. non Erichson, 1845 [126]
seminulum (Linnaeus, 1758)

Subfamily COLONINAE Horn, 1880 (1859)

COLON Herbst, 1797
KOLON Herbst, 1797 [127]

Subgenus EURYCOLON Ganglbauer, 1899
latum Kraatz, 1850
rufescens Kraatz, 1850

Subgenus MYLOECHUS Latreille, 1807
angulare Erichson, 1837
appendiculatum (Sahlberg, C.R., 1822)
calcaratum sensu auctt. Brit. non Erichson, 1837
denticulatum Kraatz, 1850
microps Czwalina, 1881
brunneum (Latreille, 1807)
dentipes (Sahlberg, C.R., 1822)
zebei Kraatz, 1854
barnevillei sensu auctt. Brit. non Kraatz, 1850

Subgenus COLON Herbst, 1797
serripes (Sahlberg, C.R., 1822)
fusculum Erichson, 1837
puncticolle sensu auctt. Brit. partim non Kraatz, 1850
viennense Herbst, 1797
puncticolle sensu auctt. Brit. partim non Kraatz, 1850

Subfamily CHOLEVINAE Kirby, 1837

Tribe ANEMADINI Hatch, 1928

NEMADUS Thomson, C.G., 1867
colonoides (Kraatz, 1851)

Tribe CHOLEVINI Kirby, 1837

NARGUS Thomson, C.G., 1867

Subgenus NARGUS Thomson, C.G., 1867
velox (Spence, 1813)

Subgenus DEMOCHRUS Thomson, C.G., 1867
anisotomoides (Spence, 1813)
wilkinii (Spence, 1813)
wilkini auctt. (misspelling)

CHOLEVA Latreille, 1796

Subgenus CHOLEVOPSIS Jeannel, 1922
spadicea (Sturm, 1839)

Subgenus CHOLEVA Latreille, 1796
agilis (Illiger, 1798)
angustata (Fabricius, 1781)
cisteloides sensu auctt. Brit. partim non (Frölich, 1799)
cisteloides (Frölich, 1799)
elongata (Paykull, 1798) [128]
fagniezi Jeannel, 1922
angustata sensu auctt. Brit. partim non (Fabricius, 1781)
glauca Britten, 1918
jeanneli Britten, 1922
sturmi sensu auctt. Brit. non Brisout de Barneville, 1863
lederiana Reitter, 1902 [129]

agilis sensu auctt. Brit. ?partim non (Illiger, 1798)
  septentrionis Jeannel, 1923
oblonga Latreille, 1807
  intermedia (Kraatz, 1852)

SCIODREPOIDES Hatch, 1933
  SCIODREPA sensu auctt. non Thomson, C.G.,
  1859
fumatus (Spence, 1813)
watsoni (Spence, 1813)

CATOPS Paykull, 1798
borealis Krogerus, 1931 [130]
chrysomeloides (Panzer, 1798)
coracinus Kellner, 1846
fuliginosus Erichson, 1837
fuscus (Panzer, 1794)
grandicollis Erichson, 1837
kirbii (Spence, 1813)
  kirbyi auctt. (misspelling)
longulus Kellner, 1846
morio (Fabricius, 1787)
nigricans (Spence, 1813)
nigriclavis Gerhardt, 1900
nigrita Erichson, 1837
tristis (Panzer, 1793) [131]
  leachi (Spence, 1813)
  montivagus Heer, 1841

CATOPIDIUS Jeannel, 1922
depressus (Murray, 1856)

PARABATHYSCIA Jeannel, 1908
  BATHYSCIA sensu auctt. partim non Schiødte,
  1849
wollastoni (Janson, E.W., 1857)

Tribe PTOMAPHAGINI Jeannel, 1911

PTOMAPHAGUS Illiger, 1798
medius Rey, 1889
  sericeus sensu auctt. Brit. partim non (Fabricius,
  1792)
  sericatus sensu auctt. Brit. non Chaudoir, 1845
subvillosus (Goeze, 1777)
  sericeus (Fabricius, 1792)
varicornis (Rosenhauer, 1847)
  variicornis auctt. (misspelling)

Subfamily PLATYPSYLLINAE Ritsema, 1869

LEPTINUS Müller, P.W.J., 1817
testaceus Müller, P.W.J., 1817

18. Family SILPHIDAE Latreille, 1806

Family author: R.G. Booth

Subfamily SILPHINAE Latreille, 1806

NECRODES Leach, 1815
littoralis (Linnaeus, 1758)

THANATOPHILUS Leach, 1815
dispar (Herbst, 1793)
rugosus (Linnaeus, 1758)
sinuatus (Fabricius, 1775)

OICEOPTOMA Leach, 1815
  OECEOPTOMA auctt. (misspelling)
  THANATOPHILUS sensu Fowler, 1888 partim non
  Leach, 1815
thoracicum (Linnaeus, 1758)

ACLYPEA Reitter, 1884
  BLITOPHAGA Reitter, 1884
opaca (Linnaeus, 1758)
undata (Müller, O.F., 1776)
  reticulata (Fabricius, 1787) non (Linnaeus, 1767)

DENDROXENA Motschulsky, 1858
  XYLODREPA Thomson, C.G., 1859
quadrimaculata (Scopoli, 1772)
  quadripunctata sensu auctt. non (Linnaeus, 1758)

SILPHA Linnaeus, 1758
  PHOSPHUGA Leach, 1817
  ABLATTARIA Reitter, 1884
atrata Linnaeus, 1758
  paedemontana Fabricius, 1775
  brunnea Herbst, 1793
  ssp. **atrata** Linnaeus, 1758
  ssp. **subrotundata** (Leach, 1817)
carinata Herbst, 1783
  griesbachiana Stephens, 1830
laevigata Fabricius, 1775
obscura Linnaeus, 1758
tristis Illiger, 1798
tyrolensis Laicharting, 1781
  nigrita Creutzer, 1799

Subfamily NICROPHORINAE Kirby, 1837

NICROPHORUS Fabricius, 1775
  NECROPHORUS auctt. (misspelling)
germanicus (Linnaeus, 1758)
humator (Gleditsch, 1767)
interruptus Stephens, 1830
  fossor Erichson, 1837
investigator Zetterstedt, 1824
  ruspator Erichson, 1837
vespillo (Linnaeus, 1758)
vespilloides Herbst, 1783
  mortuorum Fabricius, 1792
vestigator Herschel, 1807

19. Family STAPHYLINIDAE Latreille, 1802

Family author: A.G. Duff

Subfamily OMALIINAE MacLeay, 1825 [132]

Tribe ANTHOPHAGINI Thomson, C.G., 1859

ACIDOTA Stephens, 1829
crenata (Fabricius, 1792)

cruentata Mannerheim, 1830

ANTHOBIUM Leach, 1819 [133]
    LATHRIMAEUM Erichson, 1839
atrocephalum (Gyllenhal, 1827)
unicolor (Marsham, 1802)

ANTHOPHAGUS Gravenhorst, 1802

Subgenus ANTHOPHAGUS Gravenhorst, 1802
alpinus (Paykull, 1790)

Subgenus PHAGANTHUS Mulsant & Rey, 1880
caraboides (Linnaeus, 1758)

ARPEDIUM Erichson, 1939
    EUCNECOSUM Reitter, 1909 [134]
brachypterum (Gravenhorst, 1802)

DELIPHRUM Erichson, 1839
tectum (Paykull, 1789)

GEODROMICUS Redtenbacher, 1857 [135]
    PSEPHIDONUS Gistel, 1856
longipes (Mannerheim, 1830)
    globulicollis sensu auctt. Brit. non (Mannerheim, 1830)
nigrita (Müller, P.W.J., 1821)
    plagiatus sensu auctt. non (Fabricius, 1798)

LESTEVA Latreille, 1796 [136]
hanseni Lohse, 1953
    fontinalis sensu auctt. non Kiesenwetter, 1850
longoelytrata (Goeze, 1777)
monticola Kiesenwetter, 1847
pubescens Mannerheim, 1830
    luctuosa sensu auctt. Brit. non Fauvel, 1871
punctata Erichson, 1839 [137]
    villosus (Waltl, 1838)
sicula Erichson, 1840
ssp. *heeri* Fauvel, 1871 [138]

OLOPHRUM Erichson, 1839
assimile (Paykull, 1800)
consimile (Gyllenhal, 1810)
fuscum (Gravenhorst, 1806)
    nicholsoni Donisthorpe, 1910
piceum (Gyllenhal, 1810)

OROCHARES Kraatz, 1857
angustatus (Erichson, 1840)

PHILORINUM Kraatz, 1858
sordidum (Stephens, 1834)

PHYLLODREPOIDEA Ganglbauer, 1895
crenata Ganglbauer, 1895

Tribe CORYPHIINI Jakobson, 1908

CORYPHIUM Stephens, 1834
angusticolle Stephens, 1834

EUDECTUS Redtenbacher, 1857
whitei Sharp, 1871
    giraudi sensu auctt. Brit. non Redtenbacher, 1857

Tribe EUSPHALERINI Hatch, 1957

EUSPHALERUM Kraatz, 1858
    ANTHOBIUM sensu auctt. non Leach, 1819
luteum (Marsham, 1802)
    ophthalmicum (Paykull, 1800) non (Scopoli, 1763)
minutum (Fabricius, 1792)
primulae (Stephens, 1834)
sorbi (Gyllenhal, 1810) [137]
    testaceum (Gravenhorst, 1806)
sorbicola (Kangas, 1941)
    lapponicum sensu auctt. non (Mannerheim, 1830)
torquatum (Marsham, 1802)

Tribe HADROGNATHINI Portevin, 1929

HADROGNATHUS Schaum, 1852
longipalpis (Mulsant & Rey, 1851) [139]
    longipalpus auctt. (misspelling)

Tribe OMALIINI MacLeay, 1825

ACROLOCHA Thomson, C.G., 1858 [140]
    ELONIUM Leach, 1819
minuta (Olivier, 1795)
    striata (Gravenhorst, 1802) [141]
sulcula (Stephens, 1834)
    striata sensu auctt. Brit. non (Gravenhorst, 1802)

ACRULIA Thomson, C.G., 1858
inflata (Gyllenhal, 1813)

DROPEPHYLLA Mulsant & Rey, 1880 [142]
devillei Bernhauer, 1902
    grandiloqua Luze, 1910 [143]
    propinqua Bernhauer, 1943 [144]
gracilicornis (Fairmaire & Laboulbène, 1856)
heerii (Heer, 1841) [145]
ioptera (Stephens, 1832)
    linearis sensu auctt. Brit. non (Zetterstedt, 1828)
koltzei Jászay & Hlavac, 2006 [146]
vilis (Erichson, 1840)

HAPALARAEA Thomson, C.G., 1858
pygmaea (Paykull, 1800) [147]

HYPOPYCNA Mulsant & Rey, 1880
    HAPALARAEA sensu auctt. partim non Thomson, C.G., 1858
rufula (Erichson, 1840)

MICRALYMMA Westwood, 1838
marinum (Strøm, 1783)

OMALIUM Gravenhorst, 1802
allardi Fairmaire & Brisout de Barneville, 1859
caesum Gravenhorst, 1806
excavatum Stephens, 1834

exiguum Gyllenhal, 1810
italicum Bernhauer, 1902
    tricolor Mulsant & Rey, 1880 non Wollaston, 1865
laeviusculum Gyllenhal, 1827
laticolle Kraatz, 1858
    brevicolle sensu auctt. Brit. non Thomson, C.G.,
    1884
    foraminosum sensu auctt. Brit. non Mäklin, 1852
oxyacanthae Gravenhorst, 1806
riparium Thomson, C.G., 1857
rivulare (Paykull, 1789) [137]
    cursor (Müller, O.F., 1776)
rugatum Mulsant & Rey, 1880
rugulipenne Rye, 1864
septentrionis Thomson, C.G., 1856

PARAPHLOEOSTIBA Steel, 1960
gayndahensis (MacLeay, 1873) [148]

PHLOEONOMUS Heer, 1839
punctipennis Thomson, C.G., 1867
pusillus (Gravenhorst, 1806)

PHLOEOSTIBA Thomson, C.G., 1858 [149]
    PHLOEONOMUS sensu auctt. partim non Heer,
    1839
lapponica (Zetterstedt, 1838)
plana (Paykull, 1792) [137]
    flavipes (Linnaeus, 1758)

PHYLLODREPA Thomson, C.G., 1859 [150]
    HAPALARAEA sensu auctt. partim non Thomson,
    C.G., 1858
floralis (Paykull, 1789)
nigra (Gravenhorst, 1806)
puberula Bernhauer, 1903
salicis (Gyllenhal, 1810)

XYLODROMUS Heer, 1839
concinnus (Marsham, 1802) [151]
    brunnipennis sensu auctt. non (Stephens, 1834) [152]
depressus (Gravenhorst, 1802)
testaceus (Erichson, 1840) [137]
    pygmaeus (Gravenhorst, 1806)

XYLOSTIBA Ganglbauer, 1895 [153]
    PHLOEONOMUS sensu auctt. partim non Heer,
    1839
bosnica (Bernhauer, 1902) [154]
    monilicornis sensu auctt. Brit. partim non (Gyllenhal,
    1810)
monilicornis (Gyllenhal, 1810)

Subfamily PROTEININAE Erichson, 1839

MEGARTHRUS Stephens, 1829
bellevoyei (Saulcy, 1862)
    affinis Miller, 1853 non Stephens, 1834 [155]
denticollis (Beck, 1817)
depressus (Paykull, 1789) [156]
    sinuatocollis (Lacordaire, 1835) [157]
hemipterus (Illiger, 1794)
prosseni Schatzmayr, 1904
    depressus sensu auctt. non (Paykull, 1789)

METOPSIA Wollaston, 1854
clypeata (Müller, P.W.J., 1821)
    gallica (Koch, 1938)
    retusa (Stephens, 1834) [158]

PROTEINUS Latreille, 1796
atomarius Erichson, 1840 [159]
brachypterus (Fabricius, 1792)
crenulatus Pandelé, 1867
    limbatus sensu auctt. non Mäklin, 1852
laevigatus Hochhuth, 1871
    macropterus sensu auctt. non (Gravenhorst,
    1806) [160]
    serrifer Muona 1977
ovalis Stephens, 1834

Subfamily MICROPEPLINAE Leach, 1815

MICROPEPLUS Latreille, 1809
    ARRHENOPEPLUS Koch, 1937 [161]
caelatus Erichson, 1839
fulvus Erichson, 1840
porcatus (Paykull, 1789)
staphylinoides (Marsham, 1802)
tesserula Curtis, 1828

Subfamily PSELAPHINAE Latreille, 1802 [162]

Supertribe BATRISITAE Reitter, 1882

Tribe BATRISINI Reitter, 1882

BATRISODES Reitter, 1882
adnexus (Hampe, 1863)
    buqueti sensu auctt. Brit. non (Aubé, 1833)
delaporti (Aubé, 1833)
    laporti auctt. (misspelling)
venustus (Reichenbach, 1816)

Supertribe CLAVIGERITAE Leach, 1815

Tribe CLAVIGERIN Leach, 1815

CLAVIGER Preyssler, 1790
longicornis Müller, P.W.J., 1818
testaceus Preyssler, 1790

Supertribe EUPLECTITAE Streubel, 1839

Tribe EUPLECTINI Streubel, 1839

EUPLECTUS Leach, 1817 [163]
    EUPLECTOIDES Jeannel, 1954
bescidicus Reitter, 1881
    bohemicus Machulka, 1935
    brunneus sensu Pearce, 1957 non (Grimmer, 1841)
bonvouloiri Reitter, 1881 [164]
ssp. *rosae* Raffray, 1910 [165]
brunneus (Grimmer, 1841) [166]
decipiens Raffray, 1910
    duponti sensu Pearce, 1974 non Aubé, 1833
duponti Aubé, 1833
    aubeanus sensu Pearce, 1974 non Reitter, 1881

infirmus Raffray, 1910
   boeticus Jeannel, 1950
karstenii (Reichenbach, 1816)
kirbii Denny, 1825
   kirbyi auctt. (misspelling)
   nanus sensu auctt. partim non (Reichenbach, 1816)
mutator Fauvel, 1895
   fauveli Guillebeau, 1888 [167]
   falsus Bedel, 1906
   tomlini Joy, 1906
nanus (Reichenbach, 1816)
   reichenbachi Leach, 1817
   carolae Allen, 1940
piceus Motschulsky, 1835
   pearcei (Jeannel, 1954)
punctatus Mulsant, 1861 [168]
sanguineus Denny, 1825
signatus (Reichenbach, 1816)
tholini Guillebeau, 1888 [169]
   punctatus sensu auctt. partim non Mulsant, 1861

PLECTOPHLOEUS Reitter, 1891
erichsoni (Aubé, 1844) [170]
ssp. ***occidentalis*** Besuchet, 1969
nitidus (Fairmaire, 1857)

Tribe TRICHONYCHINI Reitter, 1882

Subtribe BIBLOPORINA Park, 1951

BIBLOPORUS Thomson, C.G., 1859
bicolor (Denny, 1825)
minutus Raffray, 1914
   bicolor sensu auctt. Brit. partim non (Denny, 1825)
   hoeglundi Palm, 1948
   sulcatus Jeannel, 1950

Subtribe PANAPHANTINA Jeannel, 1950

BIBLOPLECTUS Reitter, 1882 [171]
ambiguus (Reichenbach, 1816)
   championi Jeannel, 1950
delhermi Guillebeau, 1888
   pusillus sensu Jeannel, 1950 non (Denny, 1825)
   intermedius Pearce, 1951
minutissimus (Aubé, 1833)
pusillus (Denny, 1825)
   academicus Pearce, 1951
   therondi Besuchet, 1953
spinosus Raffray, 1914
   ambiguus sensu Pearce, 1951 non (Reichenbach, 1816)
   pusillus sensu Pearce, 1951 non (Denny, 1825)
tenebrosus (Reitter, 1880)
   margaretae Sharp, 1916

Subtribe TRICHONYCHINA Reitter, 1882

AMAURONYX Reitter, 1882
maerkelii (Aubé, 1844)

TRICHONYX Chaudoir, 1845
sulcicollis (Reichenbach, 1816)

Subtribe TRIMIINA Bowman, 1934

TRIMIUM Aubé, 1833
brevicorne (Reichenbach, 1816)

Supertribe GONIACERITAE Reitter, 1882 (1872)

Tribe BRACHYGLUTINI Raffray, 1904

BRACHYGLUTA Thomson, C.G., 1859
fossulata (Reichenbach, 1816)
haematica (Reichenbach, 1816)
helferi (Schmidt-Göbel, 1836)
klimschi Holdhaus, 1902 [172]
pandellei (Saulcy, 1876)
   cotus (Saulcy, 1876)
simplicior Raffray, 1904 [173]
   haematica sensu auctt. partim non (Reichenbach, 1816)
sinuata (Aubé, 1833) [173]
   haematica sensu auctt. partim non (Reichenbach, 1816)
waterhousei (Rye, 1869) [174]
   depressa sensu auctt. non (Aubé, 1833)
   simplex (Waterhouse, G.R., 1862) non (Motschulsky, 1851)

FAGNIEZIA Jeannel, 1950 [175]
   TRISSEMUS sensu auctt. partim non Jeannel, 1949
impressa (Panzer, 1803)

REICHENBACHIA Leach, 1826
juncorum (Leach, 1817)

RYBAXIS Saulcy, 1876
longicornis (Leach, 1817)
   laminata (Motschulsky, 1836) [176]

Tribe BYTHININI Raffray, 1890

BRYAXIS Kugelann, 1794
bulbifer (Reichenbach, 1816)
curtisii (Leach, 1817)
puncticollis (Denny, 1825)
   validus (Aubé, 1844)

BYTHINUS Leach, 1817
burrellii Denny, 1825
macropalpus Aubé, 1833

TYCHOBYTHINUS Ganglbauer, 1896
glabratus (Rye, 1870)
   ludyi (Reitter, 1882)

Tribe TYCHINI Raffray, 1904

TYCHUS Leach, 1817
niger (Paykull, 1800)
striola Guillebeau, 1888

Supertribe PSELAPHITAE Latreille, 1802

Tribe PSELAPHINI Latreille, 1802

PSELAPHAULAX Reitter, 1909
dresdensis (Herbst, 1792)

PSELAPHUS Herbst, 1792
heisei Herbst, 1792

Subfamily PHLOEOCHARINAE Erichson, 1839

PHLOEOCHARIS Mannerheim, 1830
subtilissima Mannerheim, 1830

Subfamily TACHYPORINAE MacLeay, 1825

Tribe MYCETOPORINI Thomson, C.G., 1859

BOLITOBIUS Leach, 1819
BRYOCHARIS Lacordaire, 1835
castaneus (Stephens, 1832)
analis sensu auctt. non (Fabricius, 1787) [177]
cingulatus (Mannerheim, 1830)

BRYOPHACIS Reitter, 1909 [178]
BRYOPORUS sensu auctt. partim non Kraatz, 1857
crassicornis (Mäklin, 1847)
maklini (Sahlberg, J., 1871)
rugipennis sensu auctt. Brit. non (Pandellé, 1869) [179]

BRYOPORUS Kraatz, 1857
cernuus (Gravenhorst, 1806)
merdarius (Olivier, 1795) non (Fabricius, 1775)

ISCHNOSOMA Stephens, 1829 [180]
MYCETOPORUS sensu auctt. partim non Mannerheim, 1830
longicorne (Mäklin, 1847)
splendidum (Gravenhorst, 1806)

LORDITHON Thomson, C.G., 1858
BOBITOBUS Tottenham, 1939
BOLITOBIUS sensu auctt. non Leach, 1819
exoletus (Erichson, 1839)
lunulatus (Linnaeus, 1761)
thoracicus (Fabricius, 1777)
trinotatus (Erichson, 1839)

MYCETOPORUS Mannerheim, 1830
ambiguus Luze, 1901 [181]
clavicornis sensu auctt. Brit. partim non (Stephens, 1832)
angularis Mulsant & Rey, 1853
baudueri Mulsant & Rey, 1875
hellieseni Strand, A., 1950
bimaculatus Lacordaire, 1835
clavicornis (Stephens, 1832)
forticornis sensu auctt. Brit. non Fauvel, 1875
despectus Strand, A., 1969
lepidus sensu auctt. partim non (Gravenhorst, 1802)

longulus sensu auctt. partim non Mannerheim, 1830
erichsonanus Façel, 1965
baudueri sensu auctt. partim non Mulsant & Rey, 1875
lepidus (Gravenhcrst, 1806)
brunneus sensu auctt. partim non (Paykull, 1789)
longulus Mannerheim, 1830
brunneus sensu auctt. partim non (Paykull, 1789)
monticola Fowler, 1888
nigricollis Stephens, 1835
splendens (Marsham, 1802) non (Fabricius, 1792)
piceolus Rey, 1883
baudueri sensu auctt. partim non Mulsant & Rey, 1875
punctus (Gravenhorst, 1806)
reichei (Pandellé, 1869) [181]
clavicornis sensu auctt. Brit. partim non (Stephens, 1832)
rufescens (Stephens, 1832)
lucidus Erichscn, 1839

PARABOLITOBIUS Li, Zhao & Sakai, 2000
BOLITOBIUS sensu auctt. partim non Leach, 1819
inclinans (Gravenhorst, 1806) [182]

Tribe TACHYPORINI MacLeay, 1825

CILEA Jacquelin du Val, 1856
LEUCOPARYFHUS Kraatz, 1857
silphoides (Linnaeus, 1767)

LAMPRINODES Luze, 1901
saginatus (Gravennorst, 1806)

SEPEDOPHILUS Gistel, 1856
CONOSOMA sensu auctt. non Kraatz, 1857
CONOSOMUS sensu auctt. non Motschulsky, 1857
bipunctatus (Gravenhorst, 1802)
constans (Fowler, 1888)
testaceus sensu auctt. partim non (Fabricius, 1792)
pubescens sensu auctt. partim non (Gravenhorst, 1802)
immaculatus (Stephens, 1832)
littoreus (Linnaeus, 1758)
lusitanicus Hammond, 1973
testaceus sensu auctt. partim non (Fabricius, 1792)
pubescens sensu auctt. partim non (Gravenhorst, 1802)
marshami (Stephens, 1832)
testaceus sensu auctt. partim non (Fabricius, 1792)
pubescens sensu auctt. partim non (Gravenhorst, 1802)
nigripennis (Stephens, 1832)
lividus (Erichson, 1839)
pedicularius (Gravenhorst, 1802)
testaceus (Fabricius, 1792) [183]
pubescens (Gravenhorst, 1802)

TACHINUS Gravenhorst, 1802 [184]
DRYMOPORUS Thomson, C.G., 1859 [185]
bipustulatus (Fabricius, 1792)
corticinus Gravenhorst, 1802
collaris Gravenhorst, 1802
elongatus Gyllenhal, 1810
flavolimbatus Pandellé, 1869

humeralis Gravenhorst, 1802
laticollis Gravenhorst, 1802
lignorum (Linnaeus, 1758)
    flavipes (Linnaeus, 1758)
marginellus (Fabricius, 1781)
pallipes (Gravenhorst, 1806)
    benicki Ullrich, 1975 [186]
proximus Kraatz, 1855
rufipennis Gyllenhal, 1810
rufipes (Linnaeus, 1758) [187]
    signatus Gravenhorst, 1802
scapularis Stephens, 1832
subterraneus (Linnaeus, 1758)

TACHYPORUS Gravenhorst, 1802
atriceps Stephens, 1832
chrysomelinus (Linnaeus, 1758)
    fasciatus Nicholson, 1911
dispar (Paykull, 1789) [188]
    chrysomelinus sensu auctt. partim non (Linnaeus, 1758)
formosus Matthews, A.H., 1838
hypnorum (Fabricius, 1775)
nitidulus (Fabricius, 1781) [189]
obtusus (Linnaeus, 1767)
    nitidicollis Stephens, 1832
pallidus Sharp, 1871
    scutellaris Rye, 1871 non Lacordaire, 1835
pusillus Gravenhorst, 1806
quadriscopulatus Pandellé, 1869
scitulus Erichson, 1839
    macropterus sensu auctt. non Stephens, 1832
solutus Erichson, 1839
tersus Erichson, 1839
transversalis Gravenhorst, 1806

Subfamily TRICHOPHYINAE Thomson, C.G., 1858

TRICHOPHYA Mannerheim, 1830
pilicornis (Gyllenhal, 1810)

Subfamily HABROCERINAE Mulsant & Rey, 1876

HABROCERUS Erichson, 1839
capillaricornis (Gravenhorst, 1806)

Subfamily ALEOCHARINAE Fleming, 1821 [190]

Tribe ACTOCHARINI Bernhauer & Schubert, 1911 [191]

ACTOCHARIS Sharp, 1870
readingii Sharp, 1870
    marina Fauvel, 1871

Tribe ALEOCHARINI Fleming, 1821

ALEOCHARA Gravenhorst, 1802 [192]

Subgenus ALEOCHARA Gravenhorst, 1802
brevipennis Gravenhorst, 1806
curtula (Goeze, 1777)
lata Gravenhorst, 1802

Subgenus BARYODMA Thomson, C.G., 1858
intricata Mannerheim, 1830

Subgenus CERANOTA Stephens, 1839
ruficornis Gravenhorst, 1802

Subgenus COPROCHARA Mulsant & Rey, 1874
bilineata Gyllenhal, 1810
binotata Kraatz, 1858 [193]
    verna sensu auctt. non Say, 1836 [194]
    pauxilla sensu auctt. non Mulsant & Rey, 1874
bipustulata (Linnaeus, 1761)
verna Say, 1836 [195]
    pauxilla Mulsant & Rey, 1874 [196]

Subgenus EMPLENOTA Casey, 1884
obscurella Gravenhorst, 1806 [197]
    algarum Fauvel, 1862
phycophila Allen, 1937 [198]

Subgenus POLYSTOMOTA Casey, 1906 [199]
    POLYSTOMA Stephens, 1835 non Zeder, 1800
grisea Kraatz, 1856
punctatella Motschulsky, 1858
    obscurella sensu auctt. non Gravenhorst, 1806

Subgenus RHEOCHARA Mulsant & Rey, 1875
spadicea (Erichson, 1837)

Subgenus XENOCHARA Mulsant & Rey, 1874
cuniculorum Kraatz, 1858
discipennis Mulsant & Rey, 1853
fumata Gravenhorst, 1802
funebris Wollaston, 1864
    diversa sensu auctt. partim non (Sahlberg, J., 1876)
    albovillosa Bernhauer, 1901 [200]
inconspicua Aubé, 1850
kamila Likovský, 1984 [201]
    diversa (Sahlberg, J., 1876) non Mulsant & Rey, 1853
lanuginosa Gravenhorst, 1802
lygaea Kraatz, 1862
maculata Brisout de Barneville, 1863
moerens Gyllenhal, 1827
moesta Gravenhorst, 1802
    crassiuscula Sahlberg, C.R., 1831
sanguinea (Linnaeus, 1758)
sparsa Heer, 1839
stichai Likovský, 1965
tristis Gravenhorst, 1806
villosa Mannerheim, 1830

TINOTUS Sharp, 1883 [202]
morion (Gravenhorst, 1802)

Tribe ATHETINI Casey, 1910

Subtribe ATHETINA Casey, 1910 [203]

ACROTONA Thomson, C.G., 1859 [204]
    ATHETA sensu auctt. partim non Thomson, C.G., 1858
    COLPODOTA Mulsant & Rey, 1873
aterrima (Gravenhorst, 1802)
benicki (Allen, 1940)
    pusilla (Brundin, 1952) [205]
convergens (Strand, 1958) [206]

obfuscata sensu Brundin (1952) non (Gravenhorst, 1802)
exigua (Erichson, 1837)
muscorum (Brisout de Barneville, 1860)
obfuscata (Gravenhorst, 1802)
parens (Mulsant & Rey, 1852)
parvula (Mannerheim, 1830)
parva sensu auctt. Brit. non (Sahlberg, C.R., 1831)
pseudotenera (Cameron, 1933) [207]
rassii Muona, 1993
pygmaea (Gravenhorst, 1802)
sylvicola (Kraatz, 1856)
silvicola auctt. (misspelling)
planipennis (Thomson, C.G., 1855) [32]
troglodytes (Motschulsky, 1858)
consanguinea (Eppelsheim, 1875) [208]

ADOTA Casey, 1909
ATHETA sensu auctt. partim non Thomson, C.G., 1858
PHYCONOMA Easton, 1971 [209]
maritima (Mannerheim, 1843)
immigrans (Easton, 1971) [210]

ALAOBIA Thomson, C.G., 1858 [211]
MYCOTA Mulsant & Rey, 1874
ATHETA 'LOHSE GROUP II' sensu auctt. partim
gagatina (Baudi, 1848)
hybrida (Sharp, 1869)
linderi (Brisout de Barneville, 1863) [212]
pallidicornis (Thomson, C.G., 1856)
humeralis (Kraatz, 1856)
scapularis (Sahlberg, C.R., 1831)
sodalis (Erichson, 1837)
subglabra (Sharp, 1869) [213]
sparreschneideri sensu auctt. Brit. non (Munster, 1923)
nondescripta (Ashe, 1957)
taxiceroides Munster, 1932 [214]
trinotata (Kraatz, 1856)

ALEVONOTA Thomson, C.G., 1858 [215]
ALEUONOTA Thomson, C.G., 1861
egregia (Rye, 1876)
gracilenta (Erichson, 1839)
rufotestacea (Kraatz, 1856)
atricapilla (Mulsant & Rey, 1852)
aurantiaca Fauvel, 1895 [216]

ALIANTA Thomson, C.G., 1858
incana (Erichson, 1837)

ALOCONOTA Thomson, C.G., 1858
ATHETA sensu auctt. partim non Thomson, C.G., 1858
GLOSSOLA Fowler, 1887

Subgenus ALOCONOTA Thomson, C.G., 1858
cambrica (Wollaston, 1855)
currax (Kraatz, 1856)
eichhoffi (Scriba, 1867)
aegyptiaca sensu (Ganglbauer, 1895) non Motschulsky, 1858
gregaria (Erichson, 1839)
insecta (Thomson, C.G., 1856)
mihoki Bernhauer, 1913 [217]

planifrons (Waterhouse, G.R., 1864)
debilicornis sensu auctt. Brit. non Erichson, 1839 [218]
subgrandis Brundin, 1954 [219]
sulcifrons (Stephens, 1832)

Subgenus DISOPORA Thomson, C.G., 1859 [220]
coulsoni (Last, 1952)
longicollis sensu auctt. Brit. non (Mulsant & Rey, 1852)
languida (Erichson, 1837) [221]
walshi (Williams, 1930)
longicollis (Mulsant & Rey, 1852) [222]
languida sensu auctt. Brit. non (Erichson, 1837)

AMIDOBIA Thomson, C.G., 1858
ATHETA sensu auctt. partim non Thomson, C.G., 1858
talpa (Heer, 1841)

AMISCHA Thomson, C.G., 1858
analis (Gravenhorst, 1802)
bifoveolata (Mannerheim, 1830)
cavifrons (Sharp, 1869) [223]
decipiens (Sharp, 1869)
forcipata Mulsant & Rey, 1873
nigrofusca (Stephens, 1829)
soror (Kraatz, 1856) [223]
similima (Sharp, 1869) [224]

ANOPLETA Mulsant & Rey, 1874 [225]
ATHETA sensu auctt. partim non Thomson, C.G., 1858
corvina (Thomson, C.G., 1856)
ellimani (Bernhauer, 1914) [226]
kochi (Roubal, 1937) [227]
verulamii Allen, 1994 [228]
puberula (Sharp, 1869)
luctuosa sensu auctt. non (Mulsant & Rey, 1853) [229]
picicornis (Mulsant & Rey, 1873)
soedermani (Bernhauer, 1931) [230]
liliputana sensu auctt. Brit. non (Brisout de Barneville, 1860)

ATHETA Thomson, C.G., 1858
MEGISTA Mulsant & Rey, 1874
?TETROPLA Mulsant & Rey, 1874 [231]
aeneicollis (Sharp, 1869)
pertyi sensu auctt. non (Heer, 1839)
aquatica (Thomson, C.G., 1852)
aquatilis (Thomson, C.G., 1867)
brunneipennis (Thomson, C.G., 1852)
valida (Kraatz, 1856)
castanoptera (Mannerheim, 1830)
ebenina (Mulsant & Rey, 1874)
graminicola (Gravenhorst, 1806)
heymesi Hubenthal 1913 [232]
hypnorum (Kiesenwetter, 1850)
incognita (Sharp, 1869)
laevicauda Sahlberg, J., 1876 [233]
triangulum (Kraatz, 1856)
xanthopus (Thomson, C.G., 1856)

'species provisionally assigned to *Atheta*' [234]
autumnalis (Erichson, 1839) [235]
basicornis (Mulsant & Rey, 1852)
boletophila (Thomson, C.G., 1856)

britanniae (Bernhauer & Scheerpeltz, 1926)
　repanda sensu auctt. non (Mulsant & Rey, 1874) [236]
　reperta (Sharp, 1913) non Casey, 1910
crassicornis (Fabricius, 1792)
　inoptata (Sharp, 1913)
diversa (Sharp, 1869)
divisa (Märkel, 1844)
euryptera (Stephens, 1832)
fungicola (Thomson, C.G., 1852)
　nitidicollis (Fairmaire & Laboulbène, 1856)
harwoodi (Williams, 1930)
intermedia (Thomson, C.G., 1852) [237]
liturata (Stephens, 1832)
nidicola (Johansen, 1914)
nigritula (Gravenhorst, 1802) [238]
oblita (Erichson, 1839)
paracrassicornis Brundin, 1954
pilicornis (Thomson, C.G., 1852)
procera (Kraatz, 1856)
ravilla (Erichson, 1839)
　angusticollis (Thomson, C.G., 1856)
strandiella (Brundin, 1954)
vaga (Heer, 1839)
　nigricornis (Thomson, C.G., 1852) non (Stephens, 1832) [239]

BADURA Mulsant & Rey, 1873 [240]
　ATHETA sensu auctt. partim non Thomson, C.G., 1858
　DIMETROTA sensu auctt. partim non Mulsant & Rey, 1873
macrocera (Thomson, C.G., 1856)
puncticollis (Benick, 1938)

BESSOBIA Thomson, C.G., 1858 [241]
　ATHETA sensu auctt. partim non Thomson, C.G., 1858
excellens (Kraatz, 1856)
fungivora (Thomson, C.G., 1867)
monticola (Thomson, C.G., 1852)
occulta (Erichson, 1837)

BOREOPHILIA Benick, 1973
　ATHETA sensu auctt. partim non Thomson, C.G., 1858
eremita (Rye, 1866) [242]
　islandica sensu auctt. Brit. non (Kraatz, 1857)
　hercynica sensu auctt. Brit. non (Renkonen, 1936)

BRUNDINIA Tottenham, 1949
　ATHETA sensu auctt. partim non Thomson, C.G., 1858
　METAXYA Mulsant & Rey, 1873 non Walker, 1856
　ACTOPHYLLA sensu auctt. partim non Bernhauer, 1909
marina (Mulsant & Rey, 1853) [243]
　granosa (Hochhuth, 1849)
meridionalis (Mulsant & Rey, 1853)

CADAVEROTA Yosii & Sawada, 1976 [244]
　ATHETA 'LOHSE GROUP II' auctt. partim
cadaverina (Brisout de Barneville, 1860)
hansseni (Strand, 1943) [245]

CALLICERUS Gravenhorst, 1802
　SEMIRIS Heer, 1839 [246]
obscurus Gravenhorst, 1802
rigidicornis (Erichson, 1839)

CERITAXA Mulsant & Rey, 1873 [247]
　ATHETA sensu auctt. partim non Thomson, C.G., 1858
dilaticornis (Kraatz, 1856)
　spissata (Mulsant & Rey, 1873) [248]
pervagata (Benick, 1974)
　dilaticornis sensu auctt. non (Kraatz, 1856)
testaceipes (Heer, 1839)
　brevicollis (Baudi, 1848)

CHAETIDA Mulsant & Rey, 1874 [249]
　ATHETA sensu auctt. partim non Thomson, C.G., 1858
longicornis (Gravenhorst, 1802)

COPROTHASSA Thomson, C.G., 1859 [250]
　ATHETA sensu auctt. partim non Thomson, C.G., 1858
melanaria (Mannerheim, 1830)
　testudinea (Erichson, 1839)

DADOBIA Thomson, C.G., 1858
immersa (Erichson, 1837)

DALOTIA Casey, 1910
　ATHETA 'LOHSE GROUP I' auctt. partim
coriaria (Kraatz, 1856) [251]

DATOMICRA Mulsant & Rey, 1874 [252]
　ATHETA sensu auctt. partim non Thomson, C.G., 1858
　DATOSTIBA Sawada, 1976
canescens (Sharp, 1869)
celata (Erichson, 1837) [253]
　arenicola (Thomson, C.G., 1868)
　germana (Sharp, 1869)
dadopora (Thomson, C.G., 1867)
　celata sensu auctt. non (Erichson, 1837)
nigra (Kraatz, 1856)
　zosterae sensu auctt. non (Thomson, C.G., 1856)
sordidula (Erichson, 1837) [254]
zosterae (Thomson, C.G., 1856) [255]
　hodierna (Sharp, 1869)
　olorophila Keys, 1933

DILACRA Thomson, C.G., 1858 [256]
　ATHETA sensu auctt. partim non Thomson, C.G., 1858
　DRALICA Mulsant & Rey, 1874
luteipes (Erichson, 1837)
vilis (Erichson, 1837) [257]

DIMETROTA Mulsant & Rey, 1873 [258]
　ATHETA sensu auctt. partim non Thomson, C.G., 1858
aeneipennis (Thomson, C.G., 1856)
　picipennis sensu auctt. non (Mannerheim, 1843) [259]
atramentaria (Gyllenhal, 1810)

cauta (Erichson, 1837)
    parvula sensu auctt. partim non (Mannerheim, 1830)
cinnamoptera (Thomson, C.G., 1856)
ischnocera (Thomson, C.G., 1870)
laevana (Mulsant & Rey, 1852)
    clintoni (Kevan, 1969) [260]
marcida (Erichson, 1837)
nigripes (Thomson, C.G., 1856)
setigera (Sharp, 1869)

DINARAEA Thomson, C.G., 1858
    ATHETA sensu auctt. partim non Thomson, C.G., 1858
aequata (Erichson, 1837)
angustula (Gyllenhal, 1810)
linearis (Gravenhorst, 1802)

DOCHMONOTA Thomson, C.G., 1859
    ATHETA sensu auctt. partim non Thomson, C.G., 1858
clancula (Erichson, 1837)

ENALODROMA Thomson, C.G., 1859
    ATHETA sensu auctt. partim non Thomson, C.G., 1858
    PTYCHANDRA Ganglbauer, 1895 non Felder, 1861
hepatica (Erichson, 1839)

GEOSTIBA Thomson, C.G., 1858 [261]
    SIPALIA sensu auctt. non Mulsant & Rey, 1853
    EVANYSTES Gistel, 1856
circellaris (Gravenhorst, 1806)

HALOBRECTA Thomson, C.G., 1858
algae (Hardy, 1851) [262]
    puncticeps (Thomson, C.G., 1852)
algophila (Fenyes, 1909) [263]
flavipes Thomson, C.G., 1861
princeps (Sharp, 1869) [264]

HYDROSMECTA Thomson, C.G., 1858
    ATHETA sensu auctt. partim non Thomson, C.G., 1858
    HYDROSMECTINA Ganglbauer, 1895 [265]
delicatissima (Bernhauer, 1908) [266]
delicatula (Sharp, 1869)
eximia (Sharp, 1869)
fragilis (Kraatz, 1854)
longula (Heer, 1839)
    thinobioides (Kraatz, 1854)
subtilissima (Kraatz, 1854)
    septentrionum (Benick, 1969) [267]

LIOGLUTA Thomson, C.G., 1858
    HYPNOTA Mulsant & Rey, 1874
alpestris (Heer, 1839)
    nitidiuscula (Sharp, 1869) [268]
    nitidula (Kraatz, 1856) [269]
granigera (Kiesenwetter, 1850) [270]
longiuscula (Gravenhorst, 1802)
    vicina (Stephens, 1832)
microptera Thomson, C.G., 1867
    oblonga (Erichson, 1839) non (Lacordaire, 1835)

oblongiuscula Sharp, 1869 [271]
pagana (Erichson, 1839)

LYPROCORRHE Thomson, C.G., 1859
    NOTOTHECTA sensu auctt. partim non Thomson, C.G., 1858
anceps (Erichson, 1837)

MICRODOTA Mulsant & Rey, 1873 [272]
    ATHETA sensu auctt. partim non Thomson, C.G., 1858
aegra (Heer, 1841)
amicula (Stephens, 1832)
atomaria (Kraatz, 1856) [273]
atricolor (Sharp, 1869)
    mortuorum sensu auctt. Brit. non (Thomson, C.G., 1867)
benickiella (Brundin 1948)
    validiuscula sensu auctt. non (Kraatz, 1856)
boreella (Brundin, 1948)
excelsa (Bernhauer, 1911)
glabricula (Thomson, C.G., 1867)
    atomaria sensu (Sharp, 1869) non (Kraatz, 1856)
    minuscula sensu auctt. non (Brisout de Barneville, 1860)
indubia (Sharp, 1869)
inquinula (Gravenhorst, 1802)
liliputana (Brisout de Barneville, 1860) [274]
    alpina (Benick, 1940)
minuscula (Brisout de Barneville, 1860)
    atomaria sensu auctt. non (Kraatz, 1856)
    perexigua (Sharp, 1869)
palleola (Erichson, 1837)
spatuloides (Benick, 1939)
    spatula sensu auctt. Brit. non (Fauvel, 1876)
subtilis (Scriba, 1866)

MOCYTA Mulsant & Rey, 1874 [275]
    ATHETA sensu auctt. partim non Thomson, C.G., 1858
amplicollis (Mulsant & Rey, 1873)
clientula (Erichson, 1839) [276]
fungi (Gravenhorst, 1806)
fussi (Bernhauer, 1908)
orbata (Erichson, 1837)
orphana (Erichson, 1837)

MYCETOTA Ádám, 1987 [277]
    ATHETA 'LOHSE GROUP I' auctt. partim
fimorum (Brisout de Barneville, 1860) [278]
laticollis (Stephens, 1832)

NEHEMITROPIA Lohse, 1971
lividipennis (Mannerheim, 1830)
    sordida (Marsham, 1802) non (Gravenhorst, 1802) [279]

NEOHILARA Lohse, 1971
    ATHETA sensu auctt. partim non Thomson, C.G., 1858
subterranea (Mulsant & Rey, 1853)

NOTOTHECTA Thomson, C.G., 1858
ATHETA sensu auctt. partim non Thomson, C.G., 1858
confusa (Märkel, 1844)
flavipes (Gravenhorst, 1806)

OREOSTIBA Ganglbauer, 1895 [280]
ATHETA 'LOHSE GROUP II' auctt. partim
tibialis (Heer, 1839)

OUSIPALIA des Gozis, 1886
ATHETA sensu auctt. partim non Thomson, C.G., 1858
caesula (Erichson, 1839)

PACHNIDA Mulsant & Rey, 1874
ALIANTA sensu auctt. partim non Thomson, C.G., 1858
nigella (Erichson, 1837)

PACHYATHETA Munster, 1925 [281]
ATHETA sensu auctt. partim non Thomson, C.G., 1858
cribrata (Kraatz, 1856)
mortuorum (Thomson, C.G., 1867) [282]
nesslingi sensu auctt. Brit. non (Bernhauer, 1928)

PARAMEOTICA Ganglbauer, 1895 [283]
ATHETA sensu auctt. partim non Thomson, C.G., 1858
DILACRA sensu auctt. partim non Thomson, C.G., 1858
difficilis (Brisout de Barneville, 1860)
laticeps sensu auctt. Brit. non (Thomson, C.G., 1856)
rigua (Williams, 1929)

PARANOPLETA Brundin in Hansen, 1954
inhabilis (Kraatz, 1856)

PHILHYGRA Mulsant & Rey, 1873 [284]
ATHETA sensu auctt. partim non Thomson, C.G., 1858
HYGROECIA Mulsant & Rey, 1873
PELURGA Mulsant & Rey, 1874
arctica (Thomson, C.G., 1856)
clavipes (Sharp, 1869)
britteni Joy, 1913
debilis (Erichson, 1837)
deformis (Kraatz, 1856)
complanata sensu auctt. non (Mannerheim, 1830)
elongatula (Gravenhorst, 1802)
balcanica (Brundin, 1944) non (Bernhauer, 1936)
fallaciosa (Sharp, 1869)
gyllenhalii (Thomson, C.G., 1856) [285]
hygrobia (Thomson, C.G., 1856)
magniceps (Sahlberg, J., 1876)
hygrotopora (Kraatz, 1856)
luridipennis (Mannerheim, 1830)
malleus (Joy, 1913)
hygrobia sensu auctt. non (Thomson, C.G., 1856)
melanocera (Thomson, C.G., 1856)
?vaga (Heer, 1839)
obtusangula (Joy, 1913)
palustris (Kiesenwetter, 1844)

parca (Mulsant & Rey, 1873)
nannion (Joy, 1931) [286]
scotica (Elliman, 1909)
caucasica (Brundin, 1944)
terminalis (Gravenhorst, 1806)
volans (Scriba, 1859)
halophila (Thomson, C.G., 1861)
tomlini (Joy, 1913)

PLATARAEA Thomson, C.G., 1858
ATHETA sensu auctt. partim non Thomson, C.G., 1858
brunnea (Fabricius, 1798)
nigriceps (Marsham, 1802)

PYCNOTA Mulsant & Rey, 1874
paradoxa (Mulsant & Rey, 1861)
nidorum (Thomson, C.G., 1868)

RHAGOCNEME Munster, 1923 [287]
ATHETA sensu auctt. partim non Thomson, C.G., 1858
subsinuata (Erichson, 1839)

SCHISTOGLOSSA Kraatz, 1856
aubei (Brisout de Barneville, 1860)
bergvalli Palm, 1968 [288]
benicki Lohse, 1981 [289]
curtipennis (Sharp, 1869)
gemina (Erichson, 1837)
viduata (Erichson, 1837)

THINOBAENA Thomson, C.G., 1859 [249]
ATHETA sensu auctt. partim non Thomson, C.G., 1858
vestita (Gravenhorst, 1806)

TRAUMOECIA Mulsant & Rey, 1874 [290]
ATHETA 'LOHSE GROUP II' auctt. partim
picipes (Thomson, C.G., 1856)
?complana (Mannerheim, 1830)

TRICHIUSA Casey, 1893
immigrata Lohse, 1984 [291]

XENOTA Mulsant & Rey, 1874 [292]
ATHETA 'LOHSE GROUP I' auctt. partim
myrmecobia (Kraatz, 1856)

Subtribe THAMIARAEINA Fenyes, 1921

THAMIARAEA Thomson, C.G., 1858
cinnamomea (Gravenhorst, 1802)
hospita (Märkel, 1844)

Tribe AUTALIINI Thomson, C.G., 1859

AUTALIA Leach, 1819
impressa (Olivier, 1795)
brevicornis Blair, 1944
longicornis Scheerpeltz, 1947
impressa sensu auctt. partim non (Olivier, 1795)
puncticollis Sharp, 1864

rivularis (Gravenhorst, 1802)

Tribe DEINOPSINI Sharp, 1883

DEINOPSIS Matthews, A.H., 1838
    DINOPSIS Agassiz, 1846 [40]
erosa (Stephens, 1832)

Tribe DIGLOTTINI Jakobson, 1909

DIGLOTTA Champion, G.C., 1899
    DIGLOSSA Haliday, 1837 non Wagler, 1832
mersa (Haliday, 1837)
    submarina (Fairmaire & Laboulbène, 1856) [293]
sinuaticollis (Mulsant & Rey, 1871) [294]
    mersa sensu Lohse, 1985 non (Haliday, 1837)
    subsinuata Lohse, 1985

Tribe FALAGRIINI Mulsant & Rey, 1873

BOHEMIELLINA Machulka, 1941
flavipennis (Cameron, 1920)
    paradoxa Machulka, 1941

BORBOROPORA Kraatz in Kraatz & Fuss, 1862
kraatzii Fuss in Kraatz & Fuss, 1862

CORDALIA Jacobs, 1925
    FALAGRIA sensu auctt. partim non Leach, 1819
obscura (Gravenhorst, 1802)

FALAGRIA Leach, 1819
caesa Erichson, 1837
    sulcata (Paykull, 1789) non (Müller, O.F., 1776)
sulcatula (Gravenhorst, 1806)

FALAGRIOMA Casey, 1906
    FALAGRIA sensu auctt. partim non Leach, 1819
thoracica (Stephens, 1832)

MYRMECOCEPHALUS MacLeay, 1873
    FALAGRIA sensu auctt. partim non Leach, 1819
concinnus (Erichson, 1839)

MYRMECOPORA Saulcy, 1865

Subgenus XENUSA Mulsant & Rey, 1875
brevipes Butler, 1909 [295]
oweni Assing, 1997 [296]
sulcata (Kiesenwetter, 1850)
    simillima (Wollaston, 1864) [297]
uvida (Erichson, 1840)

Tribe GYMNUSINI Heer, 1839

GYMNUSA Gravenhorst, 1806
brevicollis (Paykull, 1800)
variegata Kiesenwetter, 1845

Tribe HOMALOTINI Heer, 1839

Subtribe BOLITOCHARINA Thomson, C.G., 1859

BOLITOCHARA Mannerheim, 1830
bella Märkel, 1844
lucida (Gravenhorst, 1802)
mulsanti Sharp, 1875
obliqua Erichson, 1837
pulchra (Gravenhorst, 1806)
    lunulata sensu (Paykull, 1789) non (Linnaeus, 1761)

EURYUSA Erichson, 1837
optabilis Heer, 1839
sinuata Erichson, 1837

HETEROTA Mulsant & Rey, 1874
plumbea (Waterhouse, G.R., 1858)

LEPTUSA Kraatz, 1856 [298]
    PACHYGLUTA Thomson, C.G., 1858
fumida (Erichson, 1839)
    haemorrhoidalis sensu Ganglbauer, 1895 non
        (Heer, 1839)
norvegica Strand, A., 1941
pulchella (Mannerheim, 1830)
    angusta (Aubé, 1850)
ruficollis (Erichson, 1839)

PSEUDOMICRODOTA Machulka, 1935
paganettii (Bernhauer, 1909) [299]
    jelineki (Krása, 1914) [300]

PSEUDOPASILIA Ganglbauer, 1895
testacea (Brisout de Barneville, 1863) [301]

RHOPALOCERINA Reitter, 1909
clavigera (Scriba, 1859)

TACHYUSIDA Mulsant & Rey, 1871
gracilis (Erichson, 1837)

THECTUROTA Casey, 1893
    CRYPTUSA sensu auctt. non Mulsant & Rey, 1873
    PRAGENSIELLA Machulka, 1941
marchii (Dodero, 1922)
    capitalis sensu auctt. non (Mulsant & Rey, 1873)
williamsi (Bernhauer, 1936) [302]

Subtribe GYROPHAENINA Kraatz, 1856

AGARICOCHARA Kraatz, 1856
    GYROPHAENA sensu auctt. partim non
        Mannerheim, 1830
latissima (Stephens, 1832)
    laevicollis (Kraatz, 1854)

BRACHIDA Mulsant & Rey, 1871
exigua (Heer, 1839)

ENCEPHALUS Stephens, 1832
complicans Stephens, 1832

GYROPHAENA Mannerheim, 1830
affinis Mannerheim, 1830
bihamata Thomson, C.G., 1867
congrua Erichson, 1837
　fasciata sensu auctt. Brit. non (Marsham, 1802)
fasciata (Marsham, 1802) [303]
　laevipennis Kraatz, 1857
gentilis Erichson, 1839
hanseni Strand, A., 1946
joyi Wendeler, 1924
　convexicollis Joy, 1912 non Sharp, 1883
joyioides Wüsthoff, 1937
lucidula Erichson, 1837
manca Erichson, 1839
　angustata (Stephens, 1832) [304]
minima Erichson, 1837
munsteri Strand, A., 1935
nana (Paykull, 1800)
poweri Crotch, 1867
pseudonana Strand, A., 1939
pulchella Heer, 1839
rousi Dvorák, 1966 [305]
strictula Erichson, 1839
transversalis Strand, 1939 [306]
williamsi Strand, A., 1935

Subtribe HOMALOTINA Heer, 1839

ANOMOGNATHUS Solier, 1849
　THECTURA Thomson, C.G., 1859
cuspidatus (Erichson, 1839)

CYPHEA Fauvel, 1863
curtula (Erichson, 1837) [307]

HOMALOTA Mannerheim, 1830
plana (Gyllenhal, 1810)

Subtribe SILUSINA Fenyes, 1918

SILUSA Erichson, 1837
rubiginosa Erichson, 1837

Tribe HYGRONOMINI Thomson, C.G., 1859

HYGRONOMA Erichson, 1837
dimidiata (Gravenhorst, 1806)

Tribe HYPOCYPHTINI Laporte, 1835

CYPHA Leach, 1819 [308]
　HYPOCYPHTUS Gyllenhal, 1827
　HYPOCYPTUS Erichson, 1839 (misspelling)
apicalis (Brisout de Barneville, 1863) [309]
　nitida (Palm, 1935) [310]
aprilis (Rey, 1882)
　imitator (Luze, 1902) [310]
discoidea (Erichson, 1839)
laeviuscula (Mannerheim, 1830)
longicornis (Paykull, 1800)
pulicaria (Erichson, 1839)
punctum (Motschulsky, 1857)
seminulum (Erichson, 1839)
　seminula auctt. (misspelling)

hanseni (Palm, 1949) [310]
tarsalis (Luze, 1902) [311]

HOLOBUS Solier, 1849
　OLIGOTA sensu auctt. partim non Mannerheim, 1830
flavicornis (Lacordaire, 1835)

OLIGOTA Mannerheim, 1830
apicata (Erichson, 1837) [312]
granaria Erichson, 1837
inflata (Mannerheim, 1830)
parva Kraatz, 1862
　pygmaea Kraatz, 1858 non Solier, 1849
picipes (Stephens, 1832)
　atomaria sensu auctt. Brit. non Erichson, 1837
pumilio Kiesenwetter, 1858
　pusillima sensu auctt. Brit. partim non (Gravenhorst, 1806)
punctulata Heer, 1839
　ruficornis Sharp, 1870
pusillima (Gravenhorst, 1806)
　ytenensis Sharp, 1912

Tribe LOMECHUSINI Fleming, 1821

Subtribe LOMECHUSINA Fleming, 1821

LOMECHUSA Gravenhorst, 1806
　ATEMELES Dillwyn, 1829
emarginata (Paykull, 1789)
paradoxa Gravenhorst, 1806

LOMECHUSOIDES Tottenham, 1939
　LOMECHUSA sensu auctt. non Gravenhorst, 1806
strumosus (Fabricius, 1775)

Subtribe MYRMEDONIINA Thomson, C.G., 1867

DRUSILLA Leach, 1819
　ASTILBUS Dillwyn, 1829
canaliculata (Fabricius, 1787)

MYRMOECIA Mulsant & Rey, 1874
　ZYRAS sensu auctt. partim non Stephens, 1835
plicata (Erichson, 1837)

PELLA Stephens, 1835 [313]
　ZYRAS sensu auctt. partim non Stephens, 1835
　MYRMEDONIA sensu auctt. non Erichson, 1837
cognata (Märkel, 1842)
funesta (Gravenhorst, 1806)
humeralis (Gravenhorst, 1802)
laticollis (Märkel, 1845)
limbata (Paykull, 1789)
lugens (Gravenhorst, 1802)

ZYRAS Stephens, 1835
collaris (Märkel, 1842)
haworthi Stephens, 1835

Tribe MYLLAENINI Ganglbauer, 1895

MYLLAENA Erichson, 1837
brevicornis (Matthews, A.H., 1838)
dubia (Gravenhorst, 1806)
elongata (Matthews, A.H., 1838)
fowleri Matthews, A., 1883
gracilicornis Fairmaire & Brisout de Barneville, 1859
gracilis (Matthews, A.H., 1838)
infuscata Kraatz, 1853
intermedia Erichson, 1837
kraatzi Sharp, 1871
masoni Matthews, A., 1883
minuta (Gravenhorst, 1806)

Tribe OXYPODINI Thomson, C.G., 1859

Subtribe DINARDINA Mulsant & Rey, 1873

DINARDA Leach, 1819
dentata (Gravenhorst, 1806)
hagensi Wasmann, 1889
maerkeli Kiesenwetter, 1843
pygmaea Wasmann, 1894

HOMOEUSA Kraatz, 1856
acuminata (Märkel, 1842)

Subtribe MEOTICINA Seevers, 1978

MEOTICA Mulsant & Rey, 1873 [314]
anglica Benick in Muona, 1991
    indocilis sensu auctt. Brit. non (Heer, 1839)
    pallens sensu auctt. non (Redtenbacher, 1849)
exilis (Knoch in Gravenhorst, 1806)
    exiliformis Joy, 1915
exillima Sharp, 1915
filiformis (Motschulsky, 1860) [315]
    apicalis Benick, 1954
    capitalis sensu auctt. ?non Mulsant & Rey, 1873 [316]
    exilis sensu auctt. Brit. partim non (Knoch in
        Gravenhorst, 1806)
pallens (Redtenbacher, 1849) [317]
    lohsei Benick, 1953

Subtribe OXYPODINA Thomson, C.G., 1859

AMAROCHARA Thomson, C.G., 1858
bonnairei (Fauvel, 1865)
forticornis (Lacordaire, 1835)
umbrosa (Erichson, 1837)

CALODERA Mannerheim, 1830 [318]
    ITYOCARA Thomson, C.G., 1867 [319]
aethiops (Gravenhorst, 1802)
nigrita Mannerheim, 1830
protensa Mannerheim, 1830
riparia Erichson, 1837
rubens (Erichson, 1837)
rufescens Kraatz, 1856
uliginosa Erichson, 1837

COUSYA Mulsant & Rey, 1875 [320]
    OCYUSA sensu auctt. Brit. partim non Kraatz, 1856
    CHILOMORPHA Krása, 1914
defecta Mulsant & Rey, 1875
longitarsis (Thomson, C.G., 1867)
    hibernica (Rye, 1876)
    laticollis sensu auctt. Brit. non Thomson, C.G., 1870
nigrata (Fairmaire & Laboulbène, 1856)
nitidiventris (Fagel, 1958) [321]

CRATARAEA Thomson, C.G., 1858
suturalis (Mannerheim, 1830)

DEXIOGYIA Thomson, C.G., 1858
    DEXIOGYA auctt. (misspelling)
corticina (Erichson 1837)

HAPLOGLOSSA Kraatz, 1856
    MICROGLOTTA Kraatz, 1862
    MICROGLOSSA Stein, 1868 (misspelling)
gentilis (Märkel, 1844)
marginalis (Gravenhorst, 1806)
    obscura Joy, 1912
nidicola (Fairmaire, 1852)
picipennis (Gyllenhal, 1827)
villosula (Stephens, 1832)
    pulla sensu auctt. non (Gravenhorst, 1802) [322]
    puncticollis (Stephens, 1832)

HYGROPORA Kraatz, 1856
cunctans (Erichson, 1837)

ILYOBATES Kraatz, 1856
bennetti Donisthorpe, 1914
    nigricollis sensu auctt. Brit. partim non (Paykull,
        1800)
    subopacus Palm, 1935 [323]
nigricollis (Paykull, 1800)
propinquus (Aubé, 1850)

ISCHNOGLOSSA Kraatz, 1856
obscura Wunderle, 1990 [324]
prolixa (Gravenhorst, 1802)
turcica Wunderle, 1992 [325]

MNIUSA Mulsant & Rey, 1875 [326]
incrassata (Mulsant & Rey, 1852)

OCALEA Erichson, 1837
badia Erichson, 1837
latipennis Sharp, 1870
picata (Stephens, 1832)
rivularis Miller, L., 1851

OCYUSA Kraatz, 1856
    DEUBELIA Bernhauer, 1899 [327]
maura (Erichson, 1837)
picina (Aubé, 1850)

OXYPODA Mannerheim, 1830
acuminata (Stephens, 1832)
    lividipennis sensu auctt. non Mannerheim, 1830 [328]
alternans (Gravenhorst, 1802)

annularis Mannerheim, 1830
brachyptera (Stephens, 1832)
brevicornis (Stephens, 1832)
    umbrata sensu auctt. non (Gravenhorst, 1802)
carbonaria (Heer, 1841)
    sericea Heer, 1839 non Lacordaire, 1835
elongatula Aubé, 1850
exoleta Erichson, 1839
ferruginea Erichson, 1839
flavicornis Kraatz, 1856
    amoena Fairmaire & Laboulbène, 1856 [329]
formiceticola Märkel, 1841
haemorrhoa (Mannerheim, 1830)
induta Mulsant & Rey, 1861
islandica Kraatz, 1857
lentula Erichson, 1837
longipes Mulsant & Rey, 1861
lurida Wollaston, 1857
    exoleta sensu auctt. non Erichson, 1839
    perplexa sensu auctt. partim non Mulsant & Rey, 1860
mutata Sharp, 1871 [330]
    riparia Fairmaire, 1859 non Thomson, C.G., 1855
    rugulosa sensu auctt. non Kraatz, 1856
nigricornis Motschulsky, 1860
    opaca sensu auctt. Brit. partim non (Gravenhorst, 1802)
nigrocincta Mulsant & Rey, 1875
opaca (Gravenhorst, 1802)
praecox Erichson, 1839 [331]
procerula Mannerheim, 1830
    elongatula sensu auctt. Brit. partim non Aubé, 1850
recondita Kraatz, 1856
soror Thomson, C.G., 1855
spectabilis Märkel, 1844
tarda Sharp, 1871 [332]
    brachyptera sensu auctt. partim ? non (Stephens, 1832)
tirolensis Gredler, 1863
    rupicola Rye, 1866
vittata Märkel, 1842

PHLOEOPORA Erichson, 1837
    PHLOEODROMA Kraatz, 1856 [333]
concolor Kraatz, 1856
corticalis (Gravenhorst, 1802)
    angustiformis Baudi, 1869
nitidiventris Fauvel, 1904 [334]
scribae Eppelsheim, 1884
    bernhaueri Lohse, 1984 [335]
    corticalis sensu auctt. Brit. partim non (Gravenhorst, 1802)
    teres sensu auctt. Brit. ?partim non (Gravenhorst, 1802) [336]
testacea (Mannerheim, 1830)

STICHOGLOSSA Fairmaire & Laboulbène, 1856
semirufa (Erichson, 1839)

TETRALAUCOPORA Bernhauer, 1928 [337]
    CHILOPORA Kraatz, 1856 non Haime, 1854
    PAROCYUSA sensu auctt. non Bernhauer, 1902
    CHILOPORATA Strand, E., 1935
longitarsis (Erichson, 1837)
rubicunda (Erichson, 1837)

THIASOPHILA Kraatz, 1856
    THYASOPHILA Fairmaire & Laboulbène, 1856
angulata (Erichson, 1837)
inquilina (Märkel, 1842)

Subtribe TACHYUSINA Thomson, C.G., 1859

BRACHYUSA Mulsant & Rey, 1874
concolor (Erichson, 1839)

DACRILA Mulsant & Rey, 1874 [338]
fallax (Kraatz, 1856)
pruinosa (Kraatz, 1856)

DASYGNYPETA Lohse, 1974 [339]
    GNYPETA sensu auctt. partim non Thomson, C.G., 1858
velata (Erichson, 1837)

GNYPETA Thomson, C.G., 1858
caerulea (Sahlberg, C.R., 1831)
carbonaria (Mannerheim, 1830)
    labilis (Erichson, 1839)
ripicola (Kiesenwetter, 1844)
rubrior Tottenham, 1939

ISCHNOPODA Stephens, 1835 [340]
leucopus (Marsham, 1802)
scitula (Erichson, 1837) [341]
umbratica (Erichson, 1837)

TACHYUSA Erichson, 1837 [342]
concinna Heer, 1839 [343]
    coarctata sensu auctt. Brit. non Erichson, 1837
constricta Erichson, 1837
objecta (Mulsant & Rey, 1870) [344]

THINONOMA Thomson, C.G., 1859 [340]
atra (Gravenhorst, 1806)

Tribe PHYTOSINI Thomson, C.G., 1867

ARENA Fauvel, 1862
tabida (Kiesenwetter, 1850)
    octavii Fauvel, 1862

PHYTOSUS Curtis, 1838
    ACTOSUS Mulsant & Rey, 1871
balticus Kraatz, 1859
nigriventris (Chevrolat, 1843)
spinifer Curtis, 1838

Tribe PLACUSINI Mulsant & Rey, 1871

PLACUSA Erichson, 1837
complanata Erichson, 1839 [345]
depressa Mäklin, 1845
    complanata sensu auctt. Brit. partim non Erichson, 1839
    ?*humilis* Erichson, 1839
pumilio (Gravenhorst, 1802)
tachyporoides (Waltl, 1838)

denticulata Sharp, 1870

Subfamily SCAPHIDIINAE Latreille, 1806 [346]

Tribe SCAPHIDIINI Latreille, 1806

SCAPHIDIUM Olivier, 1790
quadrimaculatum Olivier, 1790

Tribe SCAPHIINI Achard, 1924

SCAPHIUM Kirby, 1837
immaculatum (Olivier, 1790)

Tribe SCAPHISOMATINI Casey, 1893

SCAPHISOMA Leach, 1815
agaricinum (Linnaeus, 1758)
assimile Erichson, 1845
boleti (Panzer, 1793)

Subfamily PIESTINAE Erichson, 1839

SIAGONIUM Kirby, 1815
quadricorne Kirby, 1815

Subfamily OXYTELINAE Fleming, 1821 [347]

Tribe COPROPHILINI Heer, 1839

COPROPHILUS Latreille, 1829 [348]
ELONIUM sensu auctt. non Leach, 1819
striatulus (Fabricius, 1792)

Tribe EUPHANIINI Reitter, 1909

DELEASTER Erichson, 1839
dichrous (Gravenhorst, 1802)

SYNTOMIUM Curtis, 1828
aeneum (Müller, P.W.J., 1821)

Tribe OXYTELINI Fleming, 1821

ANOTYLUS Thomson, C.G., 1859
clypeonitens (Pandellé, 1867)
speculifrons sensu auctt. Brit. non (Kraatz, 1857)
complanatus (Erichson, 1839)
fairmairei (Pandellé, 1867)
hamatus (Fairmaire & Laboulbène, 1856)
hammondi Schülke, 2009 [349]
complanatus sensu auctt. partim non (Erichson, 1839)
insecatus (Gravenhorst, 1806)
inustus (Gravenhorst, 1806)
maritimus Thomson, C.G., 1861
perrisi (Fauvel, 1862)
mutator (Lohse, 1963)
sculpturatus sensu auctt. partim non (Gravenhorst, 1806)
nitidulus (Gravenhorst, 1802)

rugosus (Fabricius, 1775) [137]
striatus (Strøm, 1768)
saulcyi (Pandellé, 1867)
sculpturatus (Gravenhorst, 1806)
tetracarinatus (Block, 1799)

OXYTELUS Gravenhorst, 1802
fulvipes Erichson, 1839
laqueatus (Marsham, 1802)
migrator Fauvel, 1904 [350]
piceus (Linnaeus, 1767)
sculptus Gravenhorst, 1806

PLATYSTETHUS Mannerheim, 1830
Subgenus CRAETOPYCRUS Tottenham, 1939
alutaceus Thomson, C.G., 1861
capito Heer, 1839
cornutus (Gravenhorst, 1802)
degener Mulsant & Rey, 1878
cornutus sensu auctt. Brit. partim non Gravenhorst, 1802
nitens (Sahlberg, C.R., 1832)
nodifrons Mannerheim, 1830

Subgenus PLATYSTETHUS Mannerheim, 1830
arenarius (Fourcroy, 1785)

Tribe THINOBIINI Sharp, 1887

APLODERUS Stephens, 1833
HAPLODERUS Agassiz, 1846
caelatus (Gravenhorst, 1802)

BLEDIUS Leach, 1819 [351]
Subgenus ASTYCOPS Thomson, C.G., 1859
subterraneus Erichson, 1839

Subgenus BARGUS Schiødte, 1866 [352]
annae Sharp, 1911 [353]
arcticus Sahlberg, J., 1890
denticollis sensu auctt. Brit. non Fauvel, 1872
defensus Fauvel, 1872
erraticus Erichson, 1839
filipes Sharp, 1911 [354]
fuscipes Rye, 1865
longulus Erichson, 1839
opacus (Block, 1799)
pallipes (Gravenhorst, 1806) [355]
annae sensu Pope, 1977 partim ?non Sharp, 1911
larseni Hansen, 1940
terebrans (Schiødte, 1866)

Subgenus BLEDIUS Leach, 1819
frisius Lohse, 1978 [356]
spectabilis sensu auctt. partim non Kraatz, 1857
spectabilis Kraatz, 1857
germanicus Wagner, 1935 non (Gravenhorst, 1806)
limicola Tottenham, 1940 [356][357]
tricornis (Herbst, 1784)
unicornis (Germar, 1825)

Subgenus DICARENUS Gistel, 1834
fergussoni Joy, 1912
arenarius (Paykull, 1800) non (Fourcroy, 1785)
arenoides Tottenham, 1939
subniger Schneider, O., 1900

Subgenus ELBIDUS Mulsant & Rey, 1878
bicornis (Germar, 1822) [358]
   dama Motschulsky, 1857
ssp. ***jutlandensis*** Herman, 1986 [359]
   atlanticus Lohse, 1978 non Koch, 1938
diota Schiødte, 1866

Subgenus EUCERATOBLEDIUS Znojko, 1929
furcatus (Olivier, 1811)

Subgenus HESPEROPHILUS Stephens, 1829
atricapillus (Germar, 1825) [360]
   praetermissus Williams, 1929 [361]
crassicollis Lacordaire, 1835
dissimilis Erichson, 1840
femoralis (Gyllenhal, 1827)
gallicus (Gravenhorst, 1806)
   fracticornis (Paykull, 1790) non (Müller, O.F., 1776)
   laetior Mulsant & Rey, 1878
   sharpi Fowler, 1913
lohsei Schülke, 2011 [362]
   atricapillus sensu auctt. partim non (Germar, 1825)
   praetermissus sensu auctt. Brit. partim non
     Williams, 1929
occidentalis Bondroit, 1907
   crassicollis sensu auctt. Brit. non Lacordaire, 1835

CARPELIMUS Leach, 1819

Subgenus CARPELIMUS Leach, 1819
fuliginosus (Gravenhorst, 1802)
lindrothi Palm, 1942 [363]
obesus (Kiesenwetter, 1844)
pusillus (Gravenhorst, 1802)
   lasti (Scheerpeltz, 1946)

Subgenus MYOPINUS Scheerpeltz, 1937
elongatulus (Erichson, 1839)

Subgenus PARATROGOPHLOEUS Hatch, 1957
bilineatus Stephens, 1834
erichsoni (Sharp, 1871) [364]
   bilineatus sensu auctt. partim non Stephens, 1834
rivularis (Motschulsky, 1860) [137]
   obscurus Stephens, 1834
similis Smetana, 1967 [365]

Subgenus TROGINUS Mulsant & Rey, 1878
despectus (Baudi, 1869) [366]
incongruus Steel, 1969 [367]
   zealandicus sensu auctt. Brit. partim non (Sharp, 1900)
schneideri (Ganglbauer, 1895)
   hemerinus (Joy, 1913)
zealandicus (Sharp, 1900) [364]

Subgenus TROGOPHLOEUS Mannerheim, 1830
alutaceus (Fauvel, 1898) [368]
corticinus (Gravenhorst, 1806)
   dispersepunctatus (Scheerpeltz, 1947)
foveolatus (Sahlberg, C.R., 1832)
gracilis (Mannerheim, 1830)
halophilus (Kiesenwetter, 1844)
impressus (Lacordaire, 1835)
manchuricus (Bernhauer, 1938)
ssp. ***subtilicornis*** (Roubal, 1946) [369]
   corticinus sensu auctt. Brit. partim non
     (Gravenhorst, 1806)
   strandi (Scheerpeltz, 1950)
subtilis (Erichson, 1839)

MANDA Blackwelder, 1952
ACROGNATHUS Erichson, 1839 non Agassiz, 1836
mandibularis (Gyllenhal, 1827)

OCHTHEPHILUS Mulsant & Rey, 1856 [370]
ANCYROPHORUS Kraatz, 1858
andalusiacus (Fagel, 1957)
   omalinus sensu auctt. Brit. partim non (Erichson, 1840)
angustior (Bernhauer, 1943)
   omalinus sensu auctt. Brit. partim non (Erichson, 1840)
   venustulus sensu auctt. Brit. non (Rosenhauer, 1856)
aureus (Fauvel, 1871)
omalinus (Erichson, 1840)

PLANEUSTOMUS Jacquelin du Val, 1857
flavicollis Fauvel, 1871
palpalis (Erichson, 1839)

TEROPALPUS Solier, 1849
CARPELIMUS sensu auctt. partim non Leach, 1819
unicolor (Sharp, 1900)

THINOBIUS Kiesenwetter, 1844
bicolor Joy, 1911 [371]
   brunneipennis sensu auctt. Brit. non Kraatz, 1857
   linearis sensu auctt. Brit. non Kraatz, 1857
brevipennis Kiesenwetter, 1850
ciliatus Kiesenwetter, 1844
   longipennis sensu auctt. partim non (Heer, 1841)
   praetor Smetana, 1959 [372]
crinifer Smetana, 1959 [373]
   longipennis sensu auctt. partim non (Heer, 1841)
   strandi Smetana, 1960 [374]
longipennis (Heer, 1841) [375]
major Kraatz, 1857
   angusticeps sensu auctt. non Fauvel, 1889
   macroceros Joy, 1913
   macrocerus auctt. (misspelling)
newberyi Scheerpeltz, 1925
   pallidus Newbery, 1909 non Casey, 1889

THINODROMUS Kraatz, 1857
AMISAMMUS des Gozis, 1886
CARPELIMUS sensu auctt. partim non Leach, 1819
arcuatus (Stephens, 1834)

Subfamily OXYPORINAE Fleming, 1821

OXYPORUS Fabricius, 1775
rufus (Linnaeus, 1758)

Subfamily SCYDMAENINAE Leach, 1815 [376]

Tribe EUTHEIINI Casey, 1897

EUTHEIA Stephens, 1830
EUTHIA auctt. (misspelling)
formicetorum Reitter, 1881
linearis Mulsant, 1861

clavata Reitter, 1881
plicata (Gyllenhal, 1813)
schaumii Kiesenwetter, 1858
schaumi auctt. (misspelling)
scydmaenoides Stephens, 1830

EUTHICONUS Reitter, 1881/2
conicicollis (Fairmaire & Laboulbène, 1855) [377]

Tribe CEPHENNIINI Reitter, 1882

CEPHENNIUM Müller, P.W.J. & Kunze, 1822
gallicum Ganglbauer, 1899
thoracicum sensu auctt. Brit. non Müller, P.W.J. & Kunze, 1822
edmondsi Donisthorpe, 1931
pallidum Edmonds, 1931

Tribe CYRTOSCYDMINI Schaufuss, 1889

EUCONNUS Thomson, C.G., 1859

Subgenus CLADOCONNUS Reitter, 1909
denticornis (Müller, P.W.J. & Kunze, 1822)

Subgenus EUCONNUS Thomson, C.G., 1859
fimetarius (Chaudoir, 1845)
hirticollis (Illiger, 1798)
rutilipennis (Müller, P.W.J. & Kunze, 1822)

Subgenus NAPOCHUS Thomson, C.G., 1859
duboisi Méquignon, 1929
murielae Last, 1945
pragensis (Machulka, 1923)
claviger sensu auctt. Brit. non (Müller, P.W.J. & Kunze, 1822)

Subgenus NEONAPOCHUS Machulka, 1929
maeklinii (Mannerheim, 1844)
maeklini auctt. (misspelling)

MICROSCYDMUS Saulcy & Croissandeau, 1893
EUCONNUS sensu auctt. partim non Thomson, C.G., 1859
minimus (Chaudoir, 1845)
nanus sensu auctt. Brit. partim non (Schaum, 1844)
nanus (Schaum, 1844)

NEURAPHES Thomson, C.G., 1859 [378]

Subgenus NEURAPHES Thomson, C.G., 1859
angulatus (Müller, P.W.J. & Kunze, 1822)
carinatus (Mulsant & Rey, 1861)
elongatulus (Müller, P.W.J. & Kunze, 1822)
talparum Lokay, 1921
rubicundus sensu auctt. Brit. non (Schaum, 1841)
carinatus sensu auctt. Brit. partim non (Mulsant, 1861)

Subgenus PARARAPHES Reitter, 1891
plicicollis Reitter, 1879
planifrons Blatch, 1890
praeteritus Rye, 1872
longicollis sensu auctt. Brit. non (Motschulsky, 1845)

SCYDMORAPHES Reitter, 1891
NEURAPHES sensu auctt. partim non Thomson, C.G., 1859
helvolus (Schaum 1844)
minutus sensu auctt. Brit. non (Chaudoir, 1845)
nigrescens (Reitter, 1881)
sparshalli (Denny, 1825)

STENICHNUS Thomson, C.G., 1859
CYRTOSCYDMUS Motschulsky, 1870 [379]
SCYDMAENUS sensu Fowler, 1888 non Latreille, 1802
bicolor (Denny, 1825)
exilis (Erichson, 1837)
collaris (Müller, P.W.J. & Kunze, 1822)
godarti (Latreille, 1806)
poweri (Fowler, 1834)
barnevillei (Reitter, 1884)
harwoodianus Williams, 1927
pusillus (Müller, P.W.J. & Kunze, 1822)
scutellaris sensu auctt. Brit. partim non (Müller, P.W.J. & Kunze, 1822)
stotti Donisthorpe, 1932
scutellaris (Müller, P.W.J. & Kunze, 1822)
pusillus sensu auctt. Brit. partim non (Müller, P.W.J. & Kunze, 1822)

Tribe SCYDMAENINI Leach, 1815

SCYDMAENUS Latreille, 1802
EUMICRUS Laporte, 1840

Subgenus CHOLERUS Thomson, C.G., 1859
rufus Müller, P.W.J & Kunze, 1822

Subgenus SCYDMAENUS Latreille, 1802
tarsatus Müller, P.W.J. & Kunze, 1822

Subfamily STENINAE MacLeay, 1825

DIANOUS Leach, 1819
coerulescens (Gyllenhal, 1810)

STENUS Latreille, 1796 [380]

Subgenus HEMISTENUS Motschulsky, 1860 [381]
PARASTENUS Heyden, 1905 [382]
aceris Stephens, 1833
fuscicornis Erichson, 1840
geniculatus Gravenhorst, 1806
glacialis Heer, 1839
impressus Germar, 1824
ludyi Fauvel, 1855
coarcticollis sensu auctt. non Eppelsheim, 1890
ochropus Kiesenwetter, 1858
erichsoni Rye, 1364
ossium Stephens, 1833
pallipes Gravenhorst, 1802
palustris Erichson, 1839
subaeneus Erichson, 1840

Subgenus HYPOSTENUS Rey, 1884
cicindeloides (Schaller, 1783)
fornicatus Stephens, 1833
fulvicornis Stephens, 1833
kiesenwetteri Rosenhauer, 1856
latifrons Erichson, 1339

oscillator Rye, 1870
similis (Herbst, 1784)
solutus Erichson, 1840
tarsalis Ljungh, 1810

Subgenus METATESNUS Ádám, 2001
   METASTENUS Ádám, 1987 non Walker, 1834 [383]
   HEMISTENUS sensu auctt. non Motschulsky, 1860
bifoveolatus Gyllenhal, 1827
binotatus Ljungh, 1804
brevipennis Thomson, C.G., 1851 [384]
   foveicollis Kraatz, 1857
   picipes sensu auctt. partim ?non Stephens, 1833
butrintensis Smetana, 1959
   pallitarsis sensu auctt. partim non Stephens, 1833
canescens Rosenhauer, 1856
flavipes Stephens, 1833
nitidiusculus Stephens, 1833
niveus Fauvel, 1865
pallitarsis Stephens, 1833
picipennis Erichson, 1840
picipes Stephens, 1833
pubescens Stephens, 1833
umbratilis Casey, 1884
   pseudopubescens Strand, A., 1940
   pubescens sensu auctt. Brit. partim non Stephens, 1833

Subgenus STENUS Latreille, 1796
   NESTUS Rey, 1884
argus Gravenhorst, 1806
asphaltinus Erichson, 1840
assequens Rey, 1844
   simillimus Benick, 1949
ater Mannerheim, 1830
atratulus Erichson, 1839
biguttatus (Linnaeus, 1758)
bimaculatus Gyllenhal, 1810
boops Ljungh, 1810 [385]
   buphthalmus sensu Gravenhorst, 1802 non (Schrank, 1776)
calcaratus Scriba, 1864 [386]
canaliculatus Gyllenhal, 1827
carbonarius Gyllenhal, 1827
circularis Gravenhorst, 1802
clavicornis (Scopoli, 1763)
comma LeConte, 1863
   bipunctatus Erichson, 1839 non Ljungh, 1804
contumax Assing, 1994 [387]
europaeus Puthz, 1966
   cautus sensu auctt. non Erichson, 1839
   vafellus sensu auctt. Brit. non Erichson, 1839
fossulatus Erichson, 1840
fuscipes Gravenhorst, 1802
glabellus Thomson, C.G., 1870 [388]
guttula Müller, P.W.J., 1821
guynemeri Jacquelin du Val, 1850
incanus Erichson, 1839
incrassatus Erichson, 1839
juno (Paykull, 1789)
longitarsis Thomson, C.G., 1851
lustrator Erichson, 1839
melanarius Stephens, 1833
melanopus (Marsham, 1802)
morio Gravenhorst, 1806
nanus Stephens, 1833
   declaratus Erichson, 1839
nitens Stephens, 1833
palposus Zetterstedt, 1838

proditor Erichson, 1839
providus Erichson, 1839
   rogeri Kraatz, 1857 [389]
pusillus Stephens, 1833
   exiguus Erichson, 1840 [390]
subdepressus Mulsant & Rey, 1861

Subgenus TESNUS Rey, 1884 [391]
brunnipes Stephens, 1833
crassus Stephens, 1833
formicetorum Mannerheim, 1843
nigritulus Gyllenhal, 1827
opticus Gravenhorst, 1806

Subfamily EUAESTHETINAE Thomson, C.G., 1859

EDAPHUS Motschulsky, 1857
beszedesi Reitter, 1914 [392]

EUAESTHETUS Gravenhorst, 1806
bipunctatus (Ljungh, 1804)
laeviusculus Mannerheim, 1844
ruficapillus Lacordaire, 1835

Subfamily PSEUDOPSINAE Ganglbauer, 1895

PSEUDOPSIS Newman, 1834
sulcata Newman, 1834

Subfamily PAEDERINAE Fleming, 1821 [393]

Tribe PAEDERINI Fleming, 1821

Subtribe ASTENINA Hatch, 1957

ASTENUS Dejean, 1833
   ASTENOGNATHUS Reitter, 1909 [394]
immaculatus Stephens, 1833
lyonessius (Joy, 1908)
   angustatus sensu auctt. non (Paykull, 1789)
   longelytratus Palm, 1936
   brevelytratus Coiffait, 1960
procerus (Gravenhorst, 1806)
   filiformis (Latreille, 1806) non (Fabricius, 1792)
pulchellus (Heer, 1839)
   melanurus sensu Townsend, 1948 non Küster, 1853
serpentinus (Motschulsky, 1858)
   procerus sensu auctt. Brit. partim non (Gravenhorst, 1806)
   subditus (Mulsant & Rey, 1878) [395]

Subtribe CRYPTOBIINA Casey, 1905

OCHTHEPHILUM Stephens, 1829
   CRYPTOBIUM Mannerheim, 1830
collare (Reitter, 1884) [396]
fracticorne (Paykull, 1800) [397]
   brevipenne sensu auctt. non Mulsant & Rey, 1861
jacquelini (Boieldieu, 1859)

Subtribe LATHROBIINA Laporte, 1835

ACHENIUM Leach, 1819
depressum (Gravenhorst, 1802)
humile (Nicolai, 1822)

LATHROBIUM Gravenhorst, 1802 [398]

Subgenus LATHROBIUM Gravenhorst, 1802
brunnipes (Fabricius, 1792)
dilutum Erichson, 1839
elongatum (Linnaeus, 1767)
fovulum Stephens, 1833
  britannicum (Coiffait, 1972)
fulvipenne (Gravenhorst, 1806) [399]
geminum Kraatz, 1857 [400]
  boreale Hochhuth, 1851
  volgense Hochhuth, 1851
impressum Heer, 1841
  filiforme Gravenhorst, 1806 non (Fabricius, 1792)
longulum Gravenhorst, 1802
  longipenne Fairmaire & Laboulbène, 1856
pallidipenne Hochhuth, 1851
  ripicola Czwalina, 1888 [401]
pallidum Nordmann, 1837
rufipenne Gyllenhal, 1813

LOBRATHIUM Mulsant & Rey, 1878 [402]
  LATHROBIUM sensu auctt. partim non
  Gravenhorst, 1802
multipunctum (Gravenhorst, 1802)

PLATYDOMENE Ganglbauer, 1895 [403]
angusticollis (Lacordaire, 1835)

TETARTOPEUS Czwalina, 1888 [403]
angustatus (Lacordaire, 1835)
quadratus (Paykull, 1789)
rufonitidus (Reitter, 1909)
  fennicus (Renkonen, 1938) [404]
terminatus (Gravenhorst, 1802)
zetterstedti (Rye, 1872)
  punctatus (Zetterstedt, 1828) non (Fourcroy, 1785)
  foveatus sensu auctt. non (Stephens, 1839)

Subtribe MEDONINA Casey, 1905

HYPOMEDON Mulsant & Rey, 1878 [405]
  CHLOECHARIS Lynch Arribálzaga, 1884
debilicornis (Wollaston, 1857) [406]

LITHOCHARIS Dejean, 1833
nigriceps Kraatz, 1859
ochracea (Gravenhorst, 1802) [407]

MEDON Stephens, 1833 [408]
apicalis (Kraatz, 1857)
brunneus (Erichson, 1839)
castaneus (Gravenhorst, 1802)
dilutus (Erichson, 1839)
fusculus (Mannerheim, 1830)
piceus (Kraatz, 1858)
pocofer (Peyron, 1857)
ripicola (Kraatz, 1854)

PSEUDOMEDON Mulsant & Rey, 1877 [409]
  LITHOCHARIS sensu auctt. partim non Dejean,
  1833
obscurellus (Erichson, 1840)
obsoletus (Nordmann, 1837)

SUNIUS Stephens, 1829
  HYPOMEDON sensu auctt. non Mulsant & Rey,
  1878
bicolor (Olivier, 1795)
melanocephalus (Fabricius, 1792)
propinquus (Brisout de Barneville, 1867)

Subtribe PAEDERINA Fleming, 1821

PAEDERIDUS Mulsant & Rey, 1878 [410]
  PAEDERUS sensu auctt. partim non Fabricius,
  1775
rubrothoracicus (Goeze, 1777)
ruficollis (Fabricius, 1781) [392]

PAEDERUS Fabricius, 1775
caligatus Erichson, 1840
fuscipes Curtis, 1826
littoralis Gravenhorst, 1802
riparius (Linnaeus, 1758)

Subtribe SCOPAEINA Mulsant & Rey, 1878

SCOPAEUS Erichson, 1840 [411]
gracilis (Sperk, 1835)
laevigatus (Gyllenhal, 1827)
  gracilis sensu auctt. Brit. partim non (Sperk, 1835)
minutus Erichson, 1840
  sulcicollis sensu Fowler, 1888 non (Stephens, 1833)
  gracilipes Edmonds, 1933
ryei Wollaston, 1872
  minimus sensu auctt. Brit. non Erichson, 1839 [412]
  rubidus sensu auctt. Brit. non Mulsant & Rey, 1855
sulcicollis (Stephens, 1833)
  minutus sensu auctt. partim non Erichson, 1840
  cognatus Mulsant & Rey, 1855
  abbreviatus sensu Edmonds, 1931 non Mulsant &
  Rey, 1855
  pusillus sensu Tottenham, 1949 non Kiesenwetter,
  1843

Subtribe STILICINA Casey, 1905

RUGILUS Leach, 1819
  STILICUS Berthold, 1827
angustatus (Geoffroy, 1785)
  fragilis (Gravenhorst, 1806) [413]
  scutellatus (Motschulsky, 1858)
erichsonii (Fauvel, 1867)
geniculatus (Erichson, 1839)
orbiculatus (Paykull, 1789)
rufipes Germar, 1836
similis (Erichson, 1839)
subtilis (Erichson, 1840)

Subfamily STAPHYLININAE Latreille, 1802

Tribe OTHIINI Thomson, C.G., 1859

ATRECUS Jacquelin du Val, 1856
BAPTOLINUS Kraatz, 1857
affinis (Paykull, 1789)

OTHIUS Stephens, 1829
angustus Stephens, 1833
    melanocephalus (Gravenhorst, 1806) non (Geoffroy, 1785)
laeviusculus Stephens, 1833
lapidicola Märkel & Kiesenwetter, 1848 [414]
punctulatus (Goeze, 1777)
subuliformis Stephens, 1833
    myrmecophilus Kiesenwetter, 1843 [415]

Tribe STAPHYLININI Latreille, 1802

Subtribe PHILONTHINA Kirby, 1837

BISNIUS Stephens, 1829 [416]
PHILONTHUS sensu auctt. partim non Stephens, 1829
cephalotes (Gravenhorst, 1802) [417]
fimetarius (Gravenhorst, 1802)
    rigidicornis (Gravenhorst, 1802)
nigriventris (Thomson, C.G., 1867)
parcus (Sharp, 1874)
    felix (Last, 1961)
pseudoparcus (Brunne, 1976) [418]
puella (Nordmann, 1837)
scoticus (Joy & Tomlin, 1913) [419]
sordidus (Gravenhorst, 1802)
    pachycephalus (Nordmann, 1837)
subuliformis (Gravenhorst, 1802)
    fuscus (Gravenhorst, 1802) non (Gmelin in Linnaeus, 1790)

CAFIUS Stephens, 1829
cicatricosus (Erichson, 1840)
fucicola Curtis, 1830
xantholoma (Gravenhorst, 1806)

ERICHSONIUS Fauvel, 1874
ACTOBIUS Fauvel, 1875
cinerascens (Gravenhorst, 1802)
signaticornis (Mulsant & Rey, 1853)
ytenensis (Sharp, 1913)

GABRIUS Stephens, 1829 [420]
appendiculatus Sharp, 1910
    subnigritulus sensu auctt. non Joy, 1913 [421]
astutoides (Strand, A., 1946)
    astutus sensu auctt. partim non Erichson, 1840
bishopi Sharp, 1910
breviventer (Sperk, 1835)
    coxalus (Hochhuth, 1871)
    pennatus Sharp, 1910 [422]
exiguus (Nordmann, 1837)
keysianus Sharp, 1910
nigritulus (Gravenhorst, 1802)
osseticus (Kolenati, 1846)

vernalis (Gravenhorst, 1806) non (Müller, O.F., 1776)
piliger Mulsant & Rey, 1876
    stipes Sharp, 1910
splendidulus (Gravenhorst, 1802)
trossulus (Nordmann, 1837)
velox Sharp, 1910
    primigenius Joy, 1913

GABRONTHUS Tottenham, 1955
thermarum (Aubé, 1850)

NEOBISNIUS Ganglbauer, 1895
lathrobioides (Baudi, 1848)
    cerrutii Gridelli, 1943 [423]
    rubripennis Gridelli, 1943
procerulus (Gravenhorst, 1806)
prolixus (Erichson, 1840)
villosulus (Stephens, 1833)

PHILONTHUS Stephens, 1829 [424]
SPATULONTHUS Tottenham, 1955 [425]
PARAGABRIUS Coiffait, 1963 [425]
addendus Sharp, 1867
albipes (Gravenhorst, 1802)
    alpinus sensu auctt. Brit. partim non Eppelsheim, 1875 [426]
alpinus Eppelsheim, 1875 [392]
atratus (Gravenhorst, 1802)
carbonarius (Gravenhorst, 1802) [427]
    varius (Gyllenhal, 1810)
cognatus Stephens, 1832
    fuscipennis (Mannerheim, 1830) non (Block, 1799)
concinnus (Gravenhorst, 1802)
    ochropus (Gravenhorst, 1802)
confinis Strand, A., 1941
coprophilus Jarrige, 1949
corruscus (Gravenhorst, 1802)
corvinus Erichson, 1839
cruentatus (Gmelin in Linnaeus, 1790)
debilis (Gravenhorst, 1802)
decorus (Gravenhorst, 1802)
dimidiatipennis Erichson, 1840
discoideus (Gravenhorst, 1802)
ebeninus (Gravenhorst, 1802)
fumarius (Gravenhorst, 1806)
furcifer Renkonen, 1937
intermedius (Lacordaire, 1835)
jurgans Tottenham, 1937
    varians sensu auctt. partim non (Paykull, 1789)
laminatus (Creutzer, 1799)
lepidus (Gravenhorst, 1802)
longicornis Stephens, 1832
mannerheimi Fauvel, 1869
marginatus (Müller, O.F., 1764)
micans (Gravenhorst, 1802)
micantoides Benick & Lohse, 1956
    micans sensu auctt. partim non (Gravenhorst, 1802)
nigrita (Gravenhorst, 1806)
nitidicollis (Lacordaire, 1835)
    bimaculatus (Gravenhorst, 1802) non (Schrank, 1798)
parvicornis (Gravenhorst, 1802)
    agilis (Gravenhorst, 1806) [428]
politus (Linnaeus, 1758)
    aeneus (Rossi, 1790) non (De Geer, 1774)

temporalis sensu Crowson, 1970 non Mulsant &
Rey, 1853
punctus (Gravenhorst, 1802)
quisquiliarius (Gyllenhal, 1810)
rectangulus Sharp, 1874
rotundicollis (Ménétries, 1832)
rubripennis Stephens, 1832
fulvipes sensu auctt. Brit. non (Fabricius, 1792)
varipes Mulsant & Rey, 1861 [429]
rufipes (Stephens, 1832) [430]
immundus sensu auctt. non (Gravenhorst, 1806)
sanguinolentus (Gravenhorst, 1802)
spinipes Sharp, 1874 [431]
splendens (Fabricius, 1792) [137]
niger (Müller, O.F., 1764)
succicola Thomson, C.G., 1860
chalceus sensu Ganglbauer, 1895 non Stephens,
1832
tenuicornis Mulsant & Rey, 1853
carbonarius sensu (Gyllenhal, 1810) non
(Gravenhorst, 1802)
umbratilis (Gravenhorst, 1802)
varians (Paykull, 1789) [432]
ventralis (Gravenhorst, 1802)

RABIGUS Mulsant & Rey, 1876 [433]
PHILONTHUS sensu auctt. partim non Stephens,
1829
pullus (Nordmann, 1837)

REMUS Holme, 1837 [434]
CAFIUS sensu auctt. partim non Stephens, 1829
sericeus Holme, 1837

Subtribe QUEDIINA Kraatz, 1857

ACYLOPHORUS Nordmann, 1837
glaberrimus (Herbst, 1784)

ASTRAPAEUS Gravenhorst, 1802
ulmi (Rossi, 1790) [435]

EURYPORUS Erichson, 1839
picipes (Paykull, 1800)

HETEROTHOPS Stephens, 1829
binotatus (Gravenhorst, 1802) [436]
dissimilis (Gravenhorst, 1802)
minutus Wollaston, 1860 [437]
dissimilis sensu auctt. partim non (Gravenhorst,
1802)
niger Kraatz, 1868 [438]
praevius Erichson, 1839

QUEDIUS Stephens, 1829

Subgenus DISTICHALIUS Casey, 1915
cinctus (Paykull, 1790)

Subgenus MICROSAURUS Dejean, 1833 [439]
aetolicus Kraatz, 1858
brevicornis (Thomson, C.G., 1860)
brevis Erichson, 1840
cruentus (Olivier, 1795)
fulgidus (Fabricius, 1792) [137]

?*rufitarsis* (Marsham, 1802)
assimilis (Nordmann, 1837)
invreae Gridelli, 1924
ochripennis sensu auctt. Brit. non Ménétries, 1832
assecla sensu auctt. non Mulsant & Rey, 1876
lateralis (Graverhorst, 1802)
longicornis Kraatz, 1857
lyszkowskii Lott, 2010 [440]
maurus (Sahlberg, C.R., 1830)
mesomelinus (Marsham, 1802)
microps Gravenhorst, J.L.C., 1847
infuscatus sensu Donisthorpe, 1928 non Erichson,
1840
nigrocaeruleus Fauvel, 1874
puncticollis (Thomson, C.G., 1867)
othiniensis Johansen, 1907
scitus (Gravenhorst, 1806) [137]
analis (Fabricius, 1787)
truncicola Fairmaire & Laboulbène, 1856
ventralis (Aragona, 1830) non (Gravenhorst, 1802)
xanthopus Erichson, 1839

Subgenus QUEDIONUCHUS Sharp, 1884
plagiatus Mannerheim, 1843 [137]
glaber (Müller, O.F., 1764)
flavopterus (Geoffroy, 1785)
laevigatus (Gyllenhal, 1810) non (Marsham, 1802)

Subgenus QUEDIUS Stephens, 1829
balticus Korge, 1960
molochinus sensu Last, 1952 non (Gravenhorst,
1806)
curtipennis Bernhauer, 1908
fuliginosus sensu Britten, 1944 non (Gravenhorst,
1802)
fuliginosus (Gravenhorst, 1802)
subfuliginosus Britten, 1944
levicollis (Brullé, 1832)
tristis sensu auctt. [441]
molochinus (Gravenhorst, 1806)
simplicifrons Fairmaire, 1861 [442]
hispanicus Bernhauer, 1898
pallipes sensu Gridelli, 1924 partim non Lucas,
1849
secundus Last, 1952

Subgenus RAPHIRUS Stephens, 1829
auricomus Kiesenwetter, 1850
boopoides Munster, 1923
picipennis sensu auctt. Brit. non Heer, 1839
scribae sensu auctt. Brit. non Ganglbauer, 1895
boops (Gravenhorst, 1802)
asturicus sensu auctt. non Bernhauer, 1918
arestor Tottenham, 1948
crius Tottenham, 1948
fulvicollis (Stephens, 1833)
fumatus (Stephens, 1833)
humeralis Stephens, 1832
obliteratus Erichson, 1840
suturalis sensu auctt. Brit. partim non Kiesenwetter,
1845
lucidulus Erichson, 1839 [392]
maurorufus (Gravenhorst, 1806)
nemoralis Baudi, 1848
humeralis sensu auctt. Brit. partim non Stephens,
1832
suturalis sensu auctt. Brit. partim non Kiesenwetter,
1845
nigriceps Kraatz, 1857

nitipennis (Stephens, 1833)
　attenuatus sensu (Gyllenhal, 1810) non
　　(Gravenhorst, 1806)
　hyperboreus sensu auctt. Brit. non Erichson, 1840
persimilis Mulsant & Rey, 1876
　boops sensu auctt. Brit. partim non (Gravenhorst,
　　1802)
　aridulus Jansson, 1939 [443]
　corion Tottenham, 1948 [444]
　mallius Tottenham, 1948
picipes (Mannerheim, 1830)
plancus Erichson, 1840
riparius Kellner, 1843
schatzmayri Gridelli, 1922
scintillans (Gravenhorst, 1806)
semiaeneus (Stephens, 1833)
semiobscurus (Marsham, 1802)
　rufipes (Gravenhorst, 1802) non (Linnaeus, 1758)
umbrinus Erichson, 1839
　pseudoumbrinus Lohse, 1961 [445]

VELLEIUS Leach, 1819
dilatatus (Fabricius, 1787)

Subtribe STAPHYLININA Latreille, 1802

CREOPHILUS Leach, 1819
maxillosus (Linnaeus, 1758)

DINOTHENARUS Thomson, C.G., 1858
　STAPHYLINUS sensu auctt. partim non Linnaeus,
　　1758
　TRICHODERMA sensu auctt. non Stephens, 1835
　PLATYDRACUS sensu auctt. partim non Thomson,
　　C.G., 1858
　PARABEMUS Reitter, 1909
pubescens (De Geer, 1774) [446]

EMUS Leach, 1819
hirtus (Linnaeus, 1758)

OCYPUS Leach, 1819
　GOERIUS Westwood, 1827
　STAPHYLINUS sensu auctt. partim non Linnaeus,
　　1758 [447]

Subgenus MATIDUS Motschulsky, 1860 [448]
brunnipes (Fabricius, 1781)
nitens (Schrank, 1781)
　nero (Faldermann, 1835) [449]
　similis sensu (Fabricius, 1792) non (Herbst, 1784)

Subgenus OCYPUS Leach, 1819
olens (Müller, O.F., 1764)
ophthalmicus (Scopoli, 1763)

Subgenus PSEUDOCYPUS Mulsant & Rey, 1876 [450]
aeneocephalus (De Geer, 1774)
　cupreus sensu auctt. Brit. partim non (Rossi, 1790)
fortunatarum (Wollaston, 1871)
　cupreus sensu auctt. Brit. partim non (Rossi, 1790)
fuscatus (Gravenhorst, 1802)

ONTHOLESTES Ganglbauer, 1895
murinus (Linnaeus, 1758)
tessellatus (Geoffroy, 1785)

PLATYDRACUS Thomson, C.G., 1858
　STAPHYLINUS sensu auctt. partim non Linnaeus,
　　1758
fulvipes (Scopoli, 1763)
latebricola (Gravenhorst, 1806)
stercorarius (Olivier, 1795)

STAPHYLINUS Linnaeus, 1758
caesareus Cederhjelm, 1798
dimidiaticornis Gemminger, 1851
　caesareus sensu auctt. Brit. partim non
　　(Cederhjelm, 1798)
　parumtomentosus Stein, 1903
erythropterus Linnaeus, 1758

TASGIUS Stephens, 1829 [451]
　STAPHYLINUS sensu auctt. partim non Linnaeus,
　　1758 [452]
　OCYPUS sensu auctt. partim non Leach, 1819
　ALAPSODUS Tottenham, 1939

Subgenus RAYACHEILA Motschulsky, 1845 [453]
globulifer (Geoffroy, 1785)
　siculus Stierlin, 1864 non (Aubé, 1842)
　evitendus Tottenham, 1945
melanarius (Heer, 1839)
　globulifer sensu auctt. non (Geoffroy, 1785)
morsitans (Rossi, 1790)
　compressus (Marsham, 1802) [454]
winkleri Bernhauer, 1906

Subgenus TASGIUS Stephens, 1829
ater (Gravenhorst, 1802)
pedator (Gravenhorst, 1802)

Tribe XANTHOLININI Erichson, 1839

GAUROPTERUS Thomson, C.G., 1860
fulgidus (Fabricius, 1787)

GYROHYPNUS Leach, 1819
angustatus Stephens, 1833
　scoticus (Joy, 1913) [455]
atratus (Heer, 1839)
fracticornis (Müller, O.F., 1776)
　punctulatus sensu auctt. partim non (Paykull, 1789)
punctulatus (Paykull, 1789) [456]
　liebei sensu auctt. non (Scheerpeltz, 1926)
wagneri (Scheerpeltz, 1926) [392]

HYPNOGYRA Casey, 1906 [457]
　XANTHOLINUS sensu auctt. partim non Dejean,
　　1821
　PHALACROLINUS Coiffait, 1972
angularis Ganglbauer, 1895
　glabra (Gravenhorst, 1806) non (Müller, O.F., 1776)

LEPTACINUS Erichson, 1839
batychrus (Gyllenhal, 1827)
formicetorum Märkel, 1841
intermedius Donisthorpe, 1936
　sulcifrons sensu auctt. non (Stephens, 1833)
pusillus (Stephens, 1833)
　linearis (Gravenhorst, 1802) non (Olivier, 1795)

MEGALINUS Mulsant & Rey, 1877 [458]
XANTHOLINUS sensu auctt. partim non Dejean, 1821
glabratus (Gravenhorst, 1802)

NUDOBIUS Thomson, C.G., 1860
lentus (Gravenhorst, 1806)

PHACOPHALLUS Coiffait, 1956
parumpunctatus (Gyllenhal, 1827)
pallidipennis (Motschulsky, 1858) [459]
tricolor (Kraatz, 1859)

XANTHOLINUS Dejean, 1821

Subgenus PURROLINUS Coiffait, 1956
elegans (Olivier, 1795) [460]
tricolor sensu auctt. partim non (Fabricius, 1787)
meridionalis (Lacordaire, 1835)
semirufus sensu auctt. non Reitter, 1901
jarrigei Coiffait, 1956
tricolor (Fabricius, 1787) [461]
meyeri Drugmand, 1994

Subgenus TYPHLOLINUS Reitter, 1908
laevigatus Jacobsen, 1849
distans sensu auctt. Brit. non Mulsant & Rey, 1853
cribripennis Fauvel, 1873

Subgenus XANTHOLINUS Dejean, 1821
gallicus Coiffait, 1956
linearis (Olivier, 1795)
longiventris Heer, 1839

Superfamily SCARABAEOIDEA Latreille, 1802

20. Family GEOTRUPIDAE Latreille, 1802 [462]

Family author: D.J. Mann

Subfamily BOLBOCERATINAE Mulsant, 1842

ODONTEUS Samouelle, 1819
ODONTAEUS Dejean, 1821
armiger (Scopoli, 1772)
mobilicornis (Fabricius, 1775)
testaceus (Fabricius, 1775)

Subfamily GEOTRUPINAE Latreille, 1802

TYPHAEUS Leach, 1815
TYPHOEUS auctt. (misspelling)
MINOTAURUS Mulsant, 1855
typhoeus (Linnaeus, 1758)
pumilus (Marsham, 1802)
vulgaris (Leach, 1815)
thyphoeus (Leach, 1815)

ANOPLOTRUPES Jekel, 1866
GEOTRUPES sensu auctt. partim non Latreille, 1796
stercorosus (Scriba, 1791)
sylvaticus (Panzer, 1798)

GEOTRUPES Latreille, 1797
mutator (Marsham, 1802)
spiniger (Marsham, 1802) [463]
foveatus (Marsham, 1802) [464]
puncticollis Malinowsky, 1811
sublaevigatus Stephens, 1830
stercorarius (Linnaeus, 1758)
punctatostriatus Stephens, 1830

TRYPOCOPRIS Motschulsky, 1860
GEOTRUPES sensu auctt. partim non Latreille, 1796
pyrenaeus (Charpentier, 1825)
vernalis (Linnaeus, 1758)

21. Family TROGIDAE MacLeay, 1819 [465]

Family author: D J. Mann

TROX Fabricius, 1775
perlatus (Goeze, 1777)
hispidus sensu auctt. Brit. non (Pontoppidan, 1763)
sabulosus (Linnaeus, 1758)
scaber (Linnaeus, 1767)
arenosus (Gmelin in Linnaeus, 1790)
arenarius (Paykull, 1798)
trisulcatus Curtis, 1845

22. Family LUCANIDAE Latreille, 1804 [466]

Family author: D.J. Mann

Subfamily SYNDESINAE MacLeay, 1819

SINODENDRON Hellwig, 1792
cylindricum (Linnaeus, 1758)

Subfamily LUCANINAE Latreille, 1804

DORCUS MacLeay, 1819
parallelipipedus (Linnaeus, 1758)
parallelopipedus auctt. (misspelling)

LUCANUS Scopoli, 1763
cervus (Linnaeus, 1758)
inermis (Marsham, 1802)

PLATYCERUS Geoffroy, 1762
caraboides (Linnaeus, 1758)

23. Family SCARABAEIDAE Latreille, 1802 [467]

Family author: D.J. Mann

Subfamily AEGIALIINAE Laporte, 1840

AEGIALIA Latreille, 1807

Subgenus AEGIALIA Latreille, 1807
arenaria (Fabricius, 1787)
globosa (Kugelann, 1794)

Subgenus PSAMMOPORUS Thomson, C.G., 1859
insularis Pittino, 2006 [468]
　　sabuleti sensu auctt. Brit. non (Panzer, 1797)

Subgenus RHYSOTHORAX Bedel, 1911
rufa (Fabricius, 1792) [469]
　　rufina Silfverberg, 1977

Subfamily APHODINAE Leach, 1815 [470]

Tribe APHODIINI Leach, 1815

APHODIUS Hellwig, 1798 [471] [472] [473]

Subgenus ACROSSUS Mulsant, 1842
depressus (Kugelann, 1792)
luridus (Fabricius, 1775)
　　gagates (Geoffroy, 1764)
　　variegatus (Herbst, 1783)
　　nigrosulcatus (Marsham, 1802)
rufipes (Linnaeus, 1758)
　　muticus (Stephens, 1830)

Subgenus AGOLIINUS Schmidt, 1913
lapponum Gyllenhal, 1808
　　axillaris Stephens, 1839
　　subalpinus Hardy, 1847
nemoralis Erichson 1848

Subgenus AGRILINUS Mulsant & Rey, 1870
ater (De Geer, 1774)
　　terrestris (Fabricius, 1775)
　　obscurus (Marsham, 1802) non (Fabricius, 1792)
　　terrenus Stephens, 1830
constans Duftschmid, 1805
　　nitidus Stephens, 1830
rufus (Moll, 1782) [474]
　　scybalarius (Fabricius, 1781)
　　rufescens (Fabricius, 1801)
　　castaneus (Marsham, 1802)
　　unicolor (Marsham, 1802)
　　ochraceus Stephens, 1830
sordidus (Fabricius, 1775)

Subgenus AMMOECIUS Mulsant, 1842
brevis Erichson, 1848

Subgenus APHODIUS Hellwig, 1798
fimetarius (Linnaeus, 1758)
foetens (Fabricius, 1787)
　　scrutator sensu (Marsham, 1802) non (Herbst, 1789)
　　aestivalis Stephens, 1839
foetidus (Herbst, 1783) [475]
　　scybalarius sensu auctt. non (Fabricius, 1781)
　　coprinus (Marsham, 1802)
pedellus (De Geer, 1774) [476]
　　fimetarius sensu auctt. partim non (Linnaeus, 1758)

Subgenus BODILUS Mulsant & Rey, 1870
ictericus (Laicharting, 1781)
　　nitidulus (Fabricius, 1792)

Subgenus CALAMOSTERNUS Motschulsky, 1860
granarius (Linnaeus, 1767)
　　emarginatus Stephens, 1830
　　haemorrhous Stephens, 1830
　　lucens Stephens, 1830
　　melanopus Stephens, 1830

Subgenus CHILOTHORAX Motschulsky, 1860
conspurcatus (Linnaeus, 1758)
distinctus (Müller, O.F., 1776)
　　inquinatus (Herbst, 1783)
　　attaminatus (Marsham, 1802)
　　foedatus (Marsham, 1802)
paykulli Bedel, 1907
　　inquinatus sensu (Olivier, 1789) non (Herbst, 1783)
　　tessulatus (Paykull, 1798) non (Laicharting, 1781)

Subgenus COLOBOPTERUS Mulsant, 1842
erraticus (Linnaeus, 1758)

Subgenus ESYMUS Mulsant & Rey, 1870
merdarius (Fabricius, 1775)
pusillus (Herbst, 1789)
　　phaeopterus Stephens, 1830

Subgenus EUORODALUS Dellacasa, G., 1983
coenosus (Panzer, 1798)
　　tristis Panzer, 1833

Subgenus EUPLEURUS Mulsant, 1842
subterraneus (Linnaeus, 1758)

Subgenus LABARRUS Mulsant & Rey, 1870
lividus (Olivier, 1789)
　　bilituratus (Marsham, 1802)

Subgenus LIMARUS Mulsant & Rey, 1870
zenkeri Germar, 1813

Subgenus LIOTHORAX Motschulsky, 1860
niger (Illiger, 1798) [477]
　　niger (Panzer, 1797)
plagiatus (Linnaeus, 1767)
　　elongatus Stephens, 1830

Subgenus MELINOPTERUS Mulsant, 1842
consputus Creutzer, 1799
prodromus (Brahm, 1790)
punctatosulcatus Sturm, 1805 [478]
　　sabulicola Thomson, C.G., 1868
sphacelatus (Panzer, 1798)
　　marginellus Stephens, 1830
　　punctatosulcatus sensu auctt. Brit. non Sturm, 1805

Subgenus NIMBUS Mulsant & Rey, 1870
contaminatus (Herbst, 1783)
　　ciliaris (Marsham, 1802)
obliteratus Sturm, 1823

Subgenus OTOPHORUS Mulsant, 1842
haemorrhoidalis (Linnaeus, 1758)
　　sanguinolentus (Marsham, 1802)

Subgenus PHALACRONOTHUS Motschulsky, 1860
quadrimaculatus (Linnaeus, 1761)

Subgenus PLAGIOGONUS Mulsant, 1842
arenarius (Olivier, 1789) [479]
　　putridus (Geoffroy in Fourcroy, 1785) [480]
　　rhododactylus (Marsham, 1802)

Subgenus PLANOLINUS Mulsant & Rey, 1870
borealis Gyllenhal, 1827
　　putridus sensu (Sturm, 1805) non (Geoffroy in Fourcroy, 1785)
fasciatus (Olivier, 1789) [481]
　　putridus sensu (Herbst, 1789) non (Geoffroy in Fourcroy, 1785)
　　foetidus (Fabricius, 1792) non (Herbst, 1783)
　　uliginosus Hardy, 1847
　　tenellus sensu auctt. non Say, 1823

Subgenus SIGORUS Mulsant & Rey, 1870
porcus (Fabricius, 1792)
   turpis (Marsham, 1802)
   ruficrus (Marsham, 1802)

Subgenus TEUCHESTES Mulsant, 1842
fossor (Linnaeus, 1758)

Subgenus VOLINUS Mulsant & Rey, 1870
sticticus (Panzer, 1798) [482]
   equestris (Panzer, 1798)

EUHEPTAULACUS Dellacasa, G., 1983
   HEPTAULACUS sensu auctt. non Mulsant, 1842
sus (Herbst, 1783)
villosus (Gyllenhal in Schönherr, 1806)

HEPTAULACUS Mulsant, 1842
testudinarius (Fabricius, 1775)

OXYOMUS Dejean, 1833
sylvestris (Scopoli, 1763)
   silvestris auctt. (misspelling)
   porcatus (Fabricius, 1775)
   platycephalus (Marsham, 1802)

Tribe EUPARIINI Schmidt, 1910

SAPROSITES Redtenbacher, 1858
mendax (Blackburn, 1892)
   parallelus sensu Champion, 1921 non Harold, 1867
natalensis (Peringuey, 1901) [483]

Tribe PSAMMODIINI Mulsant, 1842

Subtribe PSAMMODIINA Mulsant, 1842

BRINDALUS Landin, 1960
porcicollis (Illiger, 1803) [484]

DIASTICTUS Mulsant, 1842
vulneratus (Sturm, 1805)

PSAMMODIUS Fallén, 1807
   PSAMMOBIUS Heer, 1841
asper (Fabricius, 1775)
   sulcicollis (Illiger, 1802)

TESARIUS Rakovic, 1981
   PSAMMODIUS sensu auctt. partim non Fallén,
   1807
caelatus (LeConte, 1857)
mcclayi (Cartwright, 1955) [485]

Subtribe RHYSSEMINA Pittino & Mariani, 1986

PLEUROPHORUS Mulsant, 1842
   PSAMMOBIUS sensu auctt. partim non Heer, 1841
caesus (Creutzer in Panzer, 1796)

RHYSSEMUS Mulsant, 1842
germanus (Linnaeus, 1767)

Subfamily SCARABAEINAE Latreille, 1802

Tribe COPRINI Leach, 1815

COPRIS Geoffroy, 1762
lunaris (Linnaeus, 1758)

Tribe ONTHOPHAGINI Burmeister, 1846

ONTHOPHAGUS Latreille, 1802 [486] [487]

Subgenus ONTHOPHAGUS Latreille, 1802
taurus (Schreber, 1759)

Subgenus PALAEONTHOPHAGUS Zunino, 1979
coenobita (Herbst, 1783)
fracticornis (Preyssler, 1790) [488]
   similis sensu auctt. non (Scriba, 1790)
joannae Goljan, 1953
   ovatus sensu auctt. partim non (Linnaeus, 1767)
medius (Kugelann, 1792) [489]
   vacca sensu auctt. partim non (Linnaeus, 1767)
nuchicornis (Linnaeus, 1758)
   verticicornis (Fabricius, 1775)
   xiphias (Fabricius, 1792)
   dillwynii (Stephens, 1830)
similis (Scriba, 1790)
   fracticornis sensu auctt. Brit. non (Preyssler, 1790)
verticicornis (Laicharting, 1781) [490]
   nutans (Fabricius, 1787)

Subfamily MELOLONTHINAE Leach, 1819

Tribe HOPLIINI Latreille, 1829

HOPLIA Illiger, 1803

Subgenus DECAMERA Mulsant, 1842
philanthus (Füessly, 1775)
   farinosa sensu auctt. non (Linnaeus, 1761) [491]

Tribe MELOLONTHINI Leach, 1819

Subtribe MELOLONTHINA Leach, 1819

MELOLONTHA Fabricius, 1775 [492]
hippocastani Fabricius, 1801
melolontha (Linnaeus, 1758)
   vulgaris Fabricius, 1775

POLYPHYLLA Harris, 1842
fullo (Linnaeus, 1758)

Subtribe RHIZOTROGINA Burmeister, 1855

AMPHIMALLON Latreille, 1825
   AMPHIMALLUS Mulsant, 1842 (misspelling)
   RHIZOTROGUS sensu auctt. partim non Berthold,
   1827
fallenii (Gyllenhal, 1817) [493]
   ochraceum sensu auctt. partim non (Knoch, 1801)
solstitiale (Linnaeus, 1758)

Tribe SERICINI Kirby 1837

OMALOPLIA Schönherr, 1817
　　HOMALOPLIA Erichson, 1847
ruricola (Fabricius, 1775)
　　varius (Marsham 1802)

SERICA MacLeay, 1819
brunnea (Linnaeus, 1758)

Subfamily RUTELINAE MacLeay, 1819

Tribe ANOMALINI Streubel, 1839

ANOMALA Samouelle, 1819
　　EUCHLORA MacLeay, 1819
dubia (Scopoli, 1763)
　　aenea (De Geer, 1774)
　　frischii (Fabricius, 1775)

PHYLLOPERTHA Stephens, 1830
horticola (Linnaeus, 1758)
　　arvicola sensu (Marsham, 1802) non (Fabricius, 1775)

Subfamily CETONIINAE Leach, 1815 [494]

Tribe CETONIINI Leach, 1815

CETONIA Fabricius, 1775
aurata (Linnaeus, 1761)

PROTAETIA Burmeister, 1842
　　CETONIA sensu auctt. partim non Fabricius, 1775

Subgenus POTOSIA Mulsant & Rey, 1870
metallica (Herbst, 1782)
　　cuprea sensu auctt. Brit. non (Fabricius, 1775)
　　floricola (Herbst, 1790)

Tribe TRICHIINI Fleming, 1821

GNORIMUS Lepeletier & Audinet-Serville, 1828 [495]
　　ALEUROSTICTUS Kirby, 1827
nobilis (Linnaeus, 1758)
variabilis (Linnaeus, 1758)
　　octopunctatus (Fabricius, 1775)

TRICHIUS Fabricus, 1775
fasciatus (Linnaeus, 1758)
gallicus Dejean, 1821
　　rosaceus (Voët, 1769) [496]
　　zonatus sensu auctt. Brit. non Germar, 1831
　　abdominalis sensu auctt. Brit. non (Ménétries, 1832)

Superfamily SCIRTOIDEA Fleming, 1821

24. Family EUCINETIDAE Lacordaire, 1857

Family author: A.G. Duff

EUCINETUS Germar, 1818
meridionalis (Laporte, 1836)

25. Family CLAMBIDAE Fischer von Waldheim, 1821

Family author: C. Johnson

Subfamily CALYPTOMERINAE Crowson, 1955

CALYPTOMERUS Redtenbacher, 1849
　　CLAMBUS sensu auctt. Brit. partim non Fischer von Waldheim, 1821
　　COMAZUS Fairmaire & Laboulbène, 1854
dubius (Marsham, 1802)
　　enshamensis (Stephens, 1829)

Subfamily CLAMBINAE Fischer von Waldheim, 1821

CLAMBUS Fischer von Waldheim, 1821
armadillo (De Geer, 1774)
　　armadillus auctt. (misspelling)
　　convexus (Marsham, 1802)
　　atomarius (Sturm, 1807)
　　nanus (Stephens, 1829)
evae Endrödy-Younga, 1960
　　minutus sensu auctt. Brit. partim non (Sturm, 1807)
gibbulus (LeConte, 1850)
　　punctulum sensu auctt. Brit. partim non (Beck, 1817)
　　radula Endrödy-Younga, 1960
nigrellus Reitter, 1914
　　minutus sensu auctt. Brit. partim non (Sturm, 1807)
nigriclavis Stephens, 1832
　　minutus sensu (Stephens, 1829) non (Sturm, 1807)
pallidulus Reitter, 1911
　　minutus sensu auctt. Brit. partim non (Sturm, 1807)
pubescens Redtenbacher, 1849
punctulum (Beck, 1817)
　　punctulus auctt. (misspelling)
　　nitidus Stephens, 1832
　　borealis Strand, A., 1946
simsoni Blackburn, 1902 [497]

26. Family SCIRTIDAE Fleming, 1821

Family author: G.N. Foster

ELODES Latreille, 1796
　　HELODES sensu auctt. non Paykull, 1799
elongata Tournier, 1868 [498]
　　minuta sensu auctt. partim non (Linnaeus, 1767)
　　koelleri Klausnitzer, 1971
minuta (Linnaeus, 1767)
pseudominuta Klausnitzer, 1971 [498]

minuta sensu auctt. partim non (Linnaeus, 1767)
tricuspis Nyholm, 1985 [498]
   minuta sensu auctt. partim non (Linnaeus, 1767)
   elongata sensu auctt. partim non Tournier, 1868

ODELES Klausnitzer, 2004
marginata (Fabricius, 1798) [499]

MICROCARA Thomson, C.G., 1859
testacea (Linnaeus, 1767)
   livida (Fabricius, 1792) non (Geoffroy in Fourcroy, 1785)
   bohemani sensu auctt. Brit. non (Mannerheim, 1844)

CYPHON Paykull, 1799
coarctatus Paykull, 1799
   paykulli Guérin-Méneville, 1843
   nitidulus Thomson, C.G., 1855
hilaris Nyholm, 1944
   ochraceus sensu auctt. partim non Stephens, 1830
kongsbergensis Munster, 1924 [500]
laevipennis Tournier, 1868
   phragmiteticola Nyholm, 1955 [501]
   variabilis sensu auctt. Brit. partim non (Thunberg, 1787)
ochraceus Stephens, 1830
   pallidulus Boheman, 1849
padi (Linnaeus, 1758)
palustris Thomson, C.G., 1855
   coarctatus sensu auctt. non Paykull, 1799
pubescens (Fabricius, 1792)
   variabilis sensu auctt. Brit. partim non (Thunberg, 1787)
punctipennis Sharp, 1873
variabilis (Thunberg, 1787)

PRIONOCYPHON Redtenbacher, 1858
serricornis (Müller, P.W.J., 1821)

HYDROCYPHON Redtenbacher, 1858
deflexicollis (Müller, P.W.J., 1821)

SCIRTES Illiger, 1807
hemisphaericus (Linnaeus, 1758)
orbicularis (Panzer, 1793)

Superfamily DASCILLOIDEA Guérin-Méneville, 1843 (1834)

27. Family DASCILLIDAE Guérin-Méneville, 1843 (1834)

Family author: A.G. Duff

DASCILLUS Latreille, 1796
cervinus (Linnaeus, 1758)

Superfamily BUPRESTOIDEA Leach, 1815

28. Family BUPRESTIDAE Leach, 1815 [502]

Family author: B. Levey

Subfamily BUPRESTINAE Leach, 1815

Tribe ANTHAXIINI Gory & Laporte, 1839

ANTHAXIA Eschscholtz, 1829
Subgenus ANTHAX A Eschscholtz, 1829
nitidula (Linnaeus, 1758)
Subgenus MELANTHAXIA Rikhter, 1949
quadripunctata (Linnaeus, 1758) [503]

Tribe MELANOPHILINI Bedel, 1921

MELANOPHILA Eschscholtz, 1829
acuminata (De Geer, 1774)

Subfamily AGRILINAE Laporte, 1835

Tribe AGRILINI Laporte, 1835

AGRILUS Curtis, 1825
Subgenus AGRILUS Curtis, 1825
viridis (Linnaeus, 1758)
   littlei Curtis, 1840 [504]
Subgenus ANAMBUS Thomson, C.G., 1864
angustulus (Illiger, 1803) [505]
   pavidus (Fabricius, 1792)
   viridis sensu Stephens, 1830 partim non (Linnaeus, 1758)
biguttatus (Fabricius, 1777) [506]
   pannonicus (Piller & Mitterpacher, 1783)
cuprescens (Ménétries, 1832) [507]
cyanescens Ratzeburg, 1837 [507]
laticornis (Illiger, 1803)
   viridis sensu Stephens, 1830 partim non (Linnaeus, 1758)
sinuatus (Olivier, 1790)
sulcicollis Lacordaire, 1835 [508]

Tribe APHANISTICINI Jacquelin du Val, 1859

APHANISTICUS Latreille, 1829
emarginatus (Olivier, 1790)
pusillus (Olivier, 1790)

Tribe TRACHEINI Laporte, 1835

TRACHYS Fabricius, 1801
minuta (Linnaeus, 1758)
scrobiculatus Kiesenwetter, 1857
   pumilus sensu auctt. non (Illiger, 1803)
subglaber Rey, 1891 [509]
troglodytes Gyllenhal in Schönherr, 1817

Superfamily BYRRHOIDEA Latreille, 1804

29. Family BYRRHIDAE Latreille, 1804

Family author: R.G. Booth

Subfamily BYRRHINAE Latreille, 1804

SIMPLOCARIA Stephens, 1829
maculosa Erichson, 1847
semistriata (Fabricius, 1794) [510]
   picipes (Olivier, 1790)

MORYCHUS Erichson, 1846
aeneus (Fabricius, 1775)

CYTILUS Erichson, 1846
sericeus (Forster, 1771)
   varius (Fabricius, 1775)

BYRRHUS Linnaeus, 1767 [511]
arietinus Steffahny, 1842
   fasciatus sensu auctt. Brit. partim non (Forster, 1770)
fasciatus (Forster, 1770)
pilula (Linnaeus, 1758)
   dennii Curtis, 1826
pustulatus (Forster, 1770)
   niger (Forster, 1770)
   dorsalis Fabricius, 1787

PORCINOLUS Mulsant & Rey, 1869
   BYRRHUS sensu Fowler, 1889 partim non Linnaeus, 1767
murinus (Fabricius, 1794)
   rubidus (Kugelann, 1792) [512]
   undulatus (Kugelann, 1792) [512]

Subfamily SYNCALYPTINAE Mulsant & Rey, 1869

CHAETOPHORA Kirby & Spence, 1823
   CHAETOPHORUS Kirby & Spence, 1828
   SYNCALYPTA Dillwyn, 1829
spinosa (Rossi, 1794)

CURIMOPSIS Ganglbauer, 1902
   SYNCALYPTA sensu auctt. partim non Dillwyn, 1829
maritima (Marsham, 1802)
   striatopunctata (Steffahny, 1842)
   hirsuta (Sharp, 1871)
nigrita (Palm, 1934) [513]
setigera (Illiger, 1798)

30. Family ELMIDAE Curtis, 1830

Family author: G.N. Foster

ELMIS Latreille, 1802
   HELMIS Bedel, 1878 (misspelling)
aenea (Müller, P.W.J., 1806)

   megerlei sensu auctt. Brit. non (Duftschmid, 1805)
   maugei sensu auctt. Brit. non Bedel, 1878

ESOLUS Mulsant & Rey, 1872
   ELMIS sensu Fowler, 1889 partim non Latreille, 1802
parallelepipedus (Müller, P.W.J., 1806)
   parallelopipedus auctt. (misspelling)

LIMNIUS Illiger, 1802
   ELMIS sensu Fowler, 1889 partim non Latreille, 1802
   LATELMIS Reitter, 1883
volckmari (Panzer, 1793)
   volkmari auctt. (misspelling)

MACRONYCHUS Müller, P.W.J., 1806
quadrituberculatus Müller, P.W.J., 1806

NORMANDIA Pic, 1900
   RIOLUS sensu Fowler, 1889 partim non Mulsant & Rey, 1872
   APTYKTOPHALLUS Steffan, 1958
nitens (Müller, P.W.J., 1817)
   sodalis sensu auctt. Brit. non (Erichson, 1847)

OULIMNIUS des Gozis, 1886
   LIMNIUS sensu auctt. Brit. non Illiger, 1802
major (Rey, 1889) [514]
   falcifer Berthélemy, 1962
rivularis (Rosenhauer, 1856)
   variabilis sensu auctt. Brit. non (Stephens, 1828)
troglodytes (Gyllenhal, 1827)
tuberculatus (Müller, P.W.J., 1806)
   variabilis (Stephens, 1828)

RIOLUS Mulsant & Rey, 1872
   ELMIS sensu Fowler, 1889 partim non Latreille, 1802
cupreus (Müller, P.W.J., 1806)
subviolaceus (Müller, P.W.J., 1817)

STENELMIS Dufour, 1835
canaliculata (Gyllenhal, 1808)

31. Family DRYOPIDAE Billberg, 1820 (1817)

Family author: G.N. Foster

POMATINUS Sturm, 1853
   HELICHUS sensu auctt. Europ. non Erichson, 1847 [515]
substriatus (Müller, P.W.J., 1806)

DRYOPS Olivier, 1791

Subgenus DRYOPS Olivier, 1791
   PARNUS Fabricius, 1792
anglicanus Edwards, J., 1909
auriculatus (Fourcroy, 1785)
   prolifericornis (Fabricius, 1792)
ernesti des Gozis, 1886
   auriculatus sens (Fowler, 1889) non (Fourcroy, 1785)

griseus (Erichson, 1847)
luridus (Erichson, 1847)
   prolifericornis sensu auctt. Brit. non (Fabricius, 1792)
similaris Bollow, 1936 [516]
   griseus sensu auctt. partim non (Erichson, 1847)
striatellus (Fairmaire & Brisout de Barneville, 1859)
   algiricus sensu auctt. Brit. non Lucas, 1846

Subgenus YRDOPS Steffan, 1961
nitidulus (Heer, 1841)

32. Family LIMNICHIDAE Erichson, 1846

Family author: G.N. Foster

LIMNICHUS Dejean, 1821
pygmaeus (Sturm, 1807)

33. Family HETEROCERIDAE MacLeay, 1825

Family author: G.N. Foster

AUGYLES Schiødte, 1866 [517]
   HETEROCERUS sensu auctt. partim non Fabricius, 1792
hispidulus (Kiesenwetter, 1843)
maritimus (Guérin-Méneville, 1844)
   sericans sensu auctt. Brit. non (Kiesenwetter, 1843)
   britannicus (Kuwert, 1890)

HETEROCERUS Fabricius, 1792
fenestratus (Thunberg, 1784)
   laevigatus Panzer, 1794
   pulchellus Kiesenwetter, 1843
flexuosus Stephens, 1828
   femoralis Krynicky, 1832
fossor Kiesenwetter, 1843
   rectus Waterhouse, G.R., 1859
fusculus Kiesenwetter, 1843
marginatus (Fabricius, 1787)
obsoletus Curtis, 1828

34. Family PSEPHENIDAE Lacordaire, 1854

Family author: G.N. Foster

Subfamily EUBRIINAE Lacordaire, 1857

EUBRIA Germar, 1818
palustris Germar, 1818

35. Family PTILODACTYLIDAE Laporte, 1836

Family author: G.N. Foster

PTILODACTYLA Illiger, 1807
exotica Chapin, 1927 [518]

Superfamily ELATEROIDEA Leach, 1815

36. Family EUCNEMIDAE Eschscholtz, 1829

Family author: H. Mendel

Subfamily MELASINAE Fleming, 1821

Tribe MELASINI Fleming, 1821

MELASIS Olivier, 1790
buprestoides (Linnaeus, 1761)

ISORHIPIS Lacordaire, 1835
melasoides (Laporte, 1835) [519]

Tribe EPIPHANINI Muona, 1993

HYLIS des Gozis, 1886
   HYPOCAELUS sensu Guérin-Méneville, 1843 non Dejean, 1833
   HYPOCOELUS auctt. (misspelling)
cariniceps (Reitter, 1902)
olexai (Palm, 1955)
   procerulus sensu auctt. non (Mannerheim, 1823)

EPIPHANIS Eschscholtz, 1829
cornutus Eschscholtz, 1829

Tribe DIRHAGINI Reitter, 1911

MICRORHAGUS Dejean, 1833
   DIRHAGUS Latreille, 1834
pygmaeus (Fabricius, 1792)

Subfamily EUCNEMINAE Eschscholtz, 1829

EUCNEMIS Ahrens, 1812
capucina Ahrens, 1812

37. Family THROSCIDAE Laporte, 1840

Family author: H. Mendel

AULONOTHROSCUS Horn, 1890
   TRIXAGUS sensu auctt. partim non Kugelann, 1794
brevicollis (de Bonvouloir, 1859)

TRIXAGUS Kugelann, 1794
   THROSCUS Latreille, 1796
carinifrons (de Bonvouloir, 1859)
   seriatus Blair, 1942
dermestoides (Linnaeus, 1767)
gracilis Wollaston, 1854
   elateroides sensu auctt. Brit. non (Heer, 1841)
obtusus (Curtis, 1827)

38. Family ELATERIDAE Leach, 1815

Family author: H. Mendel

Subfamily AGRYPNINAE Candèze, 1857

AGRYPNUS Eschscholtz, 1829
ADELOCERA sensu auctt. partim non Latreille, 1829
LACON sensu auctt. non Laporte, 1836
BRACHYLACON Motschulsky, 1858
murinus (Linnaeus, 1758)

LACON Laporte, 1836
ADELOCERA sensu auctt. partim non Latreille, 1829
querceus (Herbst, 1784)
varius (Olivier, 1790)

Subfamily HYPNOIDINAE Schwarz, 1906 (1860)

HYPNOIDUS Dillwyn, 1829
CRYPTOHYPNUS Eschscholtz, 1830
riparius (Fabricius, 1792)

Subfamily DENTICOLLINAE Stein & Weise, 1877 (1848)

Tribe CTENICERINI Fleutiaux, 1936

ACTENICERUS Kiesenwetter, 1858
CORYMBITES sensu Fowler, 1889 partim non Latreille, 1834
sjaelandicus (Müller, O.F., 1764)
tessellatus sensu (Fabricius, 1775) non (Linnaeus, 1758)

ANOSTIRUS Thomson, C.G., 1859
CALOSTIRUS Thomson, C.G., 1864
CORYMBITES sensu Fowler, 1889 partim non Latreille, 1834
castaneus (Linnaeus, 1758)

CTENICERA Latreille, 1829
LUDIUS sensu Eschscholtz, 1829 non Berthold, 1827
CTENICERUS Stephens, 1830 (misspelling)
CORYMBITES Latreille, 1834
cuprea (Fabricius, 1775)
aeruginosa sensu (Fabricius, 1798) non (Olivier, 1790)
pectinicornis (Linnaeus, 1758)

CALAMBUS Thomson, C.G., 1859
SELATOSOMUS sensu auctt. partim non Stephens, 1830
bipustulatus (Linnaeus, 1767)

APLOTARSUS Stephens, 1830
SELATOSOMUS sensu auctt. partim non Stephens, 1830
angustulus (Kiesenwetter, 1858)
incanus (Gyllenhal, 1827)

quercus sensu (Gyllenhal, 1808) non (Olivier, 1790)
ochropterus (Stephens, 1830)

PARAPHOTISTUS Kishii, 1966
SELATOSOMUS sensu auctt. partim non Stephens, 1830
impressus (Fabricius, 1792)
nigricornis (Panzer, 1799)
metallicus (Paykull, 1800)

PROSTERNON Latreille, 1834
CORYMBITES sensu Fowler, 1889 non Latreille, 1834
tessellatum (Linnaeus, 1758)
holosericeum (Olivier, 1790)

SELATOSOMUS Stephens, 1830
CORYMBITES sensu Fowler, 1889 partim non Latreille, 1834
aeneus (Linnaeus, 1758)
cruciatus (Linnaeus, 1758)
melancholicus (Fabricius, 1798) [520]

Tribe DENTICOLLINI Stein & Weise, 1877 (1848)

CIDNOPUS Thomson, C.G., 1859
LIMONIUS sensu auctt. partim non Eschscholtz, 1829
aeruginosus (Olivier, 1790)
cylindricus (Rossi, 1792)

KIBUNEA Kishii, 1966
LIMONIUS sensu auctt. partim non Eschscholtz, 1829
CIDNOPUS sensu auctt. partim non Thomson, C.G., 1859
minuta (Linnaeus, 1758)

LIMONISCUS Reitter, 1905
violaceus (Müller, P.W.J., 1821)

DENTICOLLIS Piller & Mitterpacher, 1783
CAMPYLUS Fischer von Waldheim, 1823/4
linearis (Linnaeus, 1758)

ATHOUS Eschscholtz, 1829

Subgenus ATHOUS Eschscholtz, 1829
haemorrhoidalis (Fabricius, 1801)
obscurus sensu (Paykull, 1800) non (Linnaeus, 1758)
vittatus (Fabricius, 1792) [521]
advena (Scopoli, 1763) [32]

Subgenus EXANATHROTUS Méquignon, 1930
subfuscus (Müller, O.F., 1764)

Subgenus ORTHATHOUS Reitter, 1905
bicolor (Goeze, 1777)
longicollis (Olivier, 1790)
campyloides Newman, 1833
difformis Lacordaire, 1835
cavus sensu Waterhouse, G.R., 1858 non (Germar, 1817)

DIACANTHOUS Reitter, 1905
HARMINIUS sensu auctt. non Fairmaire, 1852
undulatus (De Geer, 1774)

STENAGOSTUS Thomson, C.G., 1859
ATHOUS sensu auctt. Brit. partim non Eschscholtz,
1829
rhombeus (Olivier, 1790)
villosus sensu auctt. non (Fourcroy, 1785)

HEMICREPIDIUS Germar, 1839
ATHOUS sensu auctt. Brit. partim non Eschscholtz,
1829
hirtus (Herbst, 1784)
niger sensu auctt. Brit. non (Linnaeus, 1758)

Subfamily ELATERINAE Leach, 1815

Tribe ADRASTINI Candèze, 1863

ADRASTUS Eschscholtz, 1829
pallens (Fabricius, 1792)
nitidulus (Marsham, 1802) non (Fabricius, 1787)
limbatus sensu (Stephens, 1830) non (Fabricius,
1777)
rachifer (Fourcroy, 1785)
minutus sensu (Olivier, 1790) non (Linnaeus, 1758)
pusillus (Fabricius, 1801)

SYNAPTUS Eschscholtz, 1829
filiformis (Fabricius, 1781)

Tribe AGRIOTINI Laporte, 1840

AGRIOTES Eschscholtz, 1829
acuminatus (Stephens, 1830)
sobrinus Kiesenwetter, 1858
lineatus (Linnaeus, 1767)
obscurus (Linnaeus, 1758)
pallidulus (Illiger, 1807)
sordidus (Illiger, 1807)
sputator (Linnaeus, 1758)

DALOPIUS Eschscholtz, 1829
DOLOPIUS auctt. (misspelling)
marginatus (Linnaeus, 1758)

Tribe AMPEDINI Gistel, 1848

AMPEDUS Dejean, 1833
ELATER sensu auctt. non Linnaeus, 1758
balteatus (Linnaeus, 1758)
cardinalis (Schiødte, 1865)
praeustus sensu auctt. non (Fabricius, 1792)
coccinatus (Rye, 1867)
cinnabarinus (Eschscholtz, 1829)
lythropterus Germar, 1844
elongantulus (Fabricius, 1787)
elongatulus auctt. (misspelling)
nigerrimus (Lacordaire, 1835)
aethiops sensu auctt. Brit. non (Lacordaire, 1835)
nigrinus (Herbst, 1784)
pomonae (Stephens, 1830)
pomorum (Herbst, 1784)

ferrugatus (Lacordaire, 1835)
quercicola (du Buysson, 1887)
pomonae sensu auctt. non (Stephens, 1830)
miniatus (Gorham, 1892)
rufipennis (Stephens, 1830)
sanguineus (Linnaeus, 1758)
sanguinolentus (Schrank, 1776)
tristis (Linnaeus, 1758)

BRACHYGONUS du Buysson, 1912
AMPEDUS sensu auctt. partim non Dejean, 1833
ruficeps (Mulsant & Guillebeau, 1855)
pallidulus (Reitter, 1911)

ISCHNODES Germar, 1844
sanguinicollis (Panzer, 1793)

Tribe MEGAPENTHINI Gurjeva, 1973

MEGAPENTHES Kiesenwetter, 1858
lugens (Redtenbacher, 1842)

PROCRAERUS Reitter, 1905
MEGAPENTHES sensu Fowler, 1889 partim non
Kiesenwetter, 1858
tibialis (Lacordaire, 1835)

Tribe ELATERINI Leach, 1815

ELATER Linnaeus, 1758
LUDIUS Berthold, 1827
ferrugineus Linnaeus, 1758

SERICUS Eschscholtz, 1829
SERICOSOMUS Dejean, 1833
brunneus (Linnaeus, 1758)

Tribe *incertae sedis*

PANSPAEUS Sharp, 1877
PANSPOEUS auctt. (misspelling)
guttatus Sharp, 1877 [522]

Subfamily MELANOTINAE Candèze, 1859 (1848)

MELANOTUS Eschscholtz, 1829
castanipes (Paykull, 1800) [523]
villosus sensu auctt. partim non (Geoffory in
Fourcroy, 1785)
punctolineatus (Pe erin, 1829)
niger sensu (Fabricius, 1792) non (Linnaeus, 1758)
aterrimus sensu (Stephens, 1830) non (Linnaeus,
1761)
villosus (Geoffroy in Fourcroy, 1785)
rufipes (Herbst, 1784) non (De Geer, 1774)
erythropus (Gmelin in Linnaeus, 1790)
rugosus (Marsham, 1802)

Subfamily NEGASTRIINAE Nakane & Kishii, 1956

FLEUTIAUXELLUS Méquignon, 1930
    HYPNOIDUS sensu auctt. Brit. partim non Dillwyn, 1829
maritimus (Curtis, 1840)

NEGASTRIUS Thomson, C.G., 1859
    HYPNOIDUS sensu auctt. Brit. partim non Dillwyn, 1829
arenicola (Boheman, 1852) [524]
pulchellus (Linnaeus, 1761)
sabulicola (Boheman, 1851)

OEDOSTETHUS LeConte, 1853
    HYPNOIDUS sensu auctt. Brit. partim non Dillwyn, 1829
    FLEUTIAUXELLUS sensu auctt. partim non Méquignon, 1930
quadripustulatus (Fabricius, 1792)

ZOROCHROS Thomson, C.G., 1859
    HYPNOIDUS sensu auctt. Brit. partim non Dillwyn, 1829
meridionalis (Laporte, 1840)
minimus (Lacordaire, 1835)
    dermestoides (Herbst, 1806) non (Linnaeus, 1767)
    quadriguttatus sensu auctt. Brit. non (Laporte, 1840)
    tetragraphus sensu auctt. Brit. non (Germar, 1844)
    flavipes sensu auctt. Brit. non (Aubé, 1850)

Subfamily CARDIOPHORINAE Candèze, 1859

CARDIOPHORUS Eschscholtz, 1829
asellus Erichson, 1840
gramineus (Scopoli, 1763)
    thoracicus (Fabricius, 1775)
ruficollis (Linnaeus, 1758)
vestigialis Erichson, 1840
    erichsoni du Buysson, 1901

DICRONYCHUS Brullé, 1832
    CARDIOPHORUS sensu auctt. partim non Eschscholtz, 1829
equisetioides Lohse, 1976
    equiseti sensu auctt. Brit. non (Herbst, 1784)

39. Family DRILIDAE Blanchard, 1845

Family author: A.G. Duff

DRILUS Olivier, 1790
flavescens (Fourcroy, 1785)

40. Family LYCIDAE Laporte, 1836

Family author: A.G. Duff

Subfamily EROTINAE LeConte, 1881 [525]

Tribe EROTINI LeConte, 1881

PLATYCIS Thomson, C.G., 1859
minutus (Fabricius, 1787)

EROTIDES Waterhouse, C.O., 1879
cosnardi (Chevrolat, 1829)

Tribe DICTYOPTERINI Kleine, 1928

PYROPTERUS Mulsant, 1838
nigroruber (De Geer, 1774)
    affinis (Paykull, 1799)

DICTYOPTERA Latreille, 1829
    EROS Newman, 1838
aurora (Herbst, 1784)

41. Family LAMPYRIDAE Rafinesque, 1815

Family author: A.G. Duff

LAMPYRIS Geoffroy, 1762
noctiluca (Linnaeus, 1758)

LAMPROHIZA Motschulsky, 1853
splendidula (Linnaeus, 1767) [526]

PHOSPHAENUS Laporte, 1833
hemipterus (Goeze, 1777)

42. Family CANTHARIDAE Imhoff, 1856 (1815)

Family author: K.N.A. Alexander

Subfamily CANTHARINAE Imhoff, 1856 (1815)

Tribe PODABRINI Gistel, 1856

PODABRUS Westwood, 1838
alpinus (Paykull, 1798)

Tribe CANTHARINI Imhoff, 1856 (1815)

ANCISTRONYCHA Märkel in Kiesenwetter, 1852
abdominalis (Fabricius, 1798)
    cyanea (Curtis, 1828)

CANTHARIS Linnaeus, 1758
    TELEPHORUS Schaeffer, 1766
cryptica Ashe, 1947
    pallida sensu auctt. partim non Goeze, 1777

bicolor sensu (Fowler, 1889) partim non Fabricius, 1798
decipiens Baudi, 1871
    haemorrhoidalis sensu auctt. Brit. non Fabricius, 1792
    clypeata sensu (Fowler, 1889) non Illiger, 1798
figurata Mannerheim, 1843
    scotica (Sharp, 1866)
    cruachana Chitty, 1893
fusca Linnaeus, 1758
lateralis Linnaeus, 1758
    oralis Germar, 1824
livida Linnaeus, 1758
    rufipes Herbst, 1784
    dispar Fabricius, 1792
nigra (De Geer, 1774)
    fulvicollis Fabricius, 1792 non Scopoli, 1763
    flavilabris Fallén, 1807
nigricans (Müller, O.F., 1776)
    discoidea (Stephens, 1830)
obscura Linnaeus, 1758
pallida Goeze, 1777
    bicolor Panzer, 1797 non Linnaeus, 1763
    bicolor sensu (Fowler, 1889) partim non Fabricius, 1798
paludosa Fallén, 1807
pellucida Fabricius, 1792
rufa Linnaeus, 1758
    lituratus Fallén, 1807
    darwiniana (Sharp, 1866)
rustica Fallén, 1807
thoracica (Olivier, 1790)
    bicolor Herbst, 1784 non Linnaeus, 1763
    ?caeruleocephala Thunberg, 1784

RHAGONYCHA Eschscholtz, 1830
elongata (Fallén, 1807)
fulva (Scopoli, 1763)
lignosa (Müller, O.F., 1764)
    pallida (Fabricius, 1787) non (Goeze, 1777)
limbata Thomson, C.G., 1864
    femoralis sensu auctt. Brit. non (Brullé, 1832)
lutea (Müller, O.F., 1764)
    fuscicornis (Olivier, 1790)
testacea (Linnaeus, 1758)
translucida (Krynicki, 1832)
    pilosa sensu (Stephens, 1830) non (Paykull, 1798)
    unicolor sensu Fowler, 1889 ?non (Curtis, 1840)

Subfamily SILINAE Mulsant, 1862

SILIS Charpentier, 1825
    CRUDOSILIS Kazantsev, 1995
ruficollis (Fabricius, 1775)

Subfamily MALTHININAE Kiesenwetter, 1852

Tribe MALTHININI Kiesenwetter, 1852

MALTHINUS Latreille, 1806
balteatus Suffrian, 1851
flaveolus (Herbst, 1786)
    punctatus sensu Fowler, 1889 ?non (Fourcroy, 1785)
frontalis (Marsham, 1802)
seriepunctatus Kiesenwetter, 1852

fasciatus sensu auctt. non (Olivier, 1790)

Tribe MALTHODIN Böving & Craighead, 1931

MALTHODES Kiesenwetter, 1852
crassicornis (Mäklin, 1846)
    nigellus sensu Fowler 1890 non Kiesenwetter, 1852
    brevicollis sensu auctt. Brit. non (Paykull, 1798) [527]
dispar (Germar, 1824)
fibulatus Kiesenwetter, 1852
flavoguttatus Kiesenwetter, 1852
fuscus (Waltl, 1838)
    pellucidus Kiesenwetter, 1852
guttifer Kiesenwetter, 1852
lobatus Kiesenwetter, 1852 [528]
marginatus (Latreille, 1806)
maurus (Laporte, 1840)
    misellus sensu auctt. non Kiesenwetter, 1852
minimus (Linnaeus, 1758)
    marginicollis Schilsky, 1892
mysticus Kiesenwetter, 1852
pumilus (Brébisson, 1835)
    brevicollis sensu Kiesenwetter, 1852 non (Paykull, 1798)
    atomus Thomson, C.G., 1864

Superfamily DERODONTOIDEA LeConte, 1861

43. Family DERODONTIDAE LeConte, 1861

Family author: A.G. Duff

Subfamily LARICOBIINAE Mulsant & Rey, 1864

LARICOBIUS Rosenhauer, 1846
erichsonii Rosenhauer, 1846 [529]
    erichsoni auctt. (misspelling)

Superfamily BOSTRICHOIDEA Latreille, 1802

44. Family DERMESTIDAE Latreille, 1804 [530]

Family author: A.G. Duff

Subfamily THYLODRIINAE Semenov, 1909

THYLODRIAS Motschulsky, 1839
contractus Motschulsky, 1839

Subfamily THORICTINAE Agassiz, 1846

THORICTODES Reitter, 1875
    THAUMAPHRASTUS Blaisdell, 1927
heydeni Reitter, 1875 [531]
    karanisensis (Blaisdell, 1927)

Subfamily DERMESTINAE Latreille, 1804

DERMESTES Linnaeus, 1758

Subgenus DERMESTINUS Zhantiev, 1967
carnivorus Fabricius, 1775
    carniforus Fabricius, 1775
frischii Kugelann, 1792
maculatus De Geer, 1774
    vulpinus Fabricius, 1781
murinus Linnaeus, 1758
undulatus Brahm, 1790

Subgenus DERMESTES Linnaeus, 1758
ater De Geer, 1774
    cadaverinus Fabricius, 1775
haemorrhoidalis Küster, 1852
    peruvianus sensu auctt. Brit. partim non Laporte, 1840
lardarius Linnaeus, 1758
leechi Kalik, 1952
peruvianus Laporte, 1840
    oblongus sensu auctt. Brit. non Solier in Gay, 1849

Subfamily ATTAGENINAE Laporte, 1840

ATTAGENUS Latreille, 1802

brunneus Faldermann, 1835
    elongatulus Casey, 1900
cyphonoides Reitter, 1881
    alfierii Pic, 1910
fasciatus (Thunberg, 1795)
    gloriosae (Fabricius, 1798)
pellio (Linnaeus, 1758)
smirnovi Zhantiev, 1973 [532]
trifasciatus (Fabricius, 1787)
unicolor (Brahm, 1791)
    piceus (Olivier, 1790) non (Thunberg, 1781)
    megatoma (Fabricius, 1798)

Subfamily MEGATOMINAE Leach, 1815

TROGODERMA Dejean, 1821

angustum (Solier, 1849) [533]
glabrum (Herbst, 1783)
    boron Beal, 1954
granarium Everts, 1898
    khapra Arrow, 1917
inclusum LeConte, 1854
    versicolor sensu auctt. Brit. non (Creutzer, 1799)
variabile Ballion, 1878 [534]
    parabile Beal, 1954

REESA Beal, 1967

vespulae (Milliron, 1939) [535]

GLOBICORNIS Latreille, 1829

rufitarsis (Creutzer in Panzer, 1796)
    nigripes (Fabricius, 1792) non (Olivier, 1790)
    plantaris (Curtis, 1838)

MEGATOMA Herbst, 1792

undata (Linnaeus, 1758)

ANTHRENOCERUS Arrow, 1915

australis (Hope, 1843)

ORPHINUS Motschulsky, 1858

fulvipes (Guérin-Méneville, 1838)
    brevicornis (Sharp, 1885)

CTESIAS Stephens, 1830

    TIRESIAS Stephens, 1835
serra (Fabricius, 1792)

Subfamily ANTHRENINAE Gistel, 1848

ANTHRENUS Geoffroy, 1762

Subgenus ANTHRENUS Müller, O.F., 1764
flavipes LeConte, 1854
    vorax Waterhouse, C.O. 1883
pimpinellae Fabricius, 1775
scrophulariae (Linnaeus, 1758)

Subgenus ANTHRENODES Chobaut, 1898
sarnicus Mroczkowski, 1963

Subgenus ANTHRENOPS Reitter, 1881
coloratus Reitter, 1881 [536]

Subgenus FLORILINUS Mulsant & Rey, 1868
museorum (Linnaeus, 1761)
olgae Kalik, 1946 [537]

Subgenus HELOCERUS Mulsant & Rey, 1868
fuscus Olivier, 1789
    claviger Erichson, 1846

Subgenus NATHRENUS Casey, 1900
verbasci (Linnaeus, 1767)
    varius (Fabricius, 1775)

Subfamily TRINODINAE Casey, 1900

TRINODES Dejean, 1821

hirtus (Fabricius, 1781)

45. Family BOSTRICHIDAE Latreille, 1802

Family author: A.G. Duff

Subfamily BOSTRICHINAE Latreille, 1802

BOSTRICHUS Geoffroy, 1762 [538]

    BOSTRYCHUS auctt. (misspelling)
capucinus (Linnaeus, 1758)

Subfamily DINODERINAE Thomson, C.G., 1863

RHYZOPERTHA Stephens, 1830

    RHIZOPERTHA Guérin-Méneville, 1845 (misspelling)
dominica (Fabricius, 1792)
    pusilla (Fabricius, 1798)

STEPHANOPACHYS Waterhouse, C.O., 1888

substriatus (Paykull, 1800)

Subfamily LYCTINAE Billberg, 1820

LYCTUS Fabricius, 1792
XYLOTROGUS Stephens, 1830
brunneus (Stephens, 1830)
linearis (Goeze, 1777)
fuscus (Linnaeus, 1767)
canaliculatus Fabricius, 1792

46. Family PTINIDAE Latreille, 1802

Family author: R.G. Booth

Subfamily EUCRADINAE LeConte, 1861

HEDOBIA Dejean, 1821
Subgenus PTINOMORPHUS Mulsant & Rey, 1868
imperialis (Linnaeus, 1767)

Subfamily PTININAE Latreille, 1802

Tribe GIBBIINI Jacquelin du Val, 1860

GIBBIUM Scopoli, 1777
SCOTIAS de Czenpinski, 1778
aequinoctiale Boieldieu, 1854 [539]
psylliodes sensu auctt. partim non (de Czenpinski, 1778)
psylloides (de Czenpinski, 1778)
scotias (Fabricius, 1781)

STETHOMEZIUM Hinton, 1943
squamosum Hinton, 1943

MEZIUM Samouelle, 1819
affine Boieldieu, 1856
sulcatum sensu (Marsham, 1802) non (Fabricius, 1781)

Tribe PTININI Latreille, 1802

TRIGONOGENIUS Solier, 1849
globulus Solier, 1849

SPHAERICUS Wollaston, 1854
gibboides (Boieldieu, 1854)

NIPTUS Boieldieu, 1856
hololeucus (Faldermann, 1835)

TIPNUS Thomson, C.G., 1859
NIPTUS sensu Fowler, 1890 partim non Boieldieu, 1856
unicolor (Piller & Mitterpacher, 1783)
crenatus (Fabricius, 1792)

PSEUDEUROSTUS von Heyden, L., 1906
EUROSTUS Mulsant & Rey, 1868 non Dallas, 1851
hilleri (Reitter, 1877)

PTINUS Linnaeus, 1767 [540]
BRUCHUS Geoffroy, 1762 [541]
clavipes Panzer, 1792
latro sensu auctt. partim non Fabricius, 1775
testaceus Olivier, 1790 non Thunberg, 1784
brunneus Duftschmid, 1825 non Gmelin in Linnaeus, 1790
hirtellus Sturm, 1837
female form *mobilis* Moore, 1957
dubius Sturm, 1837 [542]
exulans Erichson, 1842
fur (Linnaeus, 1758)
latro Fabricius, 1775
latefasciatus Gorham, 1883
lichenum Marsham, 1802
palliatus Perris, 1847
germanus sensu Fabricius, 1781 non Goeze, 1777
?Linnaeus, 1767
pusillus Sturm, 1837
raptor Sturm, 1837
sexpunctatus Panzer, 1792
subpilosus Sturm, 1837
tectus Boieldieu, 1856
villiger Reitter, 1884

Subfamily DRYOPHILINAE Gistel, 1848

GRYNOBIUS Thomson, C.G., 1859
PRIOBIUM sensu auctt. non Motschulsky, 1845
planus (Fabricius, 1787)
tricolor (Olivier, 1790)
castaneus (Fabricius, 1792)
excavatus (Kugelann, 1792)
eichhoffi (Seidlitz, 1889)
kiesenwetteri (Edwards, 1921)

DRYOPHILUS Chevrolat, 1832
anobioides Chevrolat, 1832
pusillus (Gyllenhal, 1808)

Subfamily ERNOBIINAE Pic, 1912

OCHINA Dejean, 1821
ptinoides (Marsham, 1802)
hederae (Müller, P.W.J., 1821)

XESTOBIUM Motschulsky, 1845
rufovillosum (De Geer, 1774)
tessellatum (de Villers, C.J., 1789)

ERNOBIUS Thomson, C.G., 1859
abietis (Fabricius, 1792) [543]
angusticollis (Ratzeburg, 1837)
parvicollis (Mulsant & Rey, 1863)
gigas (Mulsant & Rey, 1863)
reflexus sensu auctt. non (Mulsant & Rey, 1863)
subelongatus Pic, 1914
mulsantianus Sharp, 1916
mollis (Linnaeus, 1758)
reversus Sharp, 1916
nigrinus (Sturm, 1837)
pini (Sturm, 1837)
oblitus Sharp, 1916

Subfamily ANOBIINAE Fleming, 1821

STEGOBIUM Motschulsky, 1860
SITODREPA Thomson, C.G., 1863
paniceum (Linnaeus, 1758)

GASTRALLUS Jacquelin du Val, 1860
immarginatus (Müller, P.W.J., 1821)
laevigatus sensu Donisthorpe, 1936 non (Olivier, 1790)
laevigatus (Olivier, 1790) [544]

ANOBIUM Fabricius, 1775
HADROBREGMUS sensu auctt. Europ. partim non Thomson, C.G., 1859
HEMICOELUS LeConte, 1861
fulvicorne (Sturm, 1837)
inexspectatum Lohse, 1954
nitidum Fabricius, 1792 [545]
punctatum (De Geer, 1774)
domesticum (Geoffroy in Fourcroy, 1785)
striatum sensu Olivier, 1790 non Fabricius, 1787

HADROBREGMUS Thomson, C.G., 1859
ANOBIUM sensu auctt. partim non Fabricius, 1775
COELOSTETHUS LeConte, 1861
DENDROBIUM Mulsant & Rey, 1864
denticollis (Creutzer in Panzer, 1796)

PRIOBIUM Motschulsky, 1845
carpini (Herbst, 1793) [546]

Subfamily PTILININAE Shuckard, 1839

PTILINUS Geoffroy, 1762 [547]
pectinicornis (Linnaeus, 1758)

Subfamily XYLETININAE Gistel, 1848

XYLETINUS Latreille, 1809
longitarsis Jansson, 1942
ater sensu auctt. Brit. partim non (Creutzer in Panzer, 1796)

LASIODERMA Stephens, 1835
serricorne (Fabricius, 1792)

Subfamily DORCATOMINAE Thomson, C.G., 1859

MESOCOELOPUS Jacquelin du Val, 1860
collaris Mulsant & Rey, 1864 [548]

DORCATOMA Herbst, 1792
DORKATOMA Herbst, 1792 [549]
ambjoerni Baranowski, 1985 [550]
chrysomelina Sturm, 1837
dresdensis Herbst, 1792
flavicornis (Fabricius, 1792)
substriata Hummel, 1829
serra (Panzer, 1795) non Fabricius, 1792

punctulata sensu Sharp, 1914 non Mulsant & Rey, 1864

CAENOCARA Thomson, C.G., 1859
affinis (Sturm, 1837)
subglobosa sensu Donisthorpe, 1918 non Mulsant & Rey, 1864
bovistae (Hoffmann, J., 1803)

ANITYS Thomson, C.G., 1863
rubens (Hoffmann, J., 1803)

MIROSTERNOMORPHUS Español, 1977
heali Bercedo & Arnáiz, 2010 [551]

Superfamily LYMEXYLOIDEA Fleming, 1821

47. Family LYMEXYLIDAE Fleming, 1821

Family author: R.G. Booth

Subfamily HYLECOETINAE Germar, 1818

HYLECOETUS Latreille, 1806
dermestoides (Linnaeus, 1761)

Subfamily LYMEXYLINAE Fleming, 1821

LYMEXYLON Fabricius, 1775
navale (Linnaeus, 1758)

Superfamily CLEROIDEA Latreille, 1802

48. Family PHLOIOPHILIDAE Kiesenwetter, 1863

Family author: R.G. Booth

PHLOIOPHILUS Stephens, 1830
PHLOEOPHILUS auctt. (misspelling)
edwardsii Stephens, 1830
edwardsi auctt. (misspelling)

49. Family TROGOSSITIDAE Latreille, 1802

Family author: R.G. Booth

Subfamily PELTINAE Latreille, 1806

OSTOMA Laicharting, 1781
ferrugineum (Linnaeus, 1758)

THYMALUS Latreille, 1802
limbatus (Fabricius, 1787)

Subfamily LOPHOCATERINAE Crowson, 1964

LOPHOCATERES Olliff, 1883
pusillus (Klug, 1832) [552]

Subfamily TROGOSSITINAE Latreille, 1802

TENEBROIDES Piller & Mitterpacher, 1783
TENEBRIOIDES auctt. (misspelling)
mauritanicus (Linnaeus, 1758)

NEMOZOMA Latreille, 1804
NEMOSOMA Latreille, 1809 [40]
elongatum (Linnaeus, 1761)

50. Family CLERIDAE Latreille, 1802 [553]

Family author: R.G. Booth

Subfamily THANEROCLERINAE Chapin, 1924

THANEROCLERUS Lefebvre, 1838
buqueti (Lefebvre, 1835) [554]

Subfamily TILLINAE Fischer von Waldheim, 1813

TILLUS Olivier, 1790
elongatus (Linnaeus, 1758)

TILLOIDEA Laporte, 1833
unifasciata (Fabricius, 1787)

Subfamily CLERINAE Latreille, 1802

OPILO Latreille, 1802
mollis (Linnaeus, 1758)

THANASIMUS Latreille, 1806
femoralis (Zetterstedt, 1838)
rufipes (Brahm, 1797) non (De Geer, 1775)
formicarius (Linnaeus, 1758)

TRICHODES Herbst, 1792
alvearius (Fabricius, 1792)
alveolarius Latreille, 1804
interruptus Kraatz, 1894
apiarius (Linnaeus, 1758)

Subfamily TARSOSTENINAE Jacquelin du Val, 1860

PARATILLUS Gorham, 1878
carus (Newman, 1840)

TARSOSTENUS Spinola, 1844
univittatus (Rossi, 1792)

Subfamily KORYNETINAE Laporte, 1836

KORYNETES Herbst, 1792
CORYNETES Paykull, 1798 [40]
caeruleus (De Geer, 1775)
cyanellus Dejean, 1837

NECROBIA Olivier, 1795
ruficollis (Fabricius, 1775)
rufipes (De Geer, 1775)
violacea (Linnaeus, 1758)

51. Family DASYTIDAE Laporte, 1840 [555]

Family author: R.G. Booth

Subfamily RHADALINAE LeConte, 1861

APLOCNEMUS Stephens, 1830
HAPLOCNEMUS auctt. (misspelling)
impressus (Marsham, 1802)
pini (Redtenbacher, 1849)
nigricornis (Fabricius, 1792)

Subfamily DASYTINAE Laporte, 1840

DASYTES Paykull, 1799
aeratus Stephens, 1830
serricornis Stephens, 1830
aerosus Kiesenwetter, 1867
cyaneus (Fabricius, 1775) [556]
caeruleus (De Geer, 1774) non (Linnaeus, 1758)
niger (Linnaeus, 1761)
floralis sensu (Stephens, 1830) non (Olivier, 1790)
plumbeus (Müller, O.F., 1776)
flavipes (Fabricius, 1777)
tibialis Zetterstedt, 1828
flavipes sensu Joy, 1932 non (Olivier, 1790)
oculatus sensu auctt. Brit. non Kiesenwetter, 1867
virens (Marsham, 1802)
flavipes sensu (Olivier, 1790) non (Fabricius, 1777)
flavipes sensu Fowler, 1889 non (Fabricius, 1777)
puncticollis Reitter, 1888 non Germain, 1855
plumbeus sensu Joy, 1932 non (Müller, O.F., 1776)

PSILOTHRIX Küster, 1850
viridicoeruleus (Geoffroy in Fourcroy, 1785)
cyaneus (Olivier, 1790)
nobilis (Illiger, 1798)
caeruleus sensu (Stephens, 1830) non (Fabricius, 1775)
viridis sensu (Stephens, 1830) non (Rossi, 1792)

DOLICHOSOMA Stephens, 1830
lineare (Rossi, 1794)

52. Family MALACHIIDAE Fleming, 1821 [557]

Family author: R.G. Booth

COLOTES Erichson, 1840
punctatus (Erichson, 1840) [558]

EBAEUS Erichson, 1840
pedicularius (Linnaeus, 1758)
    productus sensu (Stephens, 1830) non (Olivier, 1790)

HYPEBAEUS Kiesenwetter, 1863
    EBAEUS sensu auctt. Brit. partim non Erichson, 1840
flavipes (Fabricius, 1787)
    abietinus sensu auctt. Brit. non Abeille de Perrin, 1869

AXINOTARSUS Motschulsky, 1853
marginalis (Laporte, 1840)
pulicarius (Fabricius, 1777)
ruficollis (Olivier, 1790)

TROGLOPS Erichson, 1840
cephalotes (Olivier, 1790) [559]

CORDYLEPHERUS Evers, 1985
    MALACHIUS sensu auctt. partim non Fabricius, 1775
viridis (Fabricius, 1787)

MALACHIUS Fabricius, 1775
aeneus (Linnaeus, 1758)
bipustulatus (Linnaeus, 1758)
    immaculicollis Rey, 1867
    lusitanicus var. *australis* sensu Donisthorpe, 1931 non Rey, 1867

CLANOPTILUS Motschulsky, 1854
    MALACHIUS sensu auctt. partim non Fabricius, 1775
barnevillei (Puton, 1865)
marginellus (Olivier, 1790)
    elegans sensu (Donisthorpe, 1931) non (Geoffroy in Fourcroy, 1785)
    pseudosardous (Reclairie & van der Wiel, 1932)
    angustimarginalis (Donisthorpe, 1933)
strangulatus (Abeille de Perrin, 1885)
    vulneratus Abeille de Perrin, 1891

SPHINGINUS Mulsant & Rey, 1867
lobatus (Olivier, 1790) [560]

CERAPHELES Mulsant & Rey, 1867
    DIAPHONUS Mulsant & Rey, 1867
    ANTHOCOMUS sensu Fowler, 1889 partim non Erichson, 1840
terminatus (Ménétries, 1832)

ANTHOCOMUS Erichson, 1840
fasciatus (Linnaeus, 1758)
rufus (Herbst, 1784)
    coccineus (Schaller, 1783) non (Linnaeus, 1761)

Superfamily CUCUJOIDEA Latreille, 1802

53. Family SPHINDIDAE Jacquelin du Val, 1860

Family author: R.G. Booth

SPHINDUS Dejean, 1821
dubius (Gyllenhal, 1808)

ASPIDIPHORUS Dejean, 1821
    ARPIDIPHORUS Dejean, 1821 [561]
orbiculatus (Gyllenhal, 1808)

54. Family KATERETIDAE Kirby, 1837 [562]

Family author: R.G. Booth

KATERETES Herbst, 1793
    CATERETES auctt. (misspelling)
    CERCUS Latreille, 1796
    ANISOCERA Stephens, 1832
    ANOMAEOCERA Shuckard, 1840
pedicularius (Linnaeus, 1758)
    similis (Marsham, 1802)
    unicolor (Marsham, 1802)
    spiraeae (Stephens, 1830)
pusillus (Thunberg, 1794)
    bipustulatus (Paykull, 1798) non (Thunberg, 1781) [563]
    punctulatus (Marsham, 1802)
rufilabris (Latreille, 1807)
    caricis Stephens, 1830
    junci Stephens, 1830

BRACHYPTERUS Kugelann, 1794
glaber (Newman, 1834)
    pubescens Erichson, 1843
urticae (Fabricius, 1792)
    erythropus (Marsham, 1802)

BRACHYPTEROLUS Grouvelle, 1913
    HETEROSTOMUS Jacquelin du Val, 1858 non Bigot, 1857
    BRACHYPTERUS sensu Fowler, 1889 partim non Kugelann, 1794
antirrhini (Murray, 1864)
    villiger (Reitter, 1885) [564]
linariae (Stephens, 1830)
    cornelii Spornraft, 1966
pulicarius (Linnaeus, 1758)
    gravidus (Illiger, 1798)
vestitus (Kiesenwetter, 1850)

55. Family NITIDULIDAE Latreille, 1802

Family author: R.G. Booth

Subfamily CARPOPHILINAE Erichson, 1842

Tribe CARPOPHILINI Erichson, 1842

UROPHORUS Murray, 1864
CARPOPHILUS sensu auctt. partim non Stephens, 1829
humeralis (Fabricius, 1798)

CARPOPHILUS Stephens, 1829
dimidiatus (Fabricius, 1792)
flavipes Murray, 1864
hemipterus (Linnaeus, 1758)
ligneus Murray, 1864
maculatus Murray, 1864
marginellus Motschulsky, 1858
mutilatus Erichson, 1843
nepos Murray, 1864
freemani Dobson, 1956
obsoletus Erichson, 1843
sexpustulatus (Fabricius, 1792)
bimaculatus (Marsham, 1802)
truncatus Murray, 1864 [565]
pilosellus sensu auctt. non Motschulsky, 1858
halli Dobson, 1954

Tribe EPURAEINI Kirejtshuk, 1986

EPURAEA Erichson, 1843
DADOPORA Thomson, C.G., 1859
DADOPERA Joy, 1932 (misspelling)

Subgenus EPURAEA Erichson, 1843
aestiva (Linnaeus, 1758)
depressa sensu (Illiger, 1798) non (Linnaeus, 1758)
angustula Sturm, 1844
biguttata (Thunberg, 1784)
unicolor (Olivier, 1790) [566]
obsoleta sensu Fowler, 1889 partim non (Fabricius, 1792)
binotata Reitter, 1872
nana Reitter, 1873
distincta (Grimmer, 1841)
fuscicollis (Stephens, 1835)
diffusa Brisout de Barneville, 1863
guttata (Olivier, 1811)
decemguttata sensu (Fabricius, 1792) non (Olivier, 1790)
longula Erichson, 1845
marseuli Reitter, 1872
pusilla (Illiger, 1798) non (Thunberg, 1794) [567]
melina Erichson, 1843
neglecta (Heer, 1841)
pallescens (Stephens, 1835)
aestiva sensu (Illiger, 1798) non (Linnaeus, 1758)
florea Erichson, 1845 [567]
rufomarginata (Stephens, 1830)
parvula Sturm, 1844
silacea (Herbst, 1783)
deleta Sturm, 1844 [567]
terminalis (Mannerheim, 1843)
immunda Sturm, 1844

adumbrata Mannerheim, 1852 [567]
thoracica Tournier, 1872
oblonga sensu auctt. Brit. non (Herbst, 1793)
variegata (Herbst, 1793)

Subgenus EPURAEANELLA Crotch, 1874
OMOSIPHORA Reitter, 1875
limbata (Fabricius, 1787)

Subgenus MICRURIA Reitter, 1875
MICRURULA Reitter, 1884
melanocephala (Marsham, 1802)
affinis (Stephens, 1830)
truncata (Stephens, 1830) non (Fabricius, 1792)

Subfamily MELIGETHINAE Thomson, C.G., 1859

PRIA Stephens, 1829
LARIA sensu auctt. partim non Scopoli, 1763
dulcamarae (Scopoli, 1763)
truncatella (Marsham, 1802)

MELIGETHES Stephens, 1829 [568]
aeneus (Fabricius, 1775)
caeruleus (Marsham, 1802)
latipes (Marsham, 1802)
nigrinus (Marsham, 1802)
nigricornis Stephens, 1830
urticae Stephens, 1830
atramentarius Förster, 1849
symphyti sensu auctt. Brit. partim non (Heer, 1841)
atratus (Olivier, 1790)
rufipes sensu (Marsham, 1802) non (Linnaeus, 1767)
bidens Brisout de Barneville, 1863
bidentatus sensu Joy, 1932 non Brisout de Barneville, 1863
bidentatus Brisout de Barneville, 1863
gresseri Bach, 1875
corsicus Deville, 1908
brevis Sturm, 1845
pictus Rye, 1871
brunnicornis Sturm, 1845
carinulatus Förster, 1849
erythropus sensu auctt. non (Marsham, 1802) [569]
coracinus Sturm, 1845
corvinus Erichson, 1845
difficilis (Heer, 1841)
erichsonii Brisout de Barneville, 1863
erichsoni auctt. (misspelling)
exilis Sturm, 1845
flavimanus Stephens, 1830
lumbaris Sturm, 1845
fulvipes Brisout de Barneville, 1863
rubripes Mulsant & Rey, 1863
gagathinus Erichson, 1845 [570]
gagatinus auctt. (misspelling)
lugubris sensu auctt. partim non Sturm, 1845
haemorrhoidalis Förster, 1849 [571]
incanus Sturm, 1845
kunzei Erichson, 1845
lugubris Sturm, 1845
matronalis Audisio & Spornraft, 1990 [572]
morosus Erichson, 1845
memnonius Erichson, 1845
nanus Erichson, 1845
nigrescens Stephens, 1830

xanthoceros Stephens, 1830
   picipes Sturm, 1845
obscurus Erichson, 1845
   palmatus Erichson, 1845
ochropus Sturm, 1845
ovatus Sturm, 1845
pedicularius (Gyllenhal, 1808) [573]
   viduatus (Heer, 1841) [574]
   aestimabilis Reitter, 1872
persicus (Faldermann, 1835)
   pedicularius sensu auctt. ante 1993 non (Gyllenhal, 1808) [574]
   tenebrosus Förster, 1849
planiusculus (Heer, 1841)
   murinus Erichson, 1845
rotundicollis Brisout de Barneville, 1863
ruficornis (Marsham, 1802)
   subrugosus sensu Stephens, 1830 partim non (Gyllenhal, 1808)
   flavipes Sturm, 1845 [569]
serripes (Gyllenhal, 1827)
solidus (Kugelann, 1794)
subrugosus (Gyllenhal, 1808) [575]
symphyti (Heer, 1841) [576]
umbrosus Sturm, 1845
viridescens (Fabricius, 1787)

Subfamily NITIDULINAE Latreille, 1802

NITIDULA Fabricius, 1775
bipunctata (Linnaeus, 1758)
   bipustulata (Linnaeus, 1761)
carnaria (Schaller, 1783)
   quadripustulata Fabricius, 1792
   variata Stephens, 1830
flavomaculata Rossi, 1790
   flexuosa Olivier, 1790
rufipes (Linnaeus, 1767)

OMOSITA Erichson, 1843
colon (Linnaeus, 1758)
depressa (Linnaeus, 1758)
discoidea (Fabricius, 1775)

SORONIA Erichson, 1843
grisea (Linnaeus, 1758)
punctatissima (Illiger, 1794)

AMPHOTIS Erichson, 1843
marginata (Fabricius, 1781)

CYCHRAMUS Kugelann, 1794
luteus (Fabricius, 1787)
   fungicola Heer, 1841

POCADIUS Erichson, 1843
adustus Reitter, 1888
   lanuginosus Franz, 1969 [567]
ferrugineus (Fabricius, 1775)

THALYCRA Erichson, 1843
fervida (Olivier, 1790)
   sericea (Sturm, 1839)

Subfamily CRYPTARCHINAE Thomson, C.G., 1859

CRYPTARCHA Shuckard, 1839
strigata (Fabricius, 1787)
   undata sensu (Marsham, 1802) non (Olivier, 1790)
undata (Olivier, 1790)
   imperialis (Fabricius, 1792)
   nebulosa (Marsham, 1802)

GLISCHROCHILUS Reitter, 1873
   IPS Fabricius, 1777 non De Geer, 1775

Subgenus GLISCHROCHILUS Reitter, 1873
quadripunctatus (Linnaeus, 1758)
   quadripustulatus (Linnaeus, 1761)

Subgenus LIBRODOR Reitter, 1884
hortensis (Geoffroy in Fourcroy, 1785)
   quadripunctatus sensu (Olivier, 1790) non (Linnaeus, 1758)
   olivieri Bedel, 1891
quadriguttatus (Fabricius, 1777)

PITYOPHAGUS Shuckard, 1839
ferrugineus (Linnaeus, 1761)

Subfamily CYBOCEPHALINAE Jacquelin du Val, 1858

CYBOCEPHALUS Erichson, 1844
fodori Endrödy-Younga, 1965 [577]

56. Family MONOTOMIDAE Laporte, 1840 [578]

Family author: C. Johnson

Subfamily RHIZOPHAGINAE Redtenbacher, 1845

RHIZOPHAGUS Herbst, 1793
   RYZOPHAGUS Herbst, 1793 [579]
   RHYZOPHAGUS Gyllenhal, 1813 (misspelling)

Subgenus RHIZOPHAGUS Herbst, 1793
bipustulatus (Fabricius, 1792)
   longicollis (Gyllenhal, 1827)
   gyllenhalii Thomson, C.G., 1885
dispar (Paykull, 1800)
   punctulatus Guillebeau, 1897
fenestralis (Linnaeus, 1758)
   parvulus (Paykull, 1800)
ferrugineus (Paykull, 1800)
   minor Méquignon, 1909
nitidulus (Fabricius, 1798)
oblongicollis Blatch & Horner, 1892
   punctulatus sensu auctt. Brit. non Guillebeau, 1897
   simplex sensu auctt. non Reitter, 1884
parallelocollis Gyllenhal, 1827
   terebrans sensu Stephens, 1829 non (Olivier, 1790)
perforatus Erichson, 1845
picipes (Olivier, 1790)
   politus (Hellwig, 1792)

Subgenus ANOMOPHAGUS Reitter, 1907
cribratus Gyllenhal, 1827

Subgenus EURHIZOPHAGUS Méquignon, 1909
depressus (Fabricius, 1792)
grandis Gyllenhal, 1827 [580]

CYANOSTOLUS Ganglbauer, 1899
aeneus (Richter, 1820)
    caeruleipennis (Sahlberg, C.R., 1837)
    caeruleus (Waltl, 1839)
    cyaneipennis (Hardy, 1847)

Subfamily MONOTOMINAE Laporte, 1840

MONOTOMA Herbst, 1793

Subgenus GYROCECIS Thomson, C.G., 1859
angusticollis (Gyllenhal, 1827)
    formicetorum Thomson, C.G., 1863
conicicollis Guérin-Méneville, 1837
    angusticollis sensu auctt. non (Gyllenhal, 1827)
quadrifoveolata Aubé, 1838
    subquadrifoveolata sensu Fowler, 1889 non
      Waterhouse, G.R., 1858

Subgenus MONOTOMA Herbst, 1793
bicolor Villa & Villa, 1835
brevicollis Aubé, 1838
    quadridentata Thomson, C.G., 1870
longicollis (Gyllenhal, 1827)
    gracilis Curtis, 1830
    angustata Stephens, 1830
    flavipes Kunze, 1839
picipes Herbst, 1793
    contracta Marsham, 1802
    pallida Stephens, 1830
quadricollis Aubé, 1838 [581]
    bicolor sensu auctt. partim non Villa & Villa, 1835
spinicollis Aubé, 1838
    spinigera sensu auctt. non Chaudoir, 1845
testacea Motschulsky, 1845
    quadriimpressa Motschulsky, 1845
    rufa Redtenbacher, 1849
    subquadrifoveolata Waterhouse, G.R., 1858

57. Family SILVANIDAE Kirby, 1837

Family author: R.G. Booth

Subfamily BRONTINAE Blanchard, 1845

ULEIOTA Latreille, 1796
    BRONTES Fabricius, 1801
planatus (Linnaeus, 1761)
    pallens (Fabricius, 1792)

DENDROPHAGUS Schönherr, 1809
crenatus (Paykull, 1799)

CRYPTAMORPHA Wollaston, 1854
desjardinsii (Guérin-Méneville, 1844)
    desjardinsi auctt. (misspelling)

PSAMMOECUS Latreille, 1829
    PSAMMOECHUS Boudier, 1834 (misspelling)
bipunctatus (Fabricius, 1792)

Subfamily SILVANINAE Kirby, 1837

CATHARTUS Reiche, 1854
quadricollis (Guérin-Méneville, 1844)

NAUSIBIUS Lentz, 1857
clavicornis (Kugelann, 1794)
    dentatus (Marsham, 1802)

AHASVERUS des Gozis, 1881
    CATHARTUS sensu Fowler, 1889 partim non
      Reiche, 1854
advena (Waltl, 1834)
    brunneus (Fabricius, 1792) [582]

ORYZAEPHILUS Ganglbauer, 1899
    SILVANUS sensu Fowler, 1889 partim non Latreille,
      1804
mercator (Fauvel, 1889)
surinamensis (Linnaeus, 1758)

SILVANUS Latreille, 1804
    LEPTUS Duftschmid, 1825 non Latreille, 1796
bidentatus (Fabricius, 1792)
unidentatus (Olivier, 1790)

SILVANOPRUS Reitter, 1911
    SILVANUS sensu Fowler, 1889 partim non Latreille,
      1804
fagi (Guérin-Méneville, 1844)
    similis (Erichson, 1846)

58. Family CUCUJIDAE Latreille, 1802

Family author: R.G. Booth

PEDIACUS Shuckard, 1839
depressus (Herbst, 1797)
dermestoides (Fabricius, 1792)

59. Family LAEMOPHLOEIDAE Ganglbauer, 1899

Family author: R.G. Booth

LAEMOPHLOEUS Dejean, 1835
monilis (Fabricius, 1787)

CRYPTOLESTES Ganglbauer, 1899
    LEPTUS Thomson, C.G., 1859 non Latreille, 1796
capensis (Waltl, 1834)
duplicatus (Waltl, 1839)
ferrugineus (Stephens, 1831)
pusilloides (Steel & Howe, 1952)
pusillus (Schönherr, 1817)
    minutus (Olivier, 1791) non (Geoffroy in Fourcroy,
      1785)
spartii (Curtis, 1834)
    ater (Olivier, 1795) non (Geoffroy in Fourcroy, 1785)
turcicus (Grouvelle, 1876)

NOTOLAEMUS Lefkovitch, 1959
LAEMOPHLOEUS sensu Fowler, 1889 partim non Dejean, 1835
unifasciatus (Latreille, 1804)
bimaculatus (Paykull, 1801) non (Olivier, 1791)
fasciatus (Stephens, 1839)

LEPTOPHLOEUS Casey, 1916
LAEMOPHLOEUS sensu Fowler, 1889 partim non Dejean, 1835
clematidis (Erichson, 1846)
janeti (Grouvelle, 1899)

60. Family PHALACRIDAE Leach, 1815

Family author: R.G. Booth

PHALACRUS Paykull, 1800
caricis Sturm, 1807
nigrinus sensu auctt. Brit. non (Marsham, 1802)
millefolii sensu Gyllenhal, 1813 non Paykull, 1800
championi Guillebeau, 1892
brunnipes sensu Rye, 1872 non Brisout de Barneville, 1863
corruscus (Panzer, 1796)
coruscus auctt. (misspelling)
fimetarius sensu (Paykull, 1798) non (Fabricius, 1775)
fimetarius (Fabricius, 1775)
brisouti Rye, 1872
hybridus Flach, 1888
substriatus Gyllenhal, 1813

OLIBRUS Erichson, 1845
aeneus (Fabricius, 1792)
affinis (Sturm, 1807)
particeps sensu auctt. Brit. non Mulsant, 1861
corticalis (Panzer, 1796)
flavicornis (Sturm, 1807)
helveticus Rye, 1876
liquidus Erichson, 1845
bicolor sensu auctt. Brit. non (Fabricius, 1792)
millefolii (Paykull, 1800)
pygmaeus (Sturm, 1807)

STILBUS Seidlitz, 1872
atomarius (Linnaeus, 1767)
oblongus (Erichson, 1845)
testaceus (Panzer, 1796)

61. Family CRYPTOPHAGIDAE Kirby, 1826 [583]

Family author: C. Johnson

Subfamily CRYPTOPHAGINAE Kirby, 1826

Tribe CRYPTOPHAGINI Kirby, 1826

TELMATOPHILUS Heer, 1841
brevicollis Aubé, 1862
caricis (Olivier, 1790)
obscurus (Fabricius, 1792)
schonherrii (Gyllenhal, 1808)

schoenherri auctt. (misspelling)
sparganii (Ahrens, 1812)
typhae (Fallén, 1802)
pumilus Reitter, 1875

PARAMECOSOMA Curtis, 1833
melanocephalum (Herbst, 1793)
bicolor Curtis, 1833
infuscatum Halbert, 1910

HENOTICUS Thomson, C.G., 1868
californicus (Mannerheim, 1843)
germanicus Reitter, 1906
serratus (Gyllenhal, 1808)

CRYPTOPHAGUS Herbst, 1792
KRYPTOPHAGUS Herbst, 1792 [584]
MNIONOMUS Wollaston, 1864
acutangulus Gyllenhal, 1827
waterhousei Rye, 1866
badius Sturm, 1845
cellaris (Scopoli, 1763)
crenatus (Herbst, 1792)
confusus Bruce, 1934 [585]
corticinus Thomson, C.G., 1863
dentatus (Herbst, 1793)
fumatus (Marsham, 1802)
niger sensu auctt. non Brisout de Barneville, 1863
denticulatus Heer, 1841
pilosus Gyllenhal, 1827 non Herbst, 1792
pallidus sensu auctt. Brit. non Sturm, 1845
pseudodentatus Bruce, 1934
dilutus Reitter, 1874
hexagonalis sensu auctt. ante 2006 non Tournier, 1872
subvittatus sensu auctt. Brit. non Reitter, 1888 [536]
distinguendus Sturm, 1845
umbratus Erichson, 1846
stramenti Donisthorpe, 1935
falcozi Roubal, 1927
westi Bruce, 1943
fallax Balfour-Browne, 1953
fumatus sensu auctt. non (Marsham, 1802)
spadiceus sensu auctt. non Falcoz, 1925
fuscicornis Sturm, 1845
insulicola Roubal, 1919
vagus Bruce, 1938
acuminatus Coombs & Woodroffe, 1955
intermedius Bruce, 1934
labilis Erichson, 1846
lapponicus Gyllenhal, 1827
validus Kraatz, 1856
laticollis Lucas, 1846
affinis Sturm, 1845 non Sahlberg, R.F., 1834
lycoperdi (Scopoli, 1763)
rufus (Marsham, 1802)
bituberculatus Stephens, 1830
marshami Stephens, 1830
micaceus Rey, 1889
lovendali Ganglbauer, 1899
loevendali auctt. (misspelling)
obsoletus Reitter, 1879
parallelus Brisout de Barneville, 1863
angustus sensu auctt. ante 2006 non Ganglbauer, 1899
cylindrus sensu auctt. ? Kiesenwetter, 1858
populi Paykull, 1800

grandis Kraatz, 1856
pubescens Sturm, 1845
puncticollis Lucas, 1846
    rotundatus Coombs & Woodroffe, 1955
punctipennis Brisout de Barneville, 1863
    pilosus sensu auctt. ad 2007 non Gyllenhal, 1827
    badius sensu auctt. Brit. partim non Sturm, 1845
reflexus Rey, 1889
    pallidus sensu auctt. ad 2007 non Sturm, 1845
    fowleri Joy, 1910
ruficornis Stephens, 1830
    nigritulus Reitter, 1888
saginatus Sturm, 1845
scanicus (Linnaeus, 1758)
    patruelis Sturm, 1845
    niger Brisout de Barneville, 1863
schmidtii Sturm, 1845
    schmidti auctt. (misspelling)
scutellatus Newman, 1834
    crenatus sensu auctt. Brit. non Herbst, 1792
    bicolor Sturm, 1845
setulosus Sturm, 1845
simplex Miller, 1858
subdepressus Gyllenhal, 1827
subfumatus Kraatz, 1856
uncinatus Stephens, 1830
    badius sensu auctt. Brit. partim non Sturm, 1845
    hirtulus Kraatz, 1858
    immixtus Rey, 1889
    postpositus Sahlberg, J., 1903

MICRAMBE Thomson, C.G., 1863
    CRYPTOPHAGUS sensu auctt. partim non Herbst,
    1792
    MICRAMBINUS Reitter, 1906 [586]
abietis (Paykull, 1798)
bimaculata (Panzer, 1798)
pilosula (Erichson, 1846)
    ?*villosa* (Heer, 1841) non (Beck, 1817) nec auctt.
    lindbergorum (Bruce, 1934)
ulicis (Stephens, 1830)
    vini (Panzer, 1797) non (Laicharting, 1781)
    ?*obcordata* (Marsham, 1802)
woodroffei Johnson, 2007
    ?*ocularis* (Reitter, 1872)
    villosa sensu auctt. ante 2007 non (Heer, 1841) nec
    (Beck, 1817)

ANTHEROPHAGUS Dejean, 1821
pallens (Linnaeus, 1758)
    nigricornis (Fabricius, 1787)
silaceus (Herbst, 1792)
    canescens Grouvelle, 1916
similis Curtis, 1835
    pallens sensu auctt. ad 2007 non (Linnaeus, 1758)

Tribe CAENOSCELINI Casey, 1900

CAENOSCELIS Thomson, C.G., 1863
    COENOSCELIS auctt. Brit. (misspelling)
ferruginea (Sahlberg, C.R., 1820)
    pallida (Wollaston, 1847)
sibirica Reitter, 1889 [587]
subdeplanata Brisout de Barneville, 1882

Subfamily ATOMARIINAE LeConte, 1861

Tribe HYPOCOPRINI Reitter, 1879

HYPOCOPRUS Motschulsky, 1839
latridioides Motschulsky, 1839
    lathridioides auctt. (misspelling)
    quadricollis Reitter, 1878

Tribe ATOMARIINI LeConte, 1861

ATOMARIA Stephens, 1829

Subgenus ATOMARIA Stephens, 1829
badia Erichson, 1846
    sahlbergi Sjöberg, 1947
bella Reitter, 1875
diluta Erichson, 1846
fimetarii (Fabricius, 1792)
    fimetarii (Herbst, 1793)
    fimetarius auctt. (misspelling)
impressa Erichson, 1846
linearis Stephens, 1830
lohsei Johnson & Strand, A., 1968
longicornis Thomson, C.G., 1863
    procerula sensu auctt. non Erichson, 1846
nigrirostris Stephens 1830
    umbrina sensu Erichson, 1846 non (Gyllenhal,
    1827)
    fuscicollis Mannerheim, 1852 [588]
    plicicollis Mäklin, 1863
nigriventris Stephens, 1830
    nana Erichson, 1846
pulchra Erichson, 1846
    barani Brisout de Barneville, 1863
puncticollis Thomson, C.G., 1868
    nigriventris sensu auctt. Brit. partim non Stephens,
    1830
punctithorax Reitter, 1888
    nigriventris sensu auctt. Brit. partim non Stephens,
    1830
    consanguinea Johnson, 1976
strandi Johnson, 1967
    alpina sensu auctt. Brit. non Heer, 1841
    elongatula sensu auctt. Brit. non Erichson, 1846
umbrina (Gyllenhal, 1827)
    fumata sensu auctt. Brit. non Erichson, 1846
vespertina Mäklin, 1853
    affinis sensu auctt. Brit. non (Sahlberg, R.F., 1834)
    badia sensu auctt. Brit. non Erichson, 1846
    prolixa sensu auctt. non Erichson, 1846
    pulchra sensu auctt. ad 2006 non Erichson, 1846
wollastoni Sharp, 1867

Subgenus ANCHICERA Thomson, C.G., 1863
apicalis Erichson, 1846
atra (Herbst, 1793)
atricapilla Stephens, 1830
    nigriceps Erichson, 1846
basalis Erichson, 1846
    nitidula sensu auctt. non Heer, 1841 nec (Marsham,
    1802)
clavigera Ganglbauer, 1899
fuscata (Schönherr, 1808)
    castanea Stephens, 1830
    rufipes Stephens, 1830
fuscipes (Gyllenhal, 1808)

carbonaria Stephens, 1830
concolor Märkel, 1844
gutta Newman, 1834
hislopi Wollaston, 1857
    gibbula sensu auctt. Brit. non Erichson, 1846
lewisi Reitter, 1877
mesomela (Herbst, 1792)
    mesomelaena auctt. (misspelling)
    dimidiata (Marsham, 1802)
    phaeorrhaea (Marsham, 1802)
morio Kolenati, 1846
munda Erichson, 1846
nigripennis (Kugelann, 1794)
nitidula (Marsham, 1802)
    testacea sensu auctt. Brit. non Stephens, 1830
    analis sensu auctt. Brit. non Erichson, 1846
    borealis Sjöberg, 1947
ornata Heer, 1841
    contaminata Erichson, 1846
peltata Kraatz, 1853
    nitidula Heer, 1841 non (Marsham, 1802)
pseudatra Reitter, 1888
    reitteri Løvendal, 1893
pusilla (Paykull, 1798)
    phaeogaster (Marsham, 1802)
    evanescens sensu auctt. Brit. non (Marsham, 1802)
    basella Stephens, 1830
    fulvicollis Stephens, 1830
    thoracica Stephens, 1830
    castanea sensu auctt. Brit. non Stephens, 1830
rhenonum Kraatz, 1853
    rhenana auctt. (misspelling)
    elevata Allen, 1938
    godarti sensu Sjöberg, 1947 non Guillebeau, 1885
rubella Heer, 1841
    bicolor sensu auctt. Brit. non Erichson, 1846
    berolinensis Kraatz, 1853 [588]
rubida Reitter, 1875
    ornata sensu auctt. Brit. non Heer, 1841
    cognata sensu auctt. Europ. non Erichson, 1846
    gibbula sensu Allen, 1968 non Erichson, 1846
    versicolor sensu auctt. Brit. non Erichson, 1846
    viennensis Reitter, 1875
rubricollis Brisout de Barneville, 1863
    divisa Rye, 1876
scutellaris Motschulsky, 1849
testacea Stephens, 1830
    ruficornis (Marsham, 1802) non (Gmelin in
        Linnaeus, 1790) [588]
    dorsalis Stephens, 1830
    terminata (Comolli, 1837)
turgida Erichson, 1846 [589]
zetterstedti (Zetterstedt, 1838)

OOTYPUS Ganglbauer, 1899
    EPHISTEMUS sensu Fowler, 1889 partim non
        Stephens, 1829
globosus (Waltl, 1838)
    nigriclavis sensu auctt. Brit. non (Stephens, 1829)
    palustris (Wollaston, 1846)

EPHISTEMUS Stephens, 1829
globulus (Paykull, 1798)
    gyrinoides (Marsham, 1802)
    nitens (Marsham, 1802)
    piceorhaeus (Marsham, 1802)
    pulchellus (Marsham, 1802)

dimidiatus (Sturm, 1807)
confinis Stephens, 1829
histeroides Stephens, 1835
nitidus Stephens, 1835
ovulum (Erichson, 1846)
dubius Fowler, 1889
reitteri Casey, 1900 [590]

62. Family EROTYLIDAE Latreille, 1802

Family author: R.G. Booth

Subfamily CRYPTOPHILINAE Casey, 1900 [591]

CRYPTOPHILUS Reitter, 1874
integer (Heer, 1841) [592]

Subfamily EROTYLINAE Latreille, 1802

Tribe DACNINI Gistel, 1848

DACNE Latreille, 1796
bipustulata (Thunberg, 1781)
    humeralis (Fabricius, 1787)
rufifrons (Fabricius, 1775)
    fowleri Joy, 1905

Tribe TRITOMINI Curtis, 1834

TRIPLAX Herbst, 1793
aenea (Schaller, 1783)
    bicolor (Marsham, 1802)
lacordairii Crotch, 1870
    lacordairei auctt. (misspelling)
russica (Linnaeus, 1758)
scutellaris Charpentier, 1825
    bicolor sensu Gyllenhal, 1808 non (Marsham, 1802)
    gyllenhalii Crotch, 1870

TRITOMA Fabricius, 1775 [593]
    CYRTOTRIPLAX Crotch, 1873
bipustulata Fabricius, 1775
    humeralis (Marsham, 1802)

63. Family BYTURIDAE Gistel, 1848

Family author: R.G. Booth

BYTURUS Latreille, 1796
ochraceus (Scriba, 1790)
    aestivus sensu auctt. non (Linnaeus, 1758)
    fumatus sensu auctt. partim non (Fabricius, 1775)
tomentosus (De Geer, 1774)
    flavescens (Marsham, 1802)
    urbanus (Lindemann, 1865)

64. Family BIPHYLLIDAE LeConte, 1861

Family author: R.G. Booth

BIPHYLLUS Dejean, 1821
DIPHYLLUS Berthold, 1827
lunatus (Fabricius, 1787)

DIPLOCOELUS Guérin-Méneville, 1844
fagi (Chevrolat in Guérin-Méneville, 1837)

65. Family BOTHRIDERIDAE Erichson, 1845

Family author: R.G. Booth

Subfamily TEREDINAE Seidlitz, 1888

TEREDUS Dejean, 1835
cylindricus (Olivier, 1790)
  nitidus (Fabricius, 1792)

OXYLAEMUS Erichson, 1845
cylindricus (Creutzer in Panzer, 1796)
variolosus (Dufour, 1843)

Subfamily ANOMMATINAE Ganglbauer, 1899

ANOMMATUS Wesmael, 1835
diecki Reitter, 1875 [594]
duodecimstriatus (Müller, P.W.J., 1821)
  obsoletus (Stephens, 1830)

66. Family CERYLONIDAE Billberg, 1820

Family author: R.G. Booth

Subfamily MURMIDIINAE Jacquelin du Val, 1858

MURMIDIUS Leach, 1821
ovalis (Beck, 1817)
segregatus Waterhouse, C.O., 1876

Subfamily CERYLONINAE Billberg, 1820

CERYLON Latreille, 1802
fagi Brisout de Barneville, 1867
ferrugineum Stephens, 1830
  primroseae Donisthorpe, 1942
histeroides (Fabricius, 1792)

67. Family ALEXIIDAE Imhoff, 1856

Family author: R.G. Booth

SPHAEROSOMA Stephens, 1832
ALEXIA Stephens, 1833
pilosum (Panzer, 1793)
  piliferum sensu auctt. Brit. non (Müller, P.W.J., 1821)
  quercus Stephens, 1832

68. Family ENDOMYCHIDAE Leach, 1815 [595]

Family author: R.G. Booth

Subfamily ENDOMYCHINAE Leach, 1815

ENDOMYCHUS Panzer, 1795
coccineus (Linnaeus, 1758)

Subfamily ANAMORPHINAE Strohecker, 1953

SYMBIOTES Redtenbacher, 1849
latus Redtenbacher, 1849

Subfamily MEROPHYSIINAE Seidlitz, 1872

HOLOPARAMECUS Curtis, 1833
caularum (Aubé, 1843)
depressus Curtis, 1833
  kunzei (Aubé, 1843)
singularis (Beck, 1817)

Subfamily LYCOPERDININAE Bromhead, 1838

LYCOPERDINA Latreille, 1807
bovistae (Fabricius, 1792)
succincta (Linnaeus, 1767)

Subfamily MYCETAEINAE Jacquelin du Val, 1857

MYCETAEA Stephens, 1829
subterranea (Fabricius, 1801)
  fumata sensu (Marsham, 1802) non (Linnaeus, 1767)
  hirta (Marsham, 1802) non (Herbst, 1783)
  villosa (Beck, 1817)

69. Family COCCINELLIDAE Latreille, 1807

Family author: R.G. Booth

Subfamily COCCIDULINAE Mulsant, 1846

COCCIDULA Kugelann in Illiger, 1798
rufa (Herbst, 1783)
scutellata (Herbst, 1783)

RHYZOBIUS Stephens, 1829
  RHIZOBIUS sensu auctt. non Burmeister, 1835
  RHIZOBIELLUS Oke, 1951
chrysomeloides (Herbst, 1792) [596]
litura (Fabricius, 1787)
lophanthae (Blaisdell, 1892) [597]

RODOLIA Mulsant, 1850
cardinalis (Mulsant, 1850) [598]

STETHORUS Weise, 1885
SCYMNUS sensu Fowler, 1889 partim non
    Kugelann, 1794
punctillum (Weise, 1891)
    ater sensu auctt. Brit. partim non (Kugelann, 1794)
    minimus (Rossi, 1794) non (Müller, O.F., 1776)

CLITOSTETHUS Weise, 1885
SCYMNUS sensu Fowler, 1889 partim non
    Kugelann, 1794
arcuatus (Rossi, 1794) [599]

SCYMNUS Kugelann, 1794

Subgenus SCYMNUS Kugelann, 1794
femoralis (Gyllenhal, 1827)
    rubromaculatus sensu auctt. Brit. partim non
    (Goeze, 1777)
    pygmaeus sensu auctt. Brit. partim non (Geoffroy in
    Fourcroy, 1785)
frontalis (Fabricius, 1787)
interruptus (Goeze, 1777) [600]
jakowlewi Weise, 1892 [601]
nigrinus Kugelann, 1794
rubromaculatus (Goeze, 1777) [602]
schmidti Fürsch, 1958
    pygmaeus sensu auctt. Brit. partim non (Geoffroy in
    Fourcroy, 1785)
    frontalis var. *immaculatus* sensu auctt. Brit. partim
    non Suffrian, 1843
    mimulus Capra & Fürsch, 1967

Subgenus PULLUS Mulsant, 1846
auritus Thunberg, 1795
    capitatus (Fabricius, 1798)
suturalis Thunberg, 1795

Subgenus NEOPULLUS Sasaji, 1971
haemorrhoidalis Herbst, 1797
limbatus Stephens, 1832
    testaceus sensu auctt. Europ. ante 1967 non
    Motschulsky, 1837

NEPHUS Mulsant, 1846
SCYMNUS sensu Fowler, 1889 partim non
    Kugelann, 1794
bisignatus (Boheman, 1850)
    bohemani (Stenius, 1952)
quadrimaculatus (Herbst, 1783)
    pulchellus (Herbst, 1797)
redtenbacheri (Mulsant, 1846)
    testaceus sensu auctt. Brit. non (Motschulsky, 1837)
    mulsanti (Waterhouse, G.R., 1862)
    lividus (Bold, 1872)
    limonii (Donisthorpe, 1903) [603]

CRYPTOLAEMUS Mulsant, 1853
montrouzieri Mulsant, 1853 [604]

HYPERASPIS Dejean, 1836
pseudopustulata Mulsant, 1853
    reppensis sensu auctt. Brit. non (Herbst, 1783)

Subfamily CHILOCORINAE Mulsant, 1846

PLATYNASPIS Redtenbacher, 1843
luteorubra (Goeze, 1777)

CHILOCORUS Leach, 1815
bipustulatus (Linnaeus, 1758)
renipustulatus (Scriba, 1791)
    similis (Rossi, 1790) non (Thunberg, 1781)

EXOCHOMUS Redtenbacher, 1843
quadripustulatus (Linnaeus, 1758)

Subfamily COCCINELLINAE Latreille, 1807

COCCINULA Dobzhansky, 1925
quattuordecimpustulata (Linnaeus, 1758) [605]

ANISOSTICTA Dejean, 1836
novemdecimpunctata (Linnaeus, 1758)

TYTTHASPIS Crotch, 1874
MICRASPIS Redtenbacher, 1843 non Dejean, 1836
sedecimpunctata (Linnaeus, 1761)
    sexdecimpunctata auctt. (misspelling)

MYZIA Mulsant, 1846
MYSIA Mulsant, 1846 non Lamarck, 1818
NEOMYSIA Casey, 1899
PARAMYSIA Reitter, 1911
oblongoguttata (Linnaeus, 1758)

MYRRHA Mulsant, 1846
HALYZIA sensu Fowler, 1889 partim non Mulsant,
    1846
octodecimguttata (Linnaeus, 1758)

PROPYLEA Mulsant, 1846
PROPYLAEA auctt. (misspelling)
HALYZIA sensu Fowler, 1889 partim non Mulsant,
    1846
quattuordecimpunctata (Linnaeus, 1758)
    quatuordecimpunctata auctt. (misspelling)
    conglobata sensu auctt. Brit. non (Linnaeus, 1758)

CALVIA Mulsant, 1846
HALYZIA sensu Fowler, 1889 partim non Mulsant,
    1846
quattuordecimguttata (Linnaeus, 1758)
    quatuordecimguttata auctt. (misspelling)

VIBIDIA Mulsant, 1846
HALYZIA sensu Fowler, 1889 partim non Mulsant,
    1846
duodecimguttata (Poda, 1761)

HALYZIA Mulsant, 1846
sedecimguttata (Linnaeus, 1758)
    sexdecimguttata auctt. (misspelling)

PSYLLOBORA Dejean, 1836
THEA Mulsant, 1846
HALYZIA sensu Fowler, 1889 partim non Mulsant, 1846
vigintiduopunctata (Linnaeus, 1758)

ANATIS Mulsant, 1846
ocellata (Linnaeus, 1758)

APHIDECTA Weise, 1899
APHIDEITA Weise, 1893 [32]
ADALIA sensu Fowler, 1889 partim non Mulsant, 1846
obliterata (Linnaeus, 1758)

HIPPODAMIA Dejean, 1836

Subgenus HIPPODAMIA Dejean, 1836
tredecimpunctata (Linnaeus, 1758)

Subgenus ADONIA Mulsant, 1846
variegata (Goeze, 1777)

COCCINELLA Linnaeus, 1758
hieroglyphica Linnaeus, 1758
magnifica Redtenbacher, 1843
divaricata sensu auctt. non Olivier, 1808
distincta Faldermann, 1837 non Thunberg, 1781
quinquepunctata Linnaeus, 1758
septempunctata Linnaeus, 1758
undecimpunctata Linnaeus, 1758

ADALIA Mulsant, 1846
bipunctata (Linnaeus, 1758)
decempunctata (Linnaeus, 1758)
bothnica sensu Fowler, 1889 non (Paykull, 1799)
biabilis Marriner, 1926

HARMONIA Mulsant, 1846
axyridis (Pallas, 1773) [606]
quadripunctata (Pontoppidan, 1763)

Subfamily EPILACHNINAE Mulsant, 1846

HENOSEPILACHNA Li, 1961
EPILACHNA sensu auctt. partim non Dejean, 1836
argus (Geoffory in Fourcroy, 1762) [607]

SUBCOCCINELLA Guérin-Méneville, 1844
vigintiquattuorpunctata (Linnaeus, 1758)
vigintiquatuorpunctata auctt. (misspelling)

70. Family CORYLOPHIDAE LeConte, 1852 [608]

Family author: C. Johnson

Subfamily CORYLOPHINAE LeConte, 1852

SERICODERUS Stephens, 1829
brevicornis Matthews, A., 1890 [609]
lateralis (Gyllenhal, 1827)
thoracicus Stephens, 1829

CORYLOPHUS Stephens, 1833
cassidoides (Marsham, 1802)
cassidioides auctt. (misspelling)
sublaevipennis Jacquelin du Val, 1859

Subfamily ORTHOPERINAE Jacquelin du Val, 1857

ORTHOPERUS Stephens, 1829
aequalis Sharp, 1885
nitidulus Allen, 1942
atomarius (Heer, 1841)
atomus (Gyllenhal, 1808)
?*punctum* (Marsham, 1802)
brunnipes sensu auctt. partim non (Gyllenhal, 1808)
picea Stephens, 1829
corticalis sensu Matthews, A., 1885 non (Redtenbacher, 1849)
brunnipes (Gyllenhal, 1808)
nigriclavis (Stephens, 1829)
kluki Wankowicz, 1865
corticalis (Redtenbacher, 1849) [610]
brunnipes sensu auctt. partim non (Gyllenhal, 1808)
mundus sensu auctt. non Matthews, A., 1885
punctatulus Matthews, A., 1885
improvisus Bruce, 1948
nigrescens Stephens, 1829
?*picatus* (Marsham, 1802)
atomos Stephens, 1829
truncatus Stephens, 1829
pilosiusculus sensu auctt. Brit. non Jacquelin du Val, 1859
coriaceus Mulsant & Rey, 1861
mundus Matthews, A., 1885 [611]

Subfamily RYPOBIINAE Paulian, 1950

RYPOBIUS LeConte, 1852
RHYPOBIUS auctt. (misspelling)
praetermissus Bowestead, 1999
ruficollis sensu auctt. non (Jacquelin du Val, 1854)

71. Family LATRIDIIDAE Erichson, 1842 [612]

Family author: C. Johnson

Subfamily LATRIDIINAE Erichson, 1842

LITHOSTYGNUS Broun, 1886
METOPHTHALMUS sensu auctt. non Motschulsky, 1850
serripennis Broun, 1914

STEPHOSTETHUS LeConte, 1878
LATHRIDIUS sensu auctt. Brit. partim non Illiger, 1801
alternans (Mannerheim, 1844) [613]
angusticollis (Gyllenhal, 1827)
angulatus (Mannerheim, 1844)
undulatus (Motschulsky, 1866)
kokujevi (Semenov, 1898)
kokujewi auctt. (misspelling)
caucasicus (Mannerheim, 1844) [614]
sinuatocollis sensu auctt. non (Faldermann, 1837)

campicola sensu auctt. non (Gerhardt, 1911)
lardarius (De Geer, 1775)
    dilaticollis (Motschulsky, 1866)
    pini (Motschulsky, 1866)

CARTODERE Thomson, C.G., 1859
    LATHRIDIUS sensu auctt. partim non Illiger, 1801
    CONINOMUS Thomson, C.G., 1863

Subgenus CARTODERE Thomson, C.G., 1859
constricta (Gyllenhal, 1827)
    carinata (Gyllenhal, 1827)

Subgenus ARIDIUS Motschulsky, 1866
    LATHRIDIUS sensu auctt. partim non Illiger, 1801
    CONINOMUS sensu auctt. Brit. partim non
        Thomson, C.G., 1863
bifasciata (Reitter, 1877)
nodifer (Westwood, 1839)
satelles (Blackburn, 1888)
    australica sensu auctt. non (Belon, 1887)
    norvegica (Strand, 1940)

LATRIDIUS Herbst, 1793
    LATHRIDIUS Illiger, 1801 [40]
    CONITHASSA Thomson, C.G., 1859
    ENICMUS sensu auctt. Brit. partim non Thomson,
        C.G., 1859
assimilis Mannerheim, 1844
    minutus sensu auctt. partim non (Linnaeus, 1767)
    pseudominutus (Strand, A., 1958)
consimilis Mannerheim, 1844
minutus (Linnaeus, 1767)
porcatus Herbst, 1793
    minutus sensu auctt. partim non (Linnaeus, 1767)
    ?*pulla* Marsham, 1802
    anthracinus Mannerheim, 1844
    minutissimus Motschulsky, 1858

THES Semenov, 1910
    LAR Semenov, 1904 non Gosse, 1857
    LATHRIDIUS sensu auctt. Brit. partim non Illiger,
        1801
bergrothi (Reitter, 1880)

ENICMUS Thomson, C.G., 1859
brevicornis (Mannerheim, 1844)
    carbonarius (Mannerheim, 1844)
fungicola Thomson, C.G., 1868
histrio Joy & Tomlin, 1910
    transversus sensu auctt. Brit. partim non (Olivier,
        1790)
rugosus (Herbst, 1793)
testaceus (Stephens, 1830)
    cordaticollis (Aubé, 1850)
transversus (Olivier, 1790)

DIENERELLA Reitter, 1911
    CARTODERE sensu Thomson, C.G., 1863 non
        Thomson, C.G., 1859
    MICROGRAMME Walkley, 1948

Subgenus CARTODEREMA Reitter, 1911
clathrata (Mannerheim, 1844) [615]
    elongata sensu auctt. Brit. partes maj. non (Curtis,
        1830)
    separanda sensu auctt. non (Reitter, 1887)

ruficollis (Marsham, 1802)
    collaris (Mannerheim, 1844)
    exilis (Mannerheim, 1844)
    nanula (Mannerheim, 1844)
vincenti Johnson, 2007
    elongata (Curtis, 1830) non (Gyllenhal, 1827)

Subgenus DIENERELLA Reitter, 1911
argus (Reitter, 1884)
elegans (Aubé, 1850)
    depilis (Belon, 1895)
    pilifera sensu auctt. Brit. partim non (Reitter, 1875)
filiformis (Gyllenhal, 1827)
filum (Aubé, 1850)
schueppeli (Reitter, 1880)

ADISTEMIA Fall, 1899
watsoni (Wollaston, 1871)

Subfamily CORTICARIINAE Curtis, 1829

CORTICARIA Marsham, 1802
alleni Johnson, 1974
    longicollis sensu auctt. Brit. partim non (Zetterstedt,
        1838)
    obscura sensu auctt. Brit. non Brisout de Barneville,
        1863
    depressa sensu auctt. Brit. non Thomson, C.G.,
        1871
    corsica sensu auctt. Brit. non Brisout de Barneville,
        1878
crenulata (Gyllenhal, 1827)
dubia Dajoz, 1970
    eppelsheimi sensu auctt. Brit. non Reitter, 1875
    epelsheimi auctt. Brit. (misspelling)
elongata (Gyllenhal, 1827)
fagi Wollaston, 1854
    pietschi Ganglbauer, 1899
    aequidentata Allen, 1937
ferruginea Marsham, 1802
    rufula Zetterstedt, 1838
    fenestralis sensu Reitter, 1875 non (Linnaeus,
        1758)
fulva (Comolli, 1837)
    hirtella Thomson, C.G., 1863
    flavescens Thomson, C.G., 1871
impressa (Olivier, 1790)
    denticulata (Gyllenhal, 1827)
inconspicua Wollaston, 1860
    crenicollis sensu auctt. Brit. non Mannerheim, 1844
    saginata sensu auctt. Brit. non Mannerheim,
        1844 [616]
longicollis (Zetterstedt, 1838)
longicornis (Herbst, 1783) [617]
    abietorum Motschulsky, 1867
    abietum auctt. (misspelling)
    depressa Thomson, C.G., 1871
polypori Sahlberg, J., 1900 [618]
    alleni sensu auctt. Brit. partim non Johnson, 1974
punctulata Marsham, 1802
    pubescens sensu auctt. Brit. non (Gyllenhal, 1827)
rubripes Mannerheim, 1844
    linearis (Paykull, 1798) non (Thunberg, 1784)
serrata (Paykull, 1798)
    laticollis Mannerheim, 1844
umbilicata (Beck, 1817)
    cylindrica Mannerheim, 1844

borealis Wollaston, 1855

CORTICARINA Reitter, 1880
    MELANOPHTHALMA sensu Fowler, 1889 partim
    non Motschulsky, 1866
curta (Wollaston, 1854)
    fulvipes sensu auctt. non (Comolli, 1837)
    fuscipennis sensu auctt. Brit. non (Mannerheim,
    1844)
    meridionalis (Reitter, 1875)
lambiana (Sharp, 1910)
latipennis (Sahlberg, J., 1871)
    fowleriana (Sharp, 1910)
minuta (Fabricius, 1792)
    fuscula (Gyllenhal, 1827)
similata (Gyllenhal, 1827)
    ?*fulvipes* (Comolli, 1837) non auctt.
truncatella (Mannerheim, 1844)

CORTINICARA Johnson, 1975
    CORTICARINA sensu auctt. partim non Reitter,
    1880
    MELANOPHTHALMA sensu Fowler, 1889 partim
    non Motschulsky, 1866
gibbosa (Herbst, 1793)
    pallida (Marsham, 1802)
    impressa sensu (Marsham, 1802) non (Olivier,
    1790)

MELANOPHTHALMA Motschulsky, 1866
distinguenda (Comolli, 1837)
    angulata sensu auctt. Brit. non (Wollaston, 1864)
suturalis (Mannerheim, 1844) [619]
    transversalis sensu auctt. partim non (Gyllenhal,
    1827)
transversalis (Gyllenhal, 1827)
    curticollis (Mannerheim, 1844)
    wollastoni (Waterhouse, G.R., 1859)

MIGNEAUXIA Jacquelin du Val, 1859
lederi Reitter, 1875 [620]
    ?*inflata* (Rosenhauer, 1856) non auctt.
    orientalis Reitter, 1877

Superfamily TENEBRIONOIDEA Latreille, 1802

72. Family MYCETOPHAGIDAE Leach, 1815

Family author: R.G. Booth

PSEUDOTRIPHYLLUS Reitter, 1880
    TRIPHYLLUS sensu Fowler, 1889 partim non
    Dejean, 1821
suturalis (Fabricius, 1801)
    humeralis (Stephens, 1830)

TRIPHYLLUS Dejean, 1821
bicolor (Fabricius, 1777)
    punctatus (Hellwig, 1792)
    pilosus (Herbst, 1792)
    humeralis (Marsham, 1802)
    immaculatus Roubal, 1936

LITARGUS Erichson, 1846
balteatus LeConte, 1856
    coloratus sensu auctt. Brit. non Rosenhauer, 1856
connexus (Geoffroy in Fourcroy, 1785)
    bifasciatus (Fabricius, 1787)

MYCETOPHAGUS Hellwig, 1792
    TRITOMA Geoffory, 1762 [621]
atomarius (Fabricius, 1787)
fulvicollis Fabricius, 1792
multipunctatus Fabricius, 1792
piceus (Fabricius, 1777)
populi Fabricius, 1798
quadriguttatus Müller, P.W.J., 1821
quadripustulatus (Linnaeus, 1761)

TYPHAEA Stephens, 1829
haagi Reitter, 1874 [622]
    decipiens Lohse, 1989
stercorea (Linnaeus, 1758)
    fumata (Linnaeus, 1767)

EULAGIUS Motschulsky, 1845
    ATRITOMUS Reitter, 1877
filicornis (Reitter, 1887) [623]

BERGINUS Erichson, 1846
tamarisci Wollaston, 1854 [624]

73. Family CIIDAE Leach, 1819

Family authors: G.M. Orledge & R.G. Booth

Subfamily CIINAE Leach, 1819

Tribe OROPHIINI Thomson, C.G., 1863

OCTOTEMNUS Mellié, 1847
glabriculus (Gyllenhal, 1827)

ROPALODONTUS Mellié, 1847
    RHOPALODONTUS Dohrn in Strübing, 1851 [40]
perforatus (Gyllenhal, 1813)

Tribe CIINI Leach, 1819

SULCACIS Dury, 1917
    ENTYPUS Redtenbacher, 1849 non Dahlbom, 1843
    ENNEARTHRON sensu Fowler, 1889 non Mellié,
    1847
nitidus (Fabricius, 1792)
    affinis (Gyllenhal, 1827)

STRIGOCIS Dury, 1917
    RHOPALODONTUS sensu Fowler, 1889 non Dohrn
    in Strübing, 1851
    SULCACIS sensu auctt. partim. non Dury, 1917
bicornis (Mellié, 1849)
    fronticornis sensu auctt. Brit. non (Panzer, 1809)

ORTHOCIS Casey, 1898
　　CIS sensu auctt. partim non Latreille, 1796
　　MELLIEICIS Lohse, 1964
alni (Gyllenhal, 1813)
coluber (Abeille de Perrin, 1874)
　　latifrons (Pool, 1917)

CIS Latreille, 1796
bidentatus (Olivier, 1790)
　　inermis (Marsham, 1802)
bilamellatus Wood, 1884
boleti (Scopoli, 1763)
castaneus (Herbst, 1793)
　　nitidus sensu auctt. non (Fabricius, 1792)
dentatus Mellié, 1849
fagi Waltl, 1839
　　fuscatus Mellié, 1849
　　castaneus sensu Joy, 1932 non (Herbst, 1793) nec
　　　　Mellié, 1849
　　quadridentulus sensu Donisthorpe, 1933 non Perris,
　　　　1874
festivus (Panzer, 1793)
jacquemartii Mellié, 1849
　　jacquemarti auctt. (misspelling)
lineatocribratus Mellié, 1849
micans (Fabricius, 1792)
　　hispidus (Paykull, 1798)
　　savilli Donisthorpe, 1936
punctulatus Gyllenhal, 1827
pygmaeus (Marsham, 1802)
　　nigricornis (Marsham, 1802)
　　rhododactylus (Marsham, 1802)
　　oblongus Mellié, 1849
submicans Abeille de Perrin, 1874
　　micans sensu auctt. non (Fabricius, 1792)
vestitus Mellié, 1849
　　pygmaeus sensu Nyholm, 1953 non (Marsham,
　　　　1802)
villosulus (Marsham, 1802)
　　pyrrhocephalus (Marsham, 1802)
　　flavus Stephens, 1830
　　setiger Mellié, 1849

ENNEARTHRON Mellié, 1847
cornutum (Gyllenhal, 1827)
　　concinnum (Marsham, 1802) [32]

74. Family TETRATOMIDAE Billberg, 1820

Family author: M.V.L. Barclay

Subfamily HALLOMENINAE Gistel, 1848 [625]

HALLOMENUS Panzer, 1793
binotatus (Quensel, 1790) [626]
　　humeralis Panzer, 1793

Subfamily TETRATOMINAE Billberg, 1820

TETRATOMA Fabricius, 1790
ancora Fabricius, 1790
desmarestii Latreille, 1807 [627]
　　desmaresti auctt. (misspelling)
fungorum Fabricius, 1790

75. Family MELANDRYIDAE Leach, 1815

Family author: R.G. Booth

Subfamily MELANDRYINAE Leach, 1815

ORCHESIA Latreille, 1807
　　CLINOCRARA Thomson, C.G., 1859
micans (Panzer, 1793)
minor Walker, 1836
　　tetratoma Thomson, C.G., 1864
undulata Kraatz, 1853

ANISOXYA Mulsant, 1856
fuscula (Illiger, 1798)

ABDERA Stephens, 1832
　　CARIDA Mulsant, 1856
affinis (Paykull, 1799)
biflexuosa (Curtis, 1829)
　　bifasciata sensu auctt. Brit. non (Marsham, 1802)
flexuosa (Paykull, 1799)
quadrifasciata (Curtis, 1829)
　　?bifasciata (Marsham, 1802)
triguttata (Gyllenhal, 1810)

PHLOIOTRYA Stephens, 1832
　　PHLOEOTRYA auctt. (misspelling)
vaudoueri Mulsant, 1856
　　rufipes sensu auctt. Brit. non (Gyllenhal, 1810)

XYLITA Paykull, 1798
laevigata (Hellenius, 1786)
　　buprestoides (Fabricius, 1792)

HYPULUS Paykull, 1798
quercinus (Quensel, 1790) [628]

ZILORA Mulsant, 1856
ferruginea (Paykull, 1798)

MELANDRYA Fabricius, 1801
barbata (Fabricius, 1787)
　　dubia sensu auctt. Brit. non (Schaller, 1783)
caraboides (Linnaeus, 1761)

Subfamily OSPHYINAE Mulsant, 1856 (1839)

CONOPALPUS Gyllenhal, 1810
testaceus (Olivier, 1790)

OSPHYA Illiger, 1807
bipunctata (Fabricius, 1775)

76. Family MORDELLIDAE Latreille, 1802

Family author: B. Levey

The higher classification used in this checklist follows Franciscolo (1957), the most recent authoritative treatment of the the family as a whole. The systematics of the British species of *Mordellistena* have not been fully resolved and it is likely that further species will be found here and some species currently on the British list may prove to be based on misidentifications. I have listed some currently accepted synonyms that have not appeared in the most recent British checklists, where these names were based on British material or used in past British works.

Subfamily MORDELLINAE Latreille, 1802

Tribe MORDELLINI Latreille, 1802

TOMOXIA Costa, A., 1853
bucephala Costa, A., 1853
    biguttata (Gyllenhal, 1827) non (Rossi, 1794)
    fasciata sensu (Paykull, 1800) non (Fabricius, 1775)

MORDELLA Linnaeus, 1758
holomelaena Apfelbeck, 1914
    aculeata sensu auctt. Brit. partim non Linnaeus, 1758 [629]
leucaspis Küster, 1849 [630]
    aculeata sensu auctt. Brit. partim non Linnaeus, 1758

VARIIMORDA Méquignon, 1946
    MORDELLA sensu auctt. partim non Linnaeus, 1758
villosa (Schrank, 1781)
    fasciata (Fabricius, 1775) non (Forster, 1771)

Tribe MORDELLISTENINI Ermisch, 1941

MORDELLISTENA Costa, A., 1854

Subgenus MORDELLISTENA Costa, A., 1854
brevicauda (Boheman, 1849)
humeralis (Linnaeus, 1758)
    flavescens (Marsham, 1802)
    axillaris (Gyllenhal, 1810)
neuwaldeggiana (Panzer, 1796)
    brunnea (Fabricius, 1801)
    ferruginea (Marsham, 1802)
parvula (Gyllenhal, 1827)
    inaequalis sensu auctt. Brit. non Mulsant, 1856
pseudoparvula Ermisch, 1956 [631]
    parvula sensu auctt. Brit. partim non (Gyllenhal, 1827)
    parvuloides Ermisch, 1956 [632]
    eludens Allen, 1999 [633]
pseudopumila Ermisch, 1962 [634]
    pumila sensu auctt. Brit. partim non (Gyllenhal, 1810)
pumila (Gyllenhal, 1810)
pygmaeola Ermisch, 1956 [635]
    pygmeola auctt. (misspelling)

    pumila sensu auctt. Brit. partim non (Gyllenhal, 1810)
secreta Horák, 1983 [636]
variegata (Fabricius, 1798)
    humeralis sensu auctt. Brit. partim non (Linnaeus, 1758)
    bicolor (Marsham, 1802)

Subgenus MORDELLINA Schilsky, 1908
    PSEUDOMORDELLINA Ermisch, 1951
acuticollis Schilsky, 1895 [637]
    parvula sensu auctt. Brit. partim non (Gyllenhal, 1827)
    imitatrix Allen, 1995 [638]
nanuloides Ermisch, 1967 [639]
    parvula sensu auctt. Brit. partim non (Gyllenhal, 1827)

MORDELLOCHROA Emery, 1876
abdominalis (Fabricius, 1775)
    ventralis (Fabricius, 1792)
    nigra (Marsham, 1802)

77. Family RIPIPHORIDAE Gemminger, 1870 (1855)
    RHIPIPHORIDAE auctt. (misspelling)

Family author: R.G. Booth

METOECUS Dejean, 1834
paradoxus (Linnaeus, 1761)

78. Family COLYDIIDAE Billberg, 1820

Family author: R.G. Booth

Subfamily PYCNOMERINAE Erichson, 1845

PYCNOMERUS Erichson, 1842
fuliginosus Erichson, 1842

Subfamily COLYDIINAE Billberg, 1820

Tribe ORTHOCERINI Blanchard, 1845

ORTHOCERUS Latreille, 1796
    SARROTRIUM Illiger, 1798
clavicornis (Linnaeus, 1758)
    muticus (Linnaeus, 1767)

Tribe SYNCHITINI Erichson, 1845

SYNCHITA Hellwig, 1792
humeralis (Fabricius, 1792)
    juglandis (Hellwig, 1792)
separanda (Reitter, 1882)
    humeralis sensu auctt. Brit. partes maj. non (Fabricius, 1792)
    angularis Abeille de Perrin, 1901

CICONES Curtis, 1827
undatus Guérin-Méneville, 1829 [640]
variegatus (Hellwig, 1792)

BITOMA Herbst, 1793
   MONOTOMA Panzer, 1792 [32]
   DITOMA Illiger, 1807
crenata (Fabricius, 1775)

ENDOPHLOEUS Dejean, 1834
markovichianus (Piller & Mitterpacher, 1783)
   spinosulus (Latreille, 1807)

LANGELANDIA Aubé, 1843
anophthalma Aubé, 1843

Tribe COLYDIINI Billberg, 1820

COLYDIUM Fabricius, 1792
elongatum (Fabricius, 1787)

AULONIUM Erichson, 1845
ruficorne (Olivier, 1790)
trisulcus (Geoffroy in Fourcroy, 1785)
   trisulcatum (misspelling)
   trisulcum (misspelling)

79. Family TENEBRIONIDAE Latreille, 1802

Family author: M.V.L. Barclay

Subfamily LAGRIINAE Latreille, 1825 (1820)

LAGRIA Fabricius, 1775
atripes Mulsant & Guillebeau, 1855
hirta (Linnaeus, 1758)

Subfamily TENEBRIONINAE Latreille, 1802

Tribe BOLITOPHAGINI Kirby, 1837

BOLITOPHAGUS Illiger, 1798
   BOLETOPHAGUS auctt. (misspelling)
reticulatus (Linnaeus, 1767)

ELEDONA Latreille, 1796
   HELEDONA Fowler, 1890 (misspelling)
agricola (Herbst, 1783)
   agaricola Fowler, 1890 (misspelling)

Tribe TENEBRIONINI Latreille, 1802

TENEBRIO Linnaeus, 1758
molitor Linnaeus, 1758
   laticollis Stephens, 1832
obscurus Fabricius, 1792

Tribe ALPHITOBIINI Reitter, 1917

ALPHITOBIUS Stephens, 1829
diaperinus (Panzer, 1796)
laevigatus (Fabricius, 1781)

Tribe TRIBOLIINI Gistel, 1848

TRIBOLIUM MacLeay, 1825
castaneum (Herbst, 1797)
   navale (Fabricius, 1775) [641]
confusum Jacquelin du Val, 1863
destructor Uyttenboogaart, 1934

LATHETICUS Waterhouse, C.O., 1880
oryzae Waterhouse, C.O., 1880

Tribe PALORINI Matthews, 2003

PALORUS Mulsant, 1854
   CAENOCORSE Thomson, C.G., 1859
ratzeburgii (Wissmann, 1848)
   melinus sensu Fowler, 1890 non (Herbst, 1783)
subdepressus (Wollaston, 1864)

Tribe ULOMINI Blanchard, 1845

ULOMA Dejean, 1821
culinaris (Linnaeus, 1758)

Tribe PEDININI Eschscholtz, 1829

PHYLAN Dejean, 1821
   HELIOPATHES sensu Fowler, 1890 non Dejean, 1834
gibbus (Fabricius, 1775)

Tribe MELANIMONINI Seidlitz, 1894 (1854)

MELANIMON von Steven, 1829
tibialis (Fabricius, 1781)

Tribe OPATRINI Brullé, 1832

OPATRUM Fabricius, 1775
sabulosum (Linnaeus, 1760)

Tribe HELOPINI Latreille, 1802

HELOPS Fabricius, 1775
caeruleus (Linnaeus, 1758)
   coeruleus Fowler, 1890 (misspelling)

NALASSUS Mulsant, 1854
   CYLINDRINOTUS sensu auctt. partim non Faldermann, 1837
   CYLINDRONOTUS auctt. partim (misspelling)
   HELOPS sensu Fowler, 1890 partim non Fabricius, 1775
laevioctostriatus (Goeze, 1777)
   striatus (Geoffroy in Fourcroy, 1785)

XANTHOMUS Mulsant, 1854
    CYLINDRINOTUS sensu auctt. partim non
      Faldermann, 1837
    CYLINDRONOTUS auctt. partim (misspelling)
    HELOPS sensu Fowler, 1890 partim non Fabricius,
      1775
pallidus (Curtis, 1830)

Tribe BLAPTINI Leach, 1815

BLAPS Fabricius, 1775
lethifera Marsham, 1802
    similis Latreille, 1804
mortisaga (Linnaeus, 1758)
mucronata Latreille, 1804

Subfamily DIAPERINAE Latreille, 1802

Tribe CRYPTICINI Brullé, 1832

CRYPTICUS Latreille, 1817
quisquilius (Linnaeus, 1760)

Tribe PHALERIINI Blanchard, 1845

PHALERIA Latreille, 1802
cadaverina (Fabricius, 1792)

Tribe MYRMECHIXENINI Jacquelin du Val, 1858

MYRMECHIXENUS Chevrolat, 1835
    MYRMECOXENUS Märkel, 1844
subterraneus Chevrolat, 1835
vaporariorum Guérin-Méneville, 1843

Tribe HYPOPHLAEINI Billberg, 1820

CORTICEUS Piller & Mitterpacher, 1783
    HYPOPHLAEUS Fabricius, 1790
    HYPOPHLOEUS auctt. (misspelling)
bicolor (Olivier, 1790)
fraxini (Kugelann, 1794)
linearis (Fabricius, 1790)
unicolor Piller & Mitterpacher, 1783
    castaneus (Fabricius, 1790)

Tribe SCAPHIDEMINI Reitter, 1922

SCAPHIDEMA Redtenbacher, 1849
metallicum (Fabricius, 1792)

Tribe DIAPERINI Latreille, 1802

ALPHITOPHAGUS Stephens, 1832
bifasciatus (Say, 1824)
    quadripustulatus Stephens, 1832
    unifasciatus Donisthorpe, 1925

GNATOCERUS Thunberg, 1814
    GNATHOCERUS Agassiz, 1846 [40]
cornutus (Fabricius, 1798)
maxillosus (Fabricius, 1801)

PENTAPHYLLUS Dejean, 1821
testaceus (Hellwig, 1792) [642]

PLATYDEMA Laporte & Brullé, 1831
violaceum (Fabricius, 1790)
    dytiscoides sensu Fowler, 1890 non (Rossi, 1790)

DIAPERIS Geoffroy, 1762
boleti (Linnaeus, 1758)

Subfamily ALLECULINAE Laporte, 1840

Tribe ALLECULINI Laporte, 1840

PRIONYCHUS Solier, 1835
    ERYX Stephens, 1832 non Daudin, 1803
ater (Fabricius, 1775)
melanarius (Germar, 1813)
    fairmairei sensu auctt. Brit. non Reiche, 1861

GONODERA Mulsant, 1856
    CISTELA sensu auctt. partim non Forster, 1771
luperus (Herbst, 1783)

PSEUDOCISTELA Crotch, 1873
    CISTELA Fabricius, 1775 non Geoffroy, 1762 [643]
ceramboides (Linnaeus, 1758)

ISOMIRA Mulsant, 1856
    CISTELA sensu Fowler, 1890 partim non Fabricius,
      1775
murina (Linnaeus, 1758)

MYCETOCHARA Berthold, 1827
    MYCETOCHARES Latreille, 1829 [40]
humeralis (Fabricius, 1787)
    bipustulata (Illiger, 1794)

Tribe CTENIOPODINI Solier, 1835

CTENIOPUS Solier, 1835
sulphureus (Linnaeus, 1758)
    flavus (Scopoli, 1763)

OMOPHLUS Dejean, 1834
pubescens (Linnaeus, 1758)
    betulae (Herbst, 1783)
    rufitarsis (Leske, 1785)
    armeriae (Curtis, 1836)

80. Family OEDEMERIDAE Latreille, 1810

Family author: R.G. Booth

Subfamily OEDEMERINAE Latreille, 1810

Tribe NACERDINI Mulsant, 1858

NACERDES Dejean, 1834
NACERDA Stephens, 1839 (misspelling)
ANONCODES Redtenbacher, 1845
melanura (Linnaeus, 1758)

Tribe DITYLINI Mulsant, 1858

CHRYSANTHIA Schmidt, W., 1846
CHRYSARTHRIA Schmidt, W., 1844 [32]
nigricornis Westhoff, 1881
viridis sensu Schmidt, W., 1846 non (De Geer, 1775)

Tribe ASCLERINI Gistel, 1848

ISCHNOMERA Stephens, 1832
ASCLERA Dejean, 1834
caerulea (Linnaeus, 1758)
coerulea Fowler, 1890 (misspelling)
cinerascens (Pandellé, 1867) [644]
cyanea (Fabricius, 1792) [645]
caerulea sensu auctt. partim non (Linnaeus, 1758)
graeca (Dahlgren, 1976)
sanguinicollis (Fabricius, 1787)

Tribe OEDEMERINI Latreille, 1810

OEDEMERA Olivier, 1789

Subgenus ONCOMERA Stephens, 1832
femoralis (Olivier, 1803)
femorata (Fabricius, 1792) non (Scopoli, 1763)

Subgenus OEDEMERA Olivier, 1789
lurida (Marsham, 1802)
nobilis (Scopoli, 1763)
virescens (Linnaeus, 1767)

81. Family MELOIDAE Gyllenhal, 1810

Family author: R.G. Booth

Subfamily MELOINAE Gyllenhal, 1810

LYTTA Fabricius, 1775
CANTHARIS sensu auctt. non Linnaeus, 1758
vesicatoria (Linnaeus, 1758)

MELOE Linnaeus, 1758
autumnalis Olivier, 1792
brevicollis Panzer, 1793
cicatricosus Leach, 1813
mediterraneus Müller, J., 1925 [646]
rugosus sensu auctt. partim non Marsham, 1802
proscarabaeus Linnaeus, 1758

rugosus Marsham, 1802 [647]
variegatus Donovan, 1793
violaceus Marsham, 1802

Subfamily NEMOGNATHINAE Laporte, 1840

SITARIS Latreille, 1802
APALUS sensu auctt. Brit. non Fabricius, 1775
muralis (Forster, 1771)

82. Family MYCTERIDAE Oken, 1843

Family author: R.G. Booth

MYCTERUS Clairville, 1798
curculioides (Fabricius, 1781)
curculionoides auctt. (misspelling)

83. Family PYTHIDAE Solier, 1834

Family author: R.G. Booth

PYTHO Latreille, 1796
depressus (Linnaeus, 1767)

84. Family PYROCHROIDAE Latreille, 1806

Family author: R.G. Booth

PYROCHROA Geoffroy, 1762 [648]
coccinea (Linnaeus, 1761)
serraticornis (Scopoli, 1763)

SCHIZOTUS Newman, 1838
PYROCHROA sensu Fowler, 1890 partim non Geoffroy, 1762
PYROCHROELLA Reitter, 1911
pectinicornis (Linnaeus, 1758)

85. Family SALPINGIDAE Leach, 1815

Family author: R.G. Booth

Subfamily AGLENINAE Horn, 1878

AGLENUS Erichson, 1845
brunneus (Gyllenhal, 1813)

Subfamily SALPINGINAE Leach, 1815

LISSODEMA Curtis, 1833
cursor (Gyllenhal, 1813)
kirkae Donisthorpe, 1925
denticolle (Gyllenhal, 1813)
quadripustulatum (Marsham, 1802) non (Fabricius, 1775)

RABOCERUS Mulsant, 1859
SALPINGUS sensu Fowler, 1890 partim non Illiger,
1801
foveolatus (Ljungh, 1824)
mutilatus sensu Champion, 1886 non (Beck, 1817)
bishopi Sharp, 1909
gabrieli Gerhardt, 1901
foveolatus sensu auctt. Brit. non (Ljungh, 1824)

SPHAERIESTES Stephens, 1829
SALPINGUS sensu auctt. non Illiger, 1801
SALPINGELLUS Reitter, 1911
ater (Paykull, 1798) [649]
picea Germar, 1825
stockmanni (Biström, 1977)
castaneus (Panzer, 1796)
reyi (Abeille de Perrin, 1874)
ater sensu Stephens, 1831 non (Paykull, 1798)
aeratus sensu auctt. Brit. non Mulsant, 1859

VINCENZELLUS Reitter, 1911
RHINOSIMUS sensu Fowler, 1890 partim non
Latreille, 1802
ruficollis (Panzer, 1794)
viridipennis (Latreille, 1804)

SALPINGUS Illiger, 1801
RHINOSIMUS Latreille, 1802
planirostris (Fabricius, 1787) [650]
ruficollis (Linnaeus, 1761)

Family author: R.G. Booth

Subfamily ANTHICINAE Latreille, 1819

ANTHICUS Paykull, 1798
angustatus Curtis, 1838
antherinus (Linnaeus, 1760)
bimaculatus (Illiger, 1801)
flavipes (Panzer, 1797)
scoticus Rye, 1872 [652]
tristis Schmidt, 1842
ater sensu Stephens, 1832 non (Panzer, 1796)
ssp. *schaumii* Wollaston, 1857

CORDICOLLIS Marseul, 1879
CORDICOMUS Pic, 1894
ANTHICUS sensu auctt. partim non Paykull, 1798
instabilis (Schmidt, 1842)
tibialis (Curtis, 1838) non (Waltl, 1835)

CYCLODINUS Mulsant & Rey, 1866
ANTHICUS sensu auctt. partim non Paykull, 1798
constrictus (Curtis, 1838) [653]
humilis sensu auctt. Brit. non (Germar, 1824)
albionis (Krekich-Strassoldo, 1919)
larvipennis sensu (Krekich-Strassoldo, 1919) non
(Marseul, 1879)
salinus (Crotch, 1867)
crotchi (Pic, 1893)

OMONADUS Mulsant & Rey, 1866
ANTHICUS sensu auctt. partim non Paykull, 1798
bifasciatus (Rossi, 1792)
floralis (Linnaeus, 1758)
formicarius (Goeze, 1777)
quisquilius (Thomson, C.G., 1864)

STRICTICOLLIS Marseul, 1879
STRICTICOMUS Pic, 1894
ANTHICUS sensu auctt. partim non Paykull, 1798
tobias (Marseul, 1879)

Subfamily NOTOXINAE Stephens, 1829

NOTOXUS Geoffroy, 1762 [654]
monoceros (Linnaeus, 1760)

Family author: R.G. Booth

ADERUS Stephens, 1829
XYLOPHILUS Latreille, 1825 non Mannerheim,
1823
HYLOPHILUS Berthold, 1827 non Temminck, 1822
populneus (Creutzer in Panzer, 1796)
boleti (Marsham, 1802)

EUGLENES Westwood, 1830
ADERUS sensu auctt. partim non Stephens, 1829
oculatus (Paykull, 1798)
pygmaeus sensu auctt. Brit. non (De Geer, 1774)

VANONUS Casey, 1895
ADERUS sensu auctt. partim non Stephens, 1829
brevicornis (Perris, 1869)
neglectus sensu auctt. Brit. partim non (Jacquelin
du Val, 1862)

Family author: B. Levey

The higher classification used in this checklist
follows Franciscolo (1972), the most recent
authoritative treatment of the family as a whole. I
have listed some currently accepted synonyms
that have not appeared in the most recent British
checklists, where these names were based on
British material or used in past British works. I
have not included those infrasubspecific names
which are not available under article 45 of the
Fourth Edition of the International Code of
Zoological Nomenclature.

Subfamily SCRAPTIINAE Gistel, 1848

SCRAPTIA Latreille, 1807
dubia (Olivier, 1790)
fuscula Müller, P.W.J., 1821
testacea Allen, 1940

fuscula sensu auctt. Brit. partim non Müller, P.W.J., 1821

Subfamily ANASPIDINAE Mulsant, 1856

ANASPIS Geoffroy, 1762

Subgenus ANASPIS Geoffroy, 1762
bohemica Schilsky, 1898
    hudsoni sensu Buck, 1954 partim non Donisthorpe, 1909
fasciata (Forster, 1771)
    humeralis sensu auctt. non (Fabricius, 1775)
    biguttata (Rossi, 1794)
    geoffroyi Müller, P.W.J., 1821
    quadrinotata Stephens, 1832
    quadripustulata Stephens, 1832
    scapularis Stephens, 1832
    subfasciata Stephens, 1832
frontalis (Linnaeus, 1758)
    lateralis (Fabricius, 1792)
garneysi Fowler, 1889
lurida Stephens, 1832
    testacea (Marsham, 1802) non (Fabricius, 1787)
    subtestacea Stephens, 1832
maculata (Geoffroy in Fourcroy, 1785) [655]
    melanopa (Forster, 1771)
    nigricollis (Marsham, 1802)
    obscura (Marsham, 1802)
    pallida (Marsham, 1802)
    florenceae Donisthorpe, 1928
pulicaria Costa, A., 1854
    lateralis sensu Stephens, 1832 non (Fabricius, 1792)
    forcipata Mulsant, 1856
regimbarti Schilsky, 1895
    ruficollis sensu auctt. Brit. non (Fabricius, 1792)
    fraudulenta Joy, 1912
thoracica (Linnaeus, 1758) [656]
    septentrionalis Champion, 1891
    latipalpis Schilsky, 1895
    schilskyana Csiki, 1915 [657]
    marginicollis Lindberg, 1925

Subgenus NASSIPA Emery, 1876 [658]
costai Emery, 1876
    costae auctt. (misspelling)
    fuscescens Stephens, 1832 [659]
    flava sensu auctt. non (Linnaeus, 1758)
    thoracica sensu Fowler, 1889 non (Linnaeus, 1758)
rufilabris (Gyllenhal, 1827)
    hudsoni Donisthorpe, 1909

Superfamily CHRYSOMELOIDEA Latreille, 1802

89. Family CERAMBYCIDAE Latreille, 1802 [660]

Family author: M. Rejzek

Subfamily PRIONINAE Latreille, 1802

PRIONUS Geoffroy, 1762
coriarius (Linnaeus, 1758)

Subfamily LEPTURINAE Latreille, 1802

Tribe RHAGIINI Kirby, 1837

RHAGIUM Fabricius, 1775

Subgenus RHAGIUM Fabricius, 1775
inquisitor (Linnaeus, 1758)
    indagator Fabricius, 1787

Subgenus HAGRIUM Villiers, 1978
bifasciatum Fabricius, 1775

Subgenus MEGARHAGIUM Reitter, 1913
mordax (De Geer, 1775)
    inquisitor sensu auctt. Brit. partim non (Linnaeus, 1758)

STENOCORUS Geoffroy, 1762
    STENOCHORUS auctt. (misspelling)
    TOXOTUS Dejean, 1821
meridianus (Linnaeus, 1758)

DINOPTERA Mulsant, 1863
    PACHYTA sensu auctt. partim non Dejean, 1821
    ACMAEOPS sensu auctt. non LeConte, 1850
collaris (Linnaeus, 1758)

GRAMMOPTERA Audinet-Serville, 1835
abdominalis (Stephens, 1831)
    variegata (Germar, 1824) non (Fabricius, 1775)
    analis (Herrich-Schäffer in Panzer, 1832)
ruficornis (Fabricius, 1781)
    holomelina Pool, 1905 [661]
ustulata (Schaller, 1783)
    praeusta (Fabricius, 1787)

Tribe LEPTURINI Latreille, 1802

PEDOSTRANGALIA Sokolov, 1896
    STRANGALIA sensu auctt. partim non Audinet-Serville, 1835
revestita (Linnaeus, 1767)

LEPTUROBOSCA Reitter, 1913
    LEPTURA sensu auctt. partim non Linnaeus, 1758
virens (Linnaeus, 1758)

LEPTURA Linnaeus, 1758
    STRANGALIA sensu auctt. partim non Audinet-Serville, 1835
aurulenta Fabricius, 1792
quadrifasciata Linnaeus, 1758

ANASTRANGALIA Casey, 1924
    LEPTURA sensu auctt. partim non Linnaeus, 1758
sanguinolenta (Linnaeus, 1761)

STICTOLEPTURA Casey, 1924
    LEPTURA sensu auctt. partim non Linnaeus, 1758
rubra (Linnaeus, 1758)
scutellata (Fabricius, 1781)

PARACORYMBIA Miroshnikov, 1998
LEPTURA sensu auctt. partim non Linnaeus, 1758
fulva (De Geer, 1775)

ANOPLODERA Mulsant, 1839
LEPTURA sensu auctt. partim non Linnaeus, 1758
sexguttata (Fabricius, 1775)

JUDOLIA Mulsant, 1863
PACHYTA sensu Fowler, 1890 partim non Dejean, 1821
sexmaculata (Linnaeus, 1758)

PACHYTODES Pic, 1891
PACHYTA sensu auctt. partim non Dejean, 1821
JUDOLIA sensu auctt. partim non Mulsant, 1863
cerambyciformis (Schrank, 1781)

ALOSTERNA Mulsant, 1863
GRAMMOPTERA sensu Fowler, 1890 partim non Audinet-Serville, 1835
tabacicolor (De Geer, 1775)

PSEUDOVADONIA Lobanov, Danilevsky & Murzin, 1981
LEPTURA sensu auctt. partim non Linnaeus, 1758
livida (Fabricius, 1777)

STRANGALIA Audinet-Serville, 1835
attenuata (Linnaeus, 1758)

RUTPELA Nakane & Ohbayashi, 1959
STRANGALIA sensu auctt. partim non Audinet-Serville, 1835
maculata (Poda, 1761)
armata (Herbst, 1784)

STENURELLA Villiers, 1974
STRANGALIA sensu auctt. partim non Audinet-Serville, 1835
melanura (Linnaeus, 1758)
nigra (Linnaeus, 1758)

Subfamily SPONDYLIDINAE Audinet-Serville, 1832

Tribe ASEMINI Thomson, J., 1861

ASEMUM Eschscholtz, 1830
striatum (Linnaeus, 1758)

TETROPIUM Kirby, 1837
ISARTHRON Dejean, 1835 [662]
CRIOMORPHUS Mulsant, 1839 [662]
castaneum (Linnaeus, 1758)
gabrieli Weise, 1905
fuscum sensu auctt. Brit. non (Fabricius, 1787)
crawshayi Sharp, 1905

ARHOPALUS Audinet-Serville, 1834
CRIOCEPHALUS Dejean, 1835
CRIOCEPHALUM Mulsant, 1839
ferus (Mulsant, 1839)
tristis sensu auctt. non (Fabricius, 1787)
polonicus (Motschulsky, 1845)
rusticus (Linnaeus, 1758)

Subfamily CERAMBYCINAE Latreille, 1802

Tribe CALLIDIOPINI Lacordaire, 1868

TRINOPHYLUM Bates, 1878
TRINOPHYLLUM auctt. (misspelling)
cribratum Bates, 1878

Tribe CERAMBYCINI Latreille, 1802

CERAMBYX Linnaeus, 1758
cerdo Linnaeus, 1758
heros sensu Stephens, 1831 non Scopoli, 1763
scopolii Fuessly, 1775
cerdo sensu Scopoli, 1763 non Linnaeus, 1758

Tribe GRACILIINI Mulsant, 1839

GRACILIA Audinet-Serville, 1834
minuta (Fabricius, 1781)

Tribe OBRIINI Mulsant, 1839

OBRIUM Dejean, 1821
brunneum (Fabricius, 1792)
cantharinum (Linnaeus, 1767)

Tribe PSEBIINI Lacordaire, 1868

NATHRIUS Brèthes, 1916
LEPTIDEA Mulsant, 1839 non Billberg, 1820
LEPTIDEELLA Strand, E., 1936
brevipennis (Mulsant, 1839)

Tribe MOLORCHINI Gistel, 1848

MOLORCHUS Fabricius, 1792
CAENOPTERA Thomson, C.G., 1859
minor (Linnaeus, 1758)

GLAPHYRA Newman, 1840
MOLORCHUS sensu Fowler, 1890 partim non Fabricius, 1792
CAENOPTERA sensu Joy, 1932 partim non Thomson, C.G., 1859
umbellatarum (von Schreber, 1759)

Tribe CALLICHROMATINI Swainson, 1840

AROMIA Audinet-Serville, 1833
moschata (Linnaeus, 1758)

Tribe CALLIDIINI Kirby, 1837

HYLOTRUPES Audinet-Serville, 1834
bajulus (Linnaeus, 1758)

SEMANOTUS Mulsant, 1839
russicus (Fabricius, 1776) [663]

CALLIDIUM Fabricius, 1775
violaceum (Linnaeus, 1758)

PYRRHIDIUM Fairmaire, 1864
    CALLIDIUM sensu Fowler, 1890 partim non
      Fabricius, 1775
sanguineum (Linnaeus, 1758)

PHYMATODES Mulsant, 1839 [664]
    CALLIDIUM sensu Fowler, 1890 partim non
      Fabricius, 1775
testaceus (Linnaeus, 1758)
    variabilis (Linnaeus, 1761)

POECILIUM Fairmaire, 1864
    CALLIDIUM sensu Fowler, 1890 partim non
      Fabricius, 1775
alni (Linnaeus, 1767)
lividum (Rossi, 1794) [665]

Tribe CLYTINI Mulsant, 1839

CLYTUS Laicharting, 1784
arietis (Linnaeus, 1758)
    medioniger Allen, 1959

PLAGIONOTUS Mulsant, 1842
    CLYTUS sensu Fowler, 1890 partim non
      Laicharting, 1784
arcuatus (Linnaeus, 1758)

Tribe ANAGLYPTINI Lacordaire, 1868

ANAGLYPTUS Mulsant, 1839
    ANACLYPTUS Mulsant, 1839
    CLYTUS sensu Fowler, 1890 partim non
      Laicharting, 1784
mysticus (Linnaeus, 1758)

Subfamily LAMIINAE Latreille, 1825

Tribe MESOSINI Mulsant, 1839

MESOSA Latreille, 1829
nebulosa (Fabricius, 1781)
    nubila (Gmelin in Linnaeus, 1790)

Tribe AGAPANTHIINI Mulsant, 1839

AGAPANTHIA Audinet-Serville, 1835
villosoviridescens (De Geer, 1775)
    lineatocollis (Donovan, 1797)

Tribe LAMIINI Latreille, 1825

LAMIA Fabricius, 1775
textor (Linnaeus, 1758)

Tribe POGONOCHERINI Mulsant, 1839

POGONOCHERUS Dejean, 1821
    POGONOCHAERUS auctt. (misspelling)
fasciculatus (De Geer, 1775)
hispidulus (Piller & Mitterpacher, 1783)
    bidentatus Thomson, C.G., 1866
hispidus (Linnaeus, 1758)
    dentatus (Geoffroy in Fourcroy, 1785)

Tribe ACANTHOCININI Blanchard, 1845

ACANTHOCINUS Dejean, 1821
aedilis (Linnaeus, 1758)

LEIOPUS Audinet-Serville, 1835
linnei Wallin, Nylander & Kvamme, 2009 [666]
    nebulosus sensu auctt. partim non (Linnaeus, 1758)
nebulosus (Linnaeus, 1758)

Tribe SAPERDINI Mulsant, 1839

SAPERDA Fabricius, 1775
carcharias (Linnaeus, 1758)
populnea (Linnaeus, 1758)
scalaris (Linnaeus, 1758)

STENOSTOLA Dejean, 1835
dubia (Laicharting, 1784)
    ferrea sensu auctt. Brit. non (Schrank, 1776)

Tribe PHYTOECIINI Mulsant, 1839

PHYTOECIA Dejean, 1835
cylindrica (Linnaeus, 1758)

OBEREA Dejean, 1835
oculata (Linnaeus, 1758)

Tribe TETROPINI Portevin, 1927

TETROPS Stephens, 1829
praeustus (Linnaeus, 1758)
starkii Chevrolat, 1859 [667]

90. Family MEGALOPODIDAE Latreille, 1802

Family author: M.L. Cox

**Subfamily ZEUGOPHORINAE Böving &
   Craighead, 1931**

ZEUGOPHORA Kunze, 1818
flavicollis (Marsham, 1802)
subspinosa (Fabricius, 1781)

turneri Power, 1863

91. Family ORSODACNIDAE Thomson, C.G., 1859

Family author: M.L. Cox

ORSODACNE Latreille, 1802
   ORSODACNA Latreille, 1804
cerasi (Linnaeus, 1758)
humeralis Latreille, 1804
   lineola (Panzer, 1795) non (Fabricius, 1781) [668]

92. Family CHRYSOMELIDAE Latreille, 1802

Family author: M.L. Cox

Subfamily BRUCHINAE Latreille, 1802

BRUCHUS Linnaeus, 1767 [669]
   MYLABRIS Geoffroy, 1762 [670]
   LARIA sensu auctt. non Scopoli, 1763
atomarius (Linnaeus, 1761)
   fahraei Gyllenhal, 1839
   viciae sensu Fowler, 1890 non Olivier, 1795
brachialis Fåhraeus, 1839 [671]
ervi Frölich, 1799 [672]
loti Paykull, 1800
pisorum (Linnaeus, 1758)
   pisi Linnaeus, 1767
rufimanus Boheman, 1833
   affinis sensu auctt. Brit. non Frölich, 1799
   velutinus Mulsant & Rey, 1858
rufipes Herbst, 1783
   luteicornis sensu auctt. Brit. non Illiger, 1794

BRUCHIDIUS Schilsky, 1905
   BRUCHUS sensu Fowler, 1890 partim non
      Linnaeus, 1767
cisti (Fabricius, 1775)
   unicolor (Olivier, 1795)
   canus (Germar, 1824)
   debilis (Gyllenhal, 1833)
   cisti "(Paykull, 1800)" sensu Joy, 1932
incarnatus (Boheman, 1833) [673]
olivaceus (Germar, 1824) [674]
   canus sensu auctt. Brit. non (Germar, 1824)
   unicolor sensu Schilsky, 1905 non (Olivier, 1795)
varius (Olivier, 1795) [675]
villosus (Fabricius, 1792)
   cisti sensu (Paykull, 1800) non (Fabricius, 1775)
   ater (Marsham, 1802) non (Scriba, 1790)
   fasciatus sensu auctt. non (Olivier, 1795)

ACANTHOSCELIDES Schilsky in Küster & Kraatz, 1905
obtectus (Say, 1831)
   obsoletus sensu auctt. non Say, 1831

CALLOSOBRUCHUS Pic, 1902
chinensis (Linnaeus, 1758)
   pectinicornis (Linnaeus, 1767)
maculatus (Fabricius, 1775)

quadrimaculatus (Fabricius, 1792)

Subfamily DONACIINAE Kirby, 1837

MACROPLEA Samouelle, 1819
   HAEMONIA Dejean, 1821
appendiculata (Panzer, 1794)
   equiseta (Fabricius, 1798)
mutica (Fabricius, 1792)
   curtisi (Lacordaire, 1845)

DONACIA Fabricius, 1775
aquatica (Linnaeus, 1758)
   dentipes Fabricius, 1792
bicolora Zschach, 1788
   bicolor auctt. (misspelling)
   sagittariae Fabricius, 1792
cinerea Herbst, 1784
clavipes Fabricius, 1792
   menyanthis Fabricius, 1801
   menyanthidis Gyllenhal, 1813
crassipes Fabricius, 1775
dentata Hoppe, 1795
impressa Paykull, 1799
marginata Hoppe, 1795
   limbata Panzer, 1796
   lemnae Fabricius, 1801
obscura Gyllenhal, 1813
semicuprea Panzer, 1796
simplex Fabricius, 1775
   linearis Hoppe, 1795
sparganii Ahrens, 1810
thalassina Germar, 1811
versicolorea (Brahm, 1791)
vulgaris Zschach, 1788
   typhae Ahrens, 1810

PLATEUMARIS Thomson, C.G., 1859
   DONACIA sensu Fowler, 1890 partim non Fabricius, 1775
bracata (Scopoli, 1772)
   braccata auctt. (misspelling)
discolor (Panzer, 1795)
rustica (Kunze, 1818)
   affinis (Kunze, 1818)
sericea (Linnaeus, 1758)

Subfamily CRIOCERINAE Latreille, 1804

LEMA Fabricius, 1798
cyanella (Linnaeus, 1758)
   puncticollis Curtis, 1830

OULEMA des Gozis, 1886
erichsoni (Suffrian, 1841)
melanopus (Linnaeus, 1758) [676]
   melanopa auctt.
obscura (Stephens, 1831)
   cyanella sensu auctt. partim non (Linnaeus, 1758)
   gallaeciana (Heyden, 1870)
   lichenis sensu (Weise, 1882) non (Voet, 1806)
rufocyanea (Suffrian, 1847) [677]
   melanopus sensu auctt. partim non (Linnaeus, 1758)
   duftschmidi (Redtenbacher, 1874)

septentrionis (Weise, 1880)

CRIOCERIS Geoffroy, 1762
asparagi (Linnaeus, 1758)

LILIOCERIS Reitter, 1912
    CRIOCERIS sensu Fowler, 1890 partim non Müller, O.F., 1764
lilii (Scopoli, 1763)

Subfamily CRYPTOCEPHALINAE Gyllenhal, 1813

Tribe CLYTRINI Kirby, 1837

LABIDOSTOMIS Dejean, 1836
tridentata (Linnaeus, 1758)

CLYTRA Laicharting, 1781
    CLYTHRA Fabricius, 1798 (misspelling)
laeviuscula Ratzeburg, 1837
quadripunctata (Linnaeus, 1758)

SMARAGDINA Dejean, 1836
    CYANIRIS Dejean, 1836 non Dalman, 1816
    GYNANDROPHTALMA Lacordaire, 1848
    GYNANDROPHTHALMA Jacquelin du Val, 1865
affinis (Illiger, 1794)

Tribe CRYPTOCEPHALINI Gyllenhal, 1813

CRYPTOCEPHALUS Geoffroy, 1762
aureolus Suffrian, 1847
biguttatus (Scopoli, 1763)
bilineatus (Linnaeus, 1767)
bipunctatus (Linnaeus, 1758)
coryli (Linnaeus, 1758)
decemmaculatus (Linnaeus, 1758)
exiguus Schneider, D.H., 1792
frontalis Marsham, 1802
fulvus (Goeze, 1777)
hypochaeridis (Linnaeus, 1758)
    cristula Dufour, 1843
labiatus (Linnaeus, 1761)
moraei (Linnaeus, 1758)
nitidulus Fabricius, 1787
    ochrostoma Harold, 1872
parvulus Müller, O.F., 1776
    nigrocoeruleus (Goeze, 1777)
primarius Harold, 1872
punctiger Paykull, 1799
pusillus Fabricius, 1777
querceti Suffrian, 1848
sexpunctatus (Linnaeus, 1758)
violaceus Laicharting, 1781

Subfamily LAMPROSOMATINAE Lacordaire, 1848

OOMORPHUS Curtis, 1831
    LAMPROSOMA sensu auctt. non Kirby, 1818 [678]
concolor (Sturm, 1807)

Subfamily EUMOLPINAE Hope, 1840

BROMIUS Dejean, 1836
    ADOXUS Kirby, 1837
obscurus (Linnaeus, 1758) [679]

Subfamily CHRYSOMELINAE Latreille, 1802

TIMARCHA Samouelle, 1819
goettingensis (Linnaeus, 1758)
    coriaria (Laicharting, 1781)
tenebricosa (Fabricius, 1775)

CHRYSOLINA Motschulsky, 1860 [680]
    CHRYSOMELA sensu auctt. non Linnaeus, 1758
americana (Linnaeus, 1758) [681]
banksii (Fabricius, 1775) [682]
    banksi auctt. (misspelling)
brunsvicensis (Gravenhorst, 1807)
    didymata sensu auctt. Brit. non (Scriba, 1791)
cerealis (Linnaeus, 1767)
coerulans (Scriba, 1791) [683]
    caerulans auctt. (misspelling)
fastuosa (Scopoli, 1763)
graminis (Linnaeus, 1758)
haemoptera (Linnaeus, 1758)
herbacea (Duftschmid, 1825)
    menthastri (Suffrian, 1851)
    menthrasti (Fowler, 1890) (misspelling)
hyperici (Forster, 1771)
latecincta (Demaison, 1896)
    sanguinolenta sensu auctt. Brit. partim non (Linnaeus, 1758)
    crassicornis (Helliesin, 1911) non (Fabricius, 1775)
    hellieseni Silverberg, 1977 [684]
    ssp. *intermedia* (Franz, 1938) [685]
marginata (Linnaeus, 1758)
oricalcia (Müller, O.F., 1776)
    orichalcia auctt. (misspelling)
    hobsoni (Stephens, 1831)
polita (Linnaeus, 1758)
sanguinolenta (Linnaeus, 1758)
    marginalis (Duftschmid, 1825)
staphylaea (Linnaeus, 1758)
    staphylea (Fowler, 1890) (misspelling)
sturmi (Westhoff, 1882)
    goettingensis sensu (Linnaeus, 1761) non (Linnaeus, 1758)
    violacea Weise, 1916 non (Müller, O.F., 1776)
varians (Schaller, 1783)

GASTROPHYSA Dejean, 1836
    GASTROIDEA Hope, 1840
polygoni (Linnaeus, 1758)
viridula (De Geer, 1775)

PHAEDON Latreille, 1829
armoraciae (Linnaeus, 1758)
cochleariae (Fabricius, 1792)
    regnianum Tottenham, 1941
concinnus Stephens, 1831
tumidulus (Germar, 1824)

HYDROTHASSA Thomson, C.G., 1859
glabra (Herbst, 1783)
   aucta (Fabricius, 1787)
hannoveriana (Fabricius, 1775)
   hannoverana Fowler, 1890 (misspelling)
marginella (Linnaeus, 1758)

PRASOCURIS Latreille, 1802
   HELODES Paykull, 1799 non Latreille, 1796
junci (Brahm, 1790)
phellandrii (Linnaeus, 1758)

PLAGIODERA Dejean, 1836
versicolora (Laicharting, 1781)
   armoraciae sensu (Fabricius, 1775) non (Linnaeus, 1758)

CHRYSOMELA Linnaeus, 1758
   MELASOMA Stephens, 1831
aenea Linnaeus, 1758
populi Linnaeus, 1758
tremula Fabricius, 1787
   tremulae auctt. (misspelling)
   longicollis (Suffrian, 1851)

GONIOCTENA Dejean, 1836
   PHYTODECTA Kirby, 1837
decemnotata (Marsham, 1802)
   rufipes (De Geer, 1775) non (Linnaeus, 1758)
olivacea (Forster, 1771)
pallida (Linnaeus, 1758)
viminalis (Linnaeus, 1758)

PHRATORA Dejean, 1836
   PHYLLODECTA Kirby, 1837
laticollis Suffrian, 1851
   cavifrons Thomson, C.G., 1866
polaris Schneider, J.S., 1886
vitellinae (Linnaeus, 1758)
vulgatissima (Linnaeus, 1758)

Subfamily GALERUCINAE Latreille, 1802

Tribe GALERUCINI Latreille, 1802

GALERUCELLA Crotch, 1873
calmariensis (Linnaeus, 1767)
lineola (Fabricius, 1781)
nymphaeae (Linnaeus, 1758)
pusilla (Duftschmid, 1825)
sagittariae (Gyllenhal, 1813)
   fergussoni Fowler, 1910
   grisescens sensu auctt. Brit. non (Joannis, 1866) [686]
tenella (Linnaeus, 1761)

PYRRHALTA Joannis, 1865
   GALERUCELLA sensu Fowler, 1890 partim non Crotch, 1873
viburni (Paykull, 1799)

XANTHOGALERUCA Laboissière, 1934
luteola (Müller, O.F., 1766) [687]

GALERUCA Geoffroy, 1762
   ADIMONIA Laicharting, 1781
   GALLERUCA Fabricius, 1792 (misspelling)
laticollis (Sahlberg, C.R., 1838)
   interrupta sensu auctt. non Illiger, 1802
   circumdata Duftschmid, 1825
tanaceti (Linnaeus, 1758) [688]

LOCHMAEA Weise, 1883
   LOCHMAEATA Strand, E., 1935
caprea (Linnaeus, 1758)
   capreae auctt. (misspelling)
crataegi (Forster, 1771)
suturalis (Thomson, C.G., 1866)

DIABROTICA Dejean, 1836
virgifera LeConte, 1858 [689]

PHYLLOBROTICA Dejean, 1836
quadrimaculata (Linnaeus, 1758)

LUPERUS Geoffroy, 1762
flavipes (Linnaeus, 1767)
longicornis (Fabricius, 1781)
   rufipes sensu auctt. Brit. ? (Scopoli, 1763)

CALOMICRUS Dillwyn, 1829
   LUPERUS sensu Fowler, 1890 partim non Müller, O.F., 1764
circumfusus (Marsham, 1802)
   circumfuscus auctt. (misspelling)
   nigrofasciatus sensu Weise, 1886 non (Goeze, 1777)

AGELASTICA Dejean, 1836
alni (Linnaeus, 1758)

SERMYLASSA Reitter, 1912
   SERMYLA Chapuis, 1875 non Adams, 1854
halensis (Linnaeus, 1767)

Tribe ALTICINI Newman, 1834

LUPEROMORPHA Weise, 1887
xanthodera (Fairmaire, 1888) [690]

PHYLLOTRETA Dejean, 1836
atra (Fabricius, 1775)
consobrina (Curtis, 1837)
   hintoni Donisthorpe, 1944
cruciferae (Goeze, 1777)
diademata Foudras, 1860
exclamationis (Thunberg, 1784)
flexuosa (Illiger, 1794)
   sinuata (Stephens, 1831)
nemorum (Linnaeus, 1758)
nigripes (Fabricius, 1775)
nodicornis (Marsham, 1802)
ochripes (Curtis, 1837)
punctulata (Marsham, 1802)
   aerea Allard, 1859
striolata (Fabricius, 1803)
   vittata sensu auctt. non (Fabricius, 1801)
tetrastigma (Comolli, 1837)

undulata Kutschera, 1860
vittula (Redtenbacher, 1849)

APHTHONA Dejean, 1836
?*atratula* Allard, 1859
   atrovirens sensu auctt. Brit. non Förster, 1849 [691]
atrocaerulea (Stephens, 1829)
   cyanella (Redtenbacher, 1849)
   puncticollis Allard, 1866
euphorbiae (Schrank, 1781)
   virescens Foudras, 1860
   aeneomicans sensu auctt. non Allard, 1875
herbigrada (Curtis, 1837)
lutescens (Gyllenhal, 1808)
melancholica Weise, 1888
   venustula sensu auctt. Brit. non Kutschera, 1861
nigriceps (Redtenbacher, 1842)
   pallida sensu auctt. Brit. partim non Bach, 1856
nonstriata (Goeze, 1777)
   coerulea (Fourcroy, 1785)
pallida (Bach, 1856) [692]
   nigriceps sensu auctt. Brit. partim non
      (Redtenbacher, 1842)

LONGITARSUS Berthold, 1827
absynthii Kutschera, 1862
   absinthii auctt. (misspelling)
aeneicollis (Faldermann, 1837)
   suturalis (Marsham, 1802) non (Fabricius, 1775)
   nigricollis (Foudras, 1860)
aeruginosus (Foudras, 1860)
agilis (Rye, 1868)
anchusae (Paykull, 1799)
atricillus (Linnaeus, 1761)
ballotae (Marsham, 1802)
   cerinus sensu auctt. Brit. partim non (Foudras,
      1860)
brunneus (Duftschmid, 1825)
   castaneus sensu auctt. Brit. non (Duftschmid, 1825)
curtus (Allard, 1860)
dorsalis (Fabricius, 1781) [693]
exoletus (Linnaeus, 1758)
   femoralis (Marsham, 1802)
ferrugineus (Foudras, 1860)
   waterhousei Kutschera, 1864
flavicornis (Stephens, 1831)
   tabidus sensu auctt. partim non (Fabricius, 1775)
   jacobaeae sensu auctt. partim non (Waterhouse,
      G.R., 1858)
   rufescens Fowler, 1890
fowleri Allen, 1967
   abdominalis sensu Fowler, 1890 non (Duftschmid,
      1825)
ganglbaueri Heikertinger, 1912
   piciceps sensu auctt. non (Stephens, 1831)
   senecionis Brisout de Barneville, 1873 non
      (Motschulsky, 1851)
gracilis Kutschera, 1864
   poweri (Allard, 1866)
holsaticus (Linnaeus, 1758)
jacobaeae (Waterhouse, G.R., 1858)
   tabidus sensu auctt. partim non (Fabricius, 1775)
kutscherae (Rye, 1872)
   atriceps Kutschera, 1863 non (Stephens, 1831)
longiseta Weise, 1889
   clarus Allen, 1967 [694]
luridus (Scopoli, 1763)

   brunneus sensu auctt. non (Duftschmid, 1825)
   castaneus (Duftschmid, 1825)
   fusculus sensu auctt. Brit. non Kutschera, 1864
lycopi (Foudras, 1860)
   abdominalis sensu (Allard, 1860) non (Duftschmid,
      1825)
melanocephalus (De Geer, 1775)
   piciceps (Stephens, 1831)
membranaceus (Foudras, 1860)
   cerinus sensu auctt. Brit. partim non (Foudras,
      1860)
   teucrii (Allard, 1860)
nasturtii (Fabricius, 1792)
nigerrimus (Gyllenhal, 1827)
nigrofasciatus (Goeze, 1777)
   patruelis (Allard, 1866)
   distinguendus (Rye, 1872)
obliteratoides Gruev, 1973 [695]
obliteratus (Rosenhauer, 1847)
   pulex sensu Fowler, 1890 non (Schrank, 1781)
ochroleucus (Marsham, 1802)
parvulus (Paykull, 1799)
   pumilus (Illiger, 1807)
   ater sensu Fowler, 1890 non (Fabricius, 1775)
pellucidus (Foudras, 1860)
   medicaginis (Allard, 1860)
plantagomaritimus Dollman, 1912
pratensis (Panzer, 1794)
   pusillus (Gyllenhal, 1813)
   collaris (Stephens, 1831)
   bearei Kevan, 1967 [696]
quadriguttatus (Pontoppidan, 1763)
reichei (Allard, 1860)
   fusculus Kutschera, 1864
rubiginosus (Foudras, 1860)
   flavicornis sensu (Allard, 1860) non (Stephens,
      1831)
rutilus (Illiger, 1807)
succineus (Foudras, 1860)
   laevis sensu (Allard, 1860) non (Duftschmid, 1825)
suturellus (Duftschmid, 1825)
   thoracicus (Stephens, 1831)
symphyti Heikertinger, 1912 [697]
tabidus (Fabricius, 1775)

ALTICA Geoffroy, 1762 [698]
   HALTICA Illiger, 1801 [40]
brevicollis Foudras, 1860
   coryli (Allard, 1860)
carinthiaca Weise, 1888 [699]
helianthemi (Allard, 1859) [700]
   pusilla (Duftschmid, 1825) non (Müller, O.F., 1776)
longicollis (Allard, 1860) [701]
   ericeti sensu auctt. Brit. non (Allard, 1859)
   britteni Sharp, 1914 [702]
   sandini Kemner, 1919
lythri Aubé, 1843
   tamaricis sensu auctt. Brit. partim non Schrank,
      1785
oleracea (Linnaeus, 1758)
   ampelophaga sensu auctt. Brit. non Guérin-
      Méneville, 1858
   ytenensis Sharp, 1914
palustris Weise, 1888

HERMAEOPHAGA Foudras, 1860
mercurialis (Fabricius, 1792)

BATOPHILA Foudras, 1860
aerata (Marsham, 1802)
rubi (Paykull, 1799)

LYTHRARIA Bedel, 1897
OCHROSIS sensu Fowler, 1890 non Foudras, 1860
salicariae (Paykull, 1800)

OCHROSIS Foudras, 1860
CREPIDODERA sensu Fowler, 1890 partim non
Dejean, 1836
ventralis (Illiger, 1807)

NEOCREPIDODERA Heikertinger, 1911
CREPIDODERA sensu auctt. non Dejean, 1836
ASIORESTIA Jacobson, 1925
ferruginea (Scopoli, 1763)
impressa (Fabricius, 1801)
transversa (Marsham, 1802)

DEROCREPIS Weise, 1886
CREPIDODERA sensu Fowler, 1890 partim non
Dejean, 1836
rufipes (Linnaeus, 1758)

HIPPURIPHILA Foudras, 1860
modeeri (Linnaeus, 1761)

CREPIDODERA Dejean, 1836
CHALCOIDES Foudras, 1860
aurata (Marsham, 1802) [703]
aurea (Fourcroy, 1785)
helxines sensu auctt. non (Linnaeus, 1758)
fulvicornis (Fabricius, 1792)
smaragdina Foudras, 1860
nitidula (Linnaeus, 1758)
plutus (Latreille, 1804)
chloris Foudras, 1860

EPITRIX Foudras, 1860
EPITHRIX Foudras, 1860
atropae Foudras, 1860
pubescens (Koch, J.D.W., 1803)

PODAGRICA Dejean, 1836
fuscicornis (Linnaeus, 1767)
fuscipes (Fabricius, 1775)

MANTURA Stephens, 1831
chrysanthemi (Koch, J.D.W., 1803)
matthewsii (Curtis, 1833)
obtusata (Gyllenhal, 1813)
rustica (Linnaeus, 1767)

CHAETOCNEMA Stephens, 1831 [704]
PLECTROSCELIS Dejean, 1836
aerosa (Letzner, 1847)
arida Foudras, 1860
aridula sensu auctt. Brit. partim non (Gyllenhal,
1827)
concinna (Marsham, 1802)
confusa (Boheman, 1851)
hortensis (Fourcroy, 1785)
picipes Stephens, 1831 [705]

concinna sensu auctt. partim non (Marsham, 1802)
laevicollis (Thomson, C.G., 1866) [706]
heikertingeri Lubischev, 1963 [706]
sahlbergii (Gyllenhal, 1827)
subcoerulea (Kutschera, 1864)

SPHAERODERMA Stephens, 1831
rubidum (Graëlls, 1858)
testaceum sensu auctt. partim non Fabricius, 1775
testaceum (Fabricius, 1775)
cardui (Gyllenhal, 1813)

APTEROPEDA Dejean, 1836
globosa (Illiger, 1794)
orbiculata (Marsham, 1802)
splendida Allard, 1860

MNIOPHILA Stephens, 1831
muscorum (Koch, J.D.W., 1803)

DIBOLIA Latreille, 1829
cynoglossi (Koch, J.D.W., 1803)

PSYLLIODES Latreille, 1829
affinis (Paykull, 1799)
attenuata (Koch, J.D.W., 1803)
chalcomera (Illiger, 1807)
chrysocephala (Linnaeus, 1758)
anglica (Fabricius, 1775)
luridipennis sensu auctt. partim non Kutschera,
1864
cucullata (Illiger, 1807) [707]
cuprea (Koch, J.D.W., 1803)
instabilis sensu auctt. Brit. non Foudras, 1860
dulcamarae (Koch, J.D.W., 1803)
hyoscyami (Linnaeus, 1758)
laticollis Kutschera, 1860
weberi Lohse, 1955 [708]
luridipennis Kutschera, 1864
hospes sensu auctt. non Wollaston, 1854
luteola (Müller, O.F., 1776)
marcida (Illiger, 1807)
napi (Fabricius, 1792)
picina (Marsham, 1802)
sophiae Heikertinger, 1914
cyanoptera sensu auctt. non (Illiger, 1807)

Subfamily CASSIDINAE Gyllenhal, 1813

PILEMOSTOMA Desbrochers, 1891
CASSIDA sensu Fowler, 1890 partim non Linnaeus,
1758
fastuosa (Schaller, 1783)

HYPOCASSIDA Weise, 1893
subferruginea (Schrank, 1776)

CASSIDA Linnaeus, 1758
denticollis Suffrian, 1844
chloris sensu auctt. Brit. non Suffrian, 1844
flaveola Thunberg, 1794
hemisphaerica Herbst, 1799
murraea Linnaeus, 1767
maculata Linnaeus, 1767

nebulosa Linnaeus, 1758
nobilis Linnaeus, 1758
prasina Illiger, 1798
    sanguinolenta sensu auctt. Brit. non Müller, O.F.,
    1776
    chloris Suffrian, 1844
rubiginosa Müller, O.F., 1776
    viridis sensu Scopoli, 1763 non Linnaeus, 1758
sanguinosa Suffrian, 1844
vibex Linnaeus, 1767
viridis Linnaeus, 1758
    equestris Fabricius, 1787
vittata de Villers, 1789

Superfamily CURCULIONOIDEA Latreille, 1802 [709]

Division ORTHOCERI Schönherr, 1823 [710]

93. Family NEMONYCHIDAE Bedel, 1882 [711]

Family author: M.G. Morris

Subfamily CIMBERIDINAE des Gozis, 1882 [712]

CIMBERIS des Gozis, 1881
    RHINOMACER Fabricius, 1781 [713]
attelaboides (Fabricius, 1787)

94. Family ANTHRIBIDAE Billberg, 1820 [714]

Family author: M.G. Morris

Subfamily ANTHRIBINAE Billberg, 1820

Tribe ANTHRIBINI Billberg, 1820

ANTHRIBUS Geoffroy, 1762 [715]
    BRACHYTARSUS Schönherr, 1823
fasciatus Forster, 1770
nebulosus Forster, 1770
    variegatus Fourcroy, 1785
    varius (Fabricius, 1787)

Tribe PLATYRHININI Bedel, 1882

PLATYRHINUS Clairville, 1798
    PLATYRRHINUS Fowler, 1891 (misspelling)
resinosus (Scopoli, 1763)
    latirostris (Fabricius, 1775)

Tribe PLATYSTOMINI Pierce, 1916

PLATYSTOMOS Schneider, D.H., 1791 [715]
    PLATYSTOMUS auctt. (misspelling)
    ANTHRIBUS sensu auctt. non Geoffroy, 1762
    MACROCEPHALUS Olivier, 1789 non Swederus,
    1787
albinus (Linnaeus, 1758)

Tribe DISCOPTENINI Lacordaire, 1865

PSEUDEUPARIUS Jordan, 1914 [716]
    TROPIDERES sensu auctt. non Schönherr, 1823
sepicola (Fabricius, 1792)

Tribe ZYGAENODINI Lacordaire, 1865

DISSOLEUCAS Jordan, 1925 [717]
niveirostris (Fabricius, 1798)

Subfamily CHORAGINAE Kirby, 1819

Tribe CHORAGINI Kirby, 1819

CHORAGUS Kirby, 1819
sheppardi Kirby, 1819

Tribe ARAECERINI Lacordaire, 1865

ARAECERUS Schönherr, 1823
    ARAEOCERUS auctt. (misspelling)
fasciculatus (De Geer, 1775)

Subfamily URODONTINAE Thomson, C.G., 1859 [718]

BRUCHELA Dejean, 1821
    BRUCHELLA auctt. (misspelling)
    URODON Schönherr, 1823
rufipes (Olivier, 1790) [719]

95. Family RHYNCHITIDAE Gistel, 1848

Family author: M.G. Morris

Subfamily RHYNCHITINAE Gistel, 1848

Tribe RHYNCHITINI Gistel, 1848

INVOLVULUS Schrank, 1798 [715]
caeruleus (De Geer, 1775)
cupreus (Linnaeus, 1758)

LASIORHYNCHITES Jekel, 1860 [720]
    RHYNCHITES sensu auctt. partim non Schneider,
    D.H., 1791
cavifrons (Gyllenhal, 1833)
    pubescens sensu auctt. non (Fabricius, 1775)
olivaceus (Gyllenhal, 1833)
    ophthalmicus sensu auctt. Brit. non (Stephens,
    1831)
    sericeus sensu (Fowler, 1890) non (Herbst, 1797)

NEOCOENORRHINUS Voss, 1951
RHYNCHITES sensu auctt. partim non Schneider, D.H., 1791
CAENORHINUS sensu auctt. non Thomson, C.G., 1859 [721]
COENORHINUS Voss, 1932
germanicus (Herbst, 1797)
minutus sensu (Thomson, C.G., 1866) non (Herbst, 1797)
interpunctatus (Stephens, 1831)
minutus (Herbst, 1797)
aeneovirens (Marsham, 1802)
pauxillus (Germar, 1824)

RHYNCHITES Schneider, D.H., 1791
auratus (Scopoli, 1763)
bacchus (Linnaeus, 1758)

TATIANAERHYNCHITES Legalov, 2002
NEOCOENORRHINUS sensu auctt. partim non Voss, 1951
aequatus (Linnaeus, 1767)

TEMNOCERUS Thunberg, 1815 [715]
PSELAPHORHYNCHITES Schilsky, 1903
coeruleus (Fabricius, 1798)
tomentosus (Gyllenhal, 1839)
uncinatus (Thomson, C.G., 1865)
longiceps (Thomson, C.G., 1888)
harwoodi (Joy, 1911)
nanus (Paykull, 1792)

Tribe BYCTISCINI Voss, 1923

BYCTISCUS Thomson, C.G., 1859
betulae (Linnaeus, 1758)
betuleti (Fabricius, 1792)
populi (Linnaeus, 1758)

Tribe DEPORAINI Voss, 1929

CAENORHINUS Thomson, C.G., 1859 [722]
mannerheimii (Hummel, 1823)
mannerheimi auctt. (misspelling)
megacephalus (Germar, 1824)

DEPORAUS Samouelle, 1819
betulae (Linnaeus, 1758)

96. Family ATTELABIDAE Billberg, 1820

Family author: M.G. Morris

Subfamily ATTELABINAE Billberg, 1820

ATTELABUS Linnaeus, 1758
nitens (Scopoli, 1763)

curculionoides Linnaeus, 1763

Subfamily APODERINAE Jekel, 1860

APODERUS Olivier, 1807
coryli (Linnaeus, 1758)

97. Family APIONIDAE Schönherr, 1823 [723]

Family author: M.G. Morris

Subfamily APIONINAE Schönherr, 1823

Tribe APIONINI Schönherr, 1823

APION Herbst, 1797
ERYTHRAPION Schilsky, 1906
cruentatum Walton, 1844
sanguineum sensu Stephens, 1839 non (De Geer, 1775)
desideratum Sharp, 1918
frumentarium (Linnaeus, 1758)
miniatum Germar, 1833 [724]
haematodes Kirby, 1808
frumentarium sensu (Paykull, 1792) non (Linnaeus, 1758)
brachypterum Sharp, 1918
fraudator Sharp, 1918
rubens Stephens, 1839
rubiginosum Grill, 1893
sanguineum (De Geer, 1775)

Tribe APLEMONINI Kissinger, 1968

AIZOBIUS Alonso-Zarazaga, 1991
APION sensu auctt. partim non Herbst, 1797
sedi (Germar, 1818)

HELIANTHEMAPION Wagner, 1930
APION sensu auctt. partim non Herbst, 1797
aciculare (Germar, 1817) [725]

PERAPION Wagner, 1907
APION sensu auctt. partim non Herbst, 1797

Subgenus PERAPION Wagner, 1907
affine (Kirby, 1808)
curtirostre (Germar, 1817)
brevirostre sensu (Kirby, 1808) non (Herbst, 1797)
humile (Germar, 1817)
hydrolapathi (Marsham, 1802)
marchicum (Herbst, 1797)
violaceum (Kirby, 1808)

Subgenus EROOSAPION Ehret, 1994
lemoroi (Brisout de Barneville, 1880)

PSEUDAPLEMONUS Wagner, 1930
APION sensu auctt. partim non Herbst, 1797
limonii (Kirby, 1808)

PSEUDOPERAPION Wagner, 1930
brevirostre (Herbst, 1797) [726]

Tribe ASPIDAPIINI Alonso-Zarazaga, 1990

ASPIDAPION Schilsky, 1901
    APION sensu auctt. partim non Herbst, 1797

Subgenus ASPIDAPION Schilsky, 1901
radiolus (Marsham, 1802)
soror (Rey, 1895) [727]
    foveatoscutellatum (Wagner, 1906)

Subgenus KOESTLINIA Alonso-Zarazaga, 1990
aeneum (Fabricius, 1775)

Tribe CERATAPIINI Alonso-Zarazaga, 1990

ACENTROTYPUS Alonso-Zarazaga, 1990
    APION sensu auctt. partim non Herbst, 1797
brunnipes (Boheman, 1839)
    laevigatum (Kirby, 1808) non (Paykull, 1792)

CERATAPION Schilsky, 1901 [728]
    APION sensu auctt. partim non Herbst, 1797

Subgenus CERATAPION Schilsky, 1901
armatum (Gerstaecker, 1854)
carduorum (Kirby, 1808)
    dentirostre sensu auctt. Brit. non (Gerstaecker, 1854)
    lacertense (Tottenham, 1941)
gibbirostre (Gyllenhal, 1813)
    carduorum sensu auctt. non (Kirby, 1808)

Subgenus ACANEPHODUS Alonso-Zarazaga, 1990
onopordi (Kirby, 1808)

DIPLAPION Reitter, 1916
    APION sensu auctt. partim non Herbst, 1797
confluens (Kirby, 1808)
stolidum (Germar, 1817)

OMPHALAPION Schilsky, 1901
    APION sensu auctt. partim non Herbst, 1797
beuthini (Hoffmann, Anton, 1874)
    dispar sensu auctt. Brit. non (Germar, 1817) [729]
hookerorum (Kirby, 1808)
    hookeri auctt. (misspelling) [730]
laevigatum (Paykull, 1792) [731]
    sorbi (Fabricius, 1792)

Tribe EXAPIINI Alonso-Zarazaga, 1990

EXAPION Bedel, 1887
    APION sensu auctt. partim non Herbst, 1797
    ULAPION Ehret, 1997
difficile (Herbst, 1797)
    kiesenwetteri (Desbrochers, 1870)
fuscirostre (Fabricius, 1775)
genistae (Kirby, 1811)
ulicis (Forster, 1771)

Tribe IXAPIINI Alonso-Zarazaga, 1990

IXAPION Roudier & Tempère, 1973
    APION sensu auctt. partim non Herbst, 1797
variegatum (Wencker, 1864) [732]

Tribe KALCAPIINI Alonso-Zarazaga, 1990

KALCAPION Schilsky, 1906
    APION sensu auctt. partim non Herbst, 1797
pallipes (Kirby, 1808)
semivittatum (Gyllenhal, 1833)

MELANAPION Wagner, 1930
    APION sensu auctt. partim non Herbst, 1797
minimum (Herbst, 1797)

SQUAMAPION Bokor, 1923
    APION sensu auctt. partim non Herbst, 1797
    THYMAPION Deville, 1924
atomarium (Kirby, 1808)
    minutissimum sensu auctt. Brit. non (Rosenhauer, 1856)
    serpyllicola sensu auctt. Brit. non (Wencker, 1864)
cineraceum (Wencker, 1864)
    millum sensu (Bach, 1854) non (Gyllenhal, 1833)
    annulipes sensu auctt. non (Wencker, 1864)
flavimanum (Gyllenhal, 1833)
vicinum (Kirby, 1808)

TAENIAPION Schilsky, 1906
    APION sensu auctt. partim non Herbst, 1797
urticarium (Herbst, 1784)

Tribe MALVAPIINI Alonso-Zarazaga, 1990

MALVAPION Hoffmann, Adolfe, 1958
    APION sensu auctt. partim non Herbst, 1797
malvae (Fabricius, 1775)

PSEUDAPION Schilsky, 1906
    APION sensu auctt. partim non Herbst, 1797
rufirostre (Fabricius, 1775)

RHOPALAPION Schilsky, 1906
    APION sensu auctt. partim non Herbst, 1797
longirostre (Olivier, 1807) [733]

Tribe OXYSTOMATINI Alonso-Zarazaga, 1990

Subtribe OXYSTOMATINA Alonso-Zarazaga, 1990

CYANAPION Bokor, 1923
    APION sensu auctt. partim non Herbst, 1797

Subgenus CYANAPION Bokor, 1923
spencii (Kirby, 1808)

Subgenus BOTHRYORRHYNCHAPION Bokor, 1923
afer (Gyllenhal, 1833) [734]
    afrum auctt. (misspelling)
    unicolor sensu auctt. non (Kirby, 1808)
    platalea sensu auctt. Brit. non (Germar, 1817)
gyllenhalii (Kirby, 1808)
    gyllenhali auctt. (misspelling)
    unicolor (Kirby, 1808)

EUTRICHAPION Reitter, 1916
APION sensu auctt. partim non Herbst, 1797

Subgenus EUTRICHAPION Reitter, 1916
ervi (Kirby, 1808)
viciae (Paykull, 1800)

Subgenus CNEMAPION Bokor, 1923
vorax (Herbst, 1797)

Subgenus PSILOCALYMMA Alonso-Zarazaga, 1990
punctigerum (Paykull, 1792)

HEMITRICHAPION Voss, 1959
APION sensu auctt. partim non Herbst, 1797

Subgenus TINOCYBA Alonso-Zarazaga, 1990
reflexum (Gyllenhal, 1833)
  livescerum (Gyllenhal, 1833)
waltoni (Stephens, 1839)

HOLOTRICHAPION Györffy, 1956
APION sensu auctt. partim non Herbst, 1797

Subgenus HOLOTRICHAPION Györffy, 1956
ononis (Kirby, 1808)

Subgenus APIOPS Alonso-Zarazaga, 1990
pisi (Fabricius, 1801)

Subgenus LEGARICAPION Ehret, 1990
aethiops (Herbst, 1797)

OXYSTOMA Duméril, 1806
APION sensu auctt. partim non Herbst, 1797
cerdo (Gerstaecker, 1854)
craccae (Linnaeus, 1767)
pomonae (Fabricius, 1798)
  opeticum sensu auctt. Brit. non (Bach, 1854)
subulatum (Kirby, 1808)

PIRAPION Reitter, 1916
APION sensu auctt. partim non Herbst, 1797
immune (Kirby, 1808)

Subtribe CATAPIINA Alonso-Zarazaga, 1990

CATAPION Schilsky, 1906
APION sensu auctt. partim non Herbst, 1797
curtisii (Stephens, 1831) [735]
  curtisi auctt. (misspelling)
  curtulum Desbrochers, 1870
pubescens (Kirby, 1811)
seniculus (Kirby, 1808)

Subtribe SYNAPIINA Alonso-Zarazaga, 1990

ISCHNOPTERAPION Bokor, 1923
APION sensu auctt. partim non Herbst, 1797

Subgenus ISCHNOPTERAPION Bokor, 1923
loti (Kirby, 1808)
modestum (Germar, 1817)
  sicardi (Desbrochers, 1893)

Subgenus CHLORAPION Györffy, 1956
virens (Herbst, 1797)

PROTOPIRAPION Alonso-Zarazaga, 1990
APION sensu auctt. partim non Herbst, 1797
atratulum (Germar, 1817)
  striatum (Marsham, 1802) non (Müller, O.F., 1776)

STENOPTERAPION Bokor, 1923
APION sensu auctt. partim non Herbst, 1797

Subgenus STENOPTERAPION Bokor, 1923
intermedium (Eppelsheim, 1875) [736]
meliloti (Kirby, 1808)
tenue (Kirby, 1808)

Subgenus COBOSIOTHERIUM Alonso-Zarazaga, 1990
scutellare (Kirby, 1811)

SYNAPION Schilsky, 1902
APION sensu auctt. partim non Herbst, 1797
ebeninum (Kirby, 1808)

Subtribe TRICHAPIINA Alonso-Zarazaga, 1990

BETULAPION Ehret, 1994
APION sensu auctt. partim non Herbst, 1797
TRICHAPION sensu Wagner, 1912 partim
simile (Kirby, 1811)

Tribe PIEZOTRACHELINI Voss, 1959

PROTAPION Schilsky, 1908
APION sensu auctt. partim non Herbst, 1797
apricans (Herbst, 1797)
assimile (Kirby, 1808)
  bohemani sensu (Bedel, 1885) non (Thomson, C.G., 1865)
ssp. **assimile** (Kirby, 1808)
ssp. **ryei** (Blackburn, 1874) [737]
difforme (Germar, 1818)
dissimile (Germar, 1817)
filirostre (Kirby, 1808)
  cantianum (Wagner, 1906)
fulvipes (Geoffroy in Fourcroy, 1785)
  flavipes sensu (Paykull, 1792) non (Fabricius, 1775)
  dichroum (Bedel, 1886)
laevicolle (Kirby, 1811)
nigritarse (Kirby, 1808)
ononidis (Gyllenhal, 1827)
  ononicola (Bach, 1854)
  bohemani (Thomson, C.G., 1865)
schoenherri (Boheman, 1839)
trifolii (Linnaeus, 1768)
  aestivum (Germar, 1817)
varipes (Germar, 1817)

PSEUDOPROTAPION Ehret, 1990
APION sensu auctt. partim non Herbst, 1797
astragali (Paykull, 1800)

98. Family NANOPHYIDAE Gistel, 1848

Family author: M.G. Morris

DIECKMANNIELLUS Alonso-Zarazaga, 1989
    NANOPHYES sensu auctt. partim non Schönherr, 1838
gracilis (Redtenbacher, 1849)

NANOPHYES Schönherr, 1838
    NANODES Schönherr, 1825 [738]
marmoratus (Goeze, 1777)
    lythri (Fabricius, 1787)

99. Family DRYOPHTHORIDAE Schönherr, 1825

Family author: M.G. Morris

Subfamily DRYOPHTHORINAE Schönherr, 1825

DRYOPHTHORUS Germar, 1824 [739]
corticalis (Paykull, 1792)

Subfamily RHYNCHOPHORINAE Schönherr, 1833

Tribe LITOSOMINI Lacordaire, 1865

SITOPHILUS Schönherr, 1838
    CALANDRA Schellenberg, 1798 [740]
    CALANDRA Clairville, 1798 [741]
granarius (Linnaeus, 1758)
oryzae (Linnaeus, 1763)
zeamais Motschulsky, 1855

100. Family ERIRHINIDAE Schönherr, 1825 [742]

Family author: M.G. Morris

Subfamily ERIRHININAE Schönherr, 1825

Tribe ERIRHININI Schönherr, 1825

GRYPUS Germar, 1817
    GRYPIDIUS Schönherr, 1826
equiseti (Fabricius, 1775)

NOTARIS Germar, 1817
    ERIRHINUS Schönherr, 1825 [743]
acridulus (Linnaeus, 1758)
aethiops (Fabricius, 1792)
scirpi (Fabricius, 1792) [744]
    rotundicollis (Motschulsky, 1860)

PROCAS Stephens, 1831 [745]
granulicollis Walton, 1848
picipes (Marsham, 1802)
    armillatus sensu auctt. non (Fabricius, 1801)

THRYOGENES Bedel, 1884
    ERIRHINUS sensu auctt. non Schönherr, 1825
festucae (Herbst, 1795)
fiorii Zumpt, 1928 [746]
    atrirostris Lohse, 1991
nereis (Paykull, 1800)
scirrhosus (Gyllenhal, 1836)

TOURNOTARIS Alonso-Zarazaga & Lyal, 1999
    NOTARIS sensu auctt. partim non Germar, 1817
bimaculata (Fabricius, 1787)

Tribe STENOPELMINI LeConte, 1876

STENOPELMUS Schönherr, 1836
rufinasus Gyllenhal, 1836

Tribe TANYSPHYRINI Gistel, 1848

TANYSPHYRUS Germar, 1817
lemnae (Paykull, 1792)

101. Family RAYMONDIONYMIDAE Reitter, 1913 [747]

Family author: M.G. Morris

FERRERIA Alonso-Zarazaga & Lyal, 1999
    RAYMONDIA Aubé, 1861 non Frauenfeld, 1856
    RAYMONDIONYMUS Ganglbauer, 1906 non Wollaston, 1873
marqueti (Aubé, 1863) [748]

Division GONATOCERI Schönherr, 1825

102. Family CURCULIONIDAE Latreille, 1802

Family author: M.G. Morris

Subfamily CURCULIONINAE Latreille, 1802

Tribe CURCULIONINI Latreille, 1802

ARCHARIUS Gistel, 1848
    CURCULIO sensu auctt. partim non Linnaeus, 1758
    BALANOBIUS Jekel, 1861
pyrrhoceras (Marsham, 1802)
salicivorus (Paykull, 1792)

CURCULIO Linnaeus, 1758
    BALANINUS Germar, 1817
betulae (Stephens, 1831)
    cerasorum sensu (Paykull, 1792) non Fabricius, 1775
glandium Marsham, 1802
    turbatus (Gyllenhal, 1836)
nucum Linnaeus, 1758
rubidus (Gyllenhal, 1836)
venosus (Gravenhorst, 1807)
villosus (Fabricius, 1781)
    cerasorum (Fabricius, 1775)

Tribe ACALYPTINI Thomson, C.G., 1859

ACALYPTUS Schönherr, 1833
carpini (Fabricius, 1792)

Tribe ANTHONOMINI Thomson, C.G., 1859

ANTHONOMUS Germar, 1817 [749]

Subgenus ANTHONOMUS Germar, 1817
bituberculatus Thomson, C.G., 1868
   fasciatus (Marsham, 1802) non (Scopoli, 1763)
   rosinae des Gozis, 1882
brunnipennis (Curtis, 1840)
   comari Crotch, 1869
chevrolati Desbrochers, 1868
conspersus Desbrochers, 1868
humeralis (Panzer, 1795)
   incurvus (Panzer, 1795)
pedicularius (Linnaeus, 1758)
   ulmi sensu auctt. partim non (De Geer, 1775)
piri Kollar, 1837
   cinctus Redtenbacher, 1858
pomorum (Linnaeus, 1758)
rubi (Herbst, 1795)
rufus Gyllenhal, 1836
ulmi (De Geer, 1775)
   inversus Bedel, 1884

Subgenus ANTHOMORPHUS Wiese, 1883
   PARANTHONOMUS Dietz, 1891
varians (Paykull, 1792)
   phyllocola (Herbst, 1795)

Subgenus FURCIPUS Desbrochers, 1868 [750]
rectirostris (Linnaeus, 1758) [751]

BRACHONYX Schönherr, 1825
pineti (Paykull, 1792)

BRADYBATUS Germar, 1824
fallax Gerstaecker, 1860 [752]

Tribe CIONINI Schönherr, 1825

CIONUS Clairville, 1798
alauda (Herbst, 1784)
   blattariae (Fourcroy, 1785)
hortulanus (Fourcroy, 1785)
longicollis Brisout de Barneville, 1863
nigritarsis Reitter, 1904
   thapsus sensu auctt. Brit. non (Fabricius, 1792)
scrophulariae (Linnaeus, 1758)
   woodi Donisthorpe, 1921
tuberculosus (Scopoli, 1763)
   verbasci (Fabricius, 1787)

CLEOPUS Dejean, 1821
   CIONUS sensu Fowler, 1891 partim non Clairville, 1798
pulchellus (Herbst, 1795)

Tribe ELLESCINI Thomson, C.G., 1859

Subtribe ELLESCINA Thomson, C.G., 1859

ELLESCUS Dejean, 1821 [753]
   ELLESCHUS auctt. (misspelling)
bipunctatus (Linnaeus, 1758)

Subtribe DORYTOMINA Bedel, 1886

DORYTOMUS Germar, 1817 [754]
   ETEOPHILUS Bedel, 1886
affinis (Paykull, 1800) [755]
   edoughensis Desbrochers, 1875
dejeani Faust, 1882
   costirostris (Gyllenhal, 1836) non Sahlberg, C.R., 1835
filirostris (Gyllenhal, 1836)
hirtipennis Bedel, 1884
ictor (Herbst, 1795)
   validirostris (Gyllenhal, 1836) [756]
longimanus (Forster, 1771)
   vorax (Fabricius, 1792)
majalis (Paykull, 1792)
melanophthalmus (Paykull, 1792)
   agnathus (Boheman, 1843)
rufatus (Bedel, 1888)
   pectoralis sensu Gyllenhal, 1843 non (Panzer, 1795)
   rufulus Bedel, 1884 non Mannerheim, 1853
salicinus (Gyllenhal, 1827)
salicis Walton, 1851
taeniatus (Fabricius, 1781)
   maculatus (Marsham, 1802)
   silbermanni Wencker, 1866
tortrix (Linnaeus, 1761)
tremulae (Fabricius, 1787)

Tribe MECININI Gistel, 1848 [757]

CLEOPOMIARUS Pierce, 1919
   MIARUS sensu auctt. partim non Schönherr, 1826
graminis (Gyllenhal, 1813)
   degorsi sensu auctt. ?non Abeille de Perrin, 1906 [758]
micros (Germar, 1821)
plantarum (Germar, 1824)

GYMNETRON Schönherr, 1825 [759]
   GYMNAETRON auctt. (misspelling)
beccabungae (Linnaeus, 1761)
   squamicolle Reitter, 1907
melanarium (Germar, 1821)
rostellum (Herbst, 1795)
veronicae (Germar, 1821)
   nigrum Hardy & Bold, 1852
   beccabungae sensu auctt. Brit. ante 1931 non (Linnaeus, 1761)
villosulum Gyllenhal, 1838

MECINUS Germar, 1821
circulatus (Marsham, 1802)
collaris Germar, 1821
janthinus Germar, 1821

labilis (Herbst, 1795) [760]
pascuorum (Gyllenhal, 1813) [760]
   marshalli Donisthorpe, 1943
   melas sensu Donisthorpe, 1948 non Boheman, 1838
pyraster (Herbst, 1795)

MIARUS Schönherr, 1826
campanulae (Linnaeus, 1767)

RHINUSA Stephens, 1829
   GYMNETRON sensu auctt. partim non Schönherr, 1825
antirrhini (Paykull, 1800)
collina (Gyllenhal, 1813)
linariae (Panzer, 1795)

Tribe RHAMPHINI Rafinesque, 1815 [761]

ISOCHNUS Thomson, C.G., 1859
   RHYNCHAENUS sensu auctt. partim non Clairville, 1798
foliorum (Müller, O.F., 1764)
   saliceti (Paykull, 1792)
sequensi (Stierlin, 1894) [762]
   populi (Fabricius, 1792) non (Linnaeus, 1758)
   populicola (Silfverberg, 1977)

ORCHESTES Illiger, 1798
   RHYNCHAENUS sensu auctt. partim non Clairville, 1798

Subgenus ORCHESTES Illiger, 1798
alni (Linnaeus, 1758)
   saltator (Fourcroy, 1785)
   ferrugineus (Marsham, 1802)
calceatus (Germar, 1821) [763]
   scutellaris (Fabricius, 1801) [32]
iota (Fabricius, 1787)
pilosus (Fabricius, 1781)
   ilicis (Fabricius, 1787)
   sparsus sensu auctt. Brit. partim non (Fåhraeus, 1843)
quercus (Linnaeus, 1758)
rusci (Herbst, 1795)
signifer (Creutzer, 1799)
   avellanae (Donovan, 1797) non (Paykull, 1792) [764]
   erythropus sensu auctt. Brit. non (Germar, 1821) [765]
   sparsus sensu auctt. Brit. partim non (Fåhraeus, 1843)
testaceus (Müller, O.F., 1776)
   alni sensu (Hoffmann, 1958) non (Linnaeus, 1758)

Subgenus SALIUS Schrank, 1798
   EUTHORON Thomson, C.G., 1859
fagi (Linnaeus, 1758)

PSEUDORCHESTES Bedel, 1894
   RHYNCHAENUS sensu auctt. partim non Clairville, 1798
pratensis (Germar, 1821)

RHAMPHUS Clairville, 1798
   RAMPHUS auctt. (misspelling)
oxyacanthae (Marsham, 1802)
pulicarius (Herbst, 1795)

flavicornis Clairville, 1798
subaeneus Illiger, 1807 [766]

RHYNCHAENUS Clairville, 1798
lonicerae (Herbst, 1795) [767]
   xylostei Clairville, 1798

TACHYERGES Schönherr, 1825
   RHYNCHAENUS sensu auctt. partim non Clairville, 1798
decoratus (Germar, 1821)
pseudostigma (Tempère, 1982) [768]
salicis (Linnaeus, 1758)
stigma (Germar, 1821)

Tribe SMICRONYCHINI Seidlitz, 1891

SMICRONYX Schönherr, 1843
coecus (Reich, 1797)
jungermanniae (Reich, 1797)
reichi (Gyllenhal, 1836)
   seriepilosus Tournier, 1874

Tribe STOREINI Lacordaire, 1863

PACHYTYCHIUS Jekel, 1861
haematocephalus (Gyllenhal, 1836)

Tribe STYPHLINI Jekel, 1861

ORTHOCHAETES Germar, 1824
insignis (Aubé, 1863)
setiger (Beck, 1817)

PSEUDOSTYPHLUS Tournier, 1874
pillumus (Gyllenhal, 1836)
   pilumnus auctt. (misspelling)

Tribe TYCHIINI Gistel, 1848

SIBINIA Germar, 1817 [769]
arenariae Stephens, 1831
primita (Herbst, 1795) [770]
   signata (Gyllenhal, 1813) ? Panzer, 1795
   plantaginis sensu auctt. Brit. non Eppelsheim, 1875
pyrrhodactyla (Marsham, 1802)
   potentillae Germar, 1824 [771]
   lloydi (Donisthorpe, 1929) [772]
sodalis Germar, 1824

TYCHIUS Germar, 1817
   MICCOTROGUS Schönherr, 1825 [773]
breviusculus Desbrochers, 1873 [774]
   haematopus sensu auctt. non Gyllenhal, 1836
crassirostris Kirsch, 1871 [775]
junceus (Reich, 1797)
   flavicollis Stephens, 1831
   haematopus Gyllenhal, 1836
lineatulus Stephens, 1831
meliloti Stephens, 1831
parallelus (Panzer, 1794)
   venustus sensu auctt. non (Fabricius, 1781)

picirostris (Fabricius, 1787)
   cinerascens (Marsham, 1802)
   posticus (Gyllenhal, 1836)
polylineatus (Germar, 1824)
pusillus Germar, 1842
   pygmaeus Brisout de Barneville, H., 1860
quinquepunctatus (Linnaeus, 1758)
schneideri (Herbst, 1795)
squamulatus Gyllenhal, 1836
   flavicollis sensu auctt. non Stephens, 1831 [776]
stephensi Gyllenhal, 1836
   tomentosus (Herbst, 1795) non (Olivier, 1790)
tibialis Boheman, 1843

Subfamily BAGOINAE Thomson, C.G., 1859

BAGOUS Germar, 1817 [777]

Subgenus BAGOUS Germar, 1817
   PROBAGOUS Sharp, 1917
argillaceus Gyllenhal, 1836
binodulus (Herbst, 1795)
brevis Gyllenhal, 1836
collignensis (Herbst, 1797)
   frit sensu auctt. partim non (Herbst, 1795)
   claudicans Boheman, 1845
czwalinae Seidlitz, 1891
   czwalinai auctt. (misspelling)
   heasleri Newbery, 1902
diglyptus Boheman, 1845
frit (Herbst, 1795)
limosus (Gyllenhal, 1827)
longitarsis Thomson, C.G., 1868
   arduus Sharp, 1917 [778]
   tomlini Sharp, 1917
lutulosus (Gyllenhal, 1827)
nodulosus Gyllenhal, 1836
subcarinatus Gyllenhal, 1836
   frit sensu Newbery, 1902 non (Herbst, 1795)
tempestivus (Herbst, 1795)
   cnemerythrus (Marsham, 1802)
   angustulus Thomson, C.G., 1870
Subgenus ABAGOUS Sharp, 1917
glabrirostris (Herbst, 1795)
lutosus (Gyllenhal, 1813)
lutulentus (Gyllenhal, 1813)
   binotatus Stephens, 1831
   nigritarsis Thomson, C.G., 1865
puncticollis Boheman, 1854
   collignensis sensu auctt. partim non (Herbst, 1797)
robustus Brisout de Barneville, H., 1863
   rudis Sharp, 1917 [779]
Subgenus CYPRUS Schönherr, 1825
tubulus Caldara & O'Brien, 1994
   cylindrus (Paykull, 1800) non (Fabricius, 1781)
   attenuatus (Ahrens, 1812) non (Fabricius, 1801)
   angustus Silfverberg, 1977 non (Tanner, 1954) [780]
Subgenus EPHIMEROPUS Hochhuth, 1847
petro (Herbst, 1795)
   aubei (Cussac, 1851)
Subgenus HYDRONOMUS Schönherr, 1825
alismatis (Marsham, 1802)

Subfamily BARIDINAE Schönherr, 1836 [781]

Tribe BARIDINI Schönherr, 1836

AULACOBARIS Desbrochers, 1892
   BARIS sensu auctt. partim non Germar, 1817
lepidii (Germar, 1824)
picicornis (Marsham, 1802)

BARIS Germar, 1817 [782]
analis (Olivier, 1790)

COSMOBARIS Casey, 1920
   BARIS sensu auctt. partim non Germar, 1817
scolopacea (Germar, 1818)

MELANOBARIS Alonso-Zarazaga & Lyal, 1999
   BARIS sensu auctt. partim non Germar, 1817
laticollis (Marsham, 1802)

Tribe APOSTASIMERINI Schönherr, 1844

LIMNOBARIS Bedel, 1885 [783]
dolorosa (Goeze, 1777)
   pilistriata (Stephens, 1831)
t-album (Linnaeus, 1758)
   ssp. *atriplicis* (Fabricius, 1777)

Subfamily CEUTORHYNCHINAE Gistel, 1848 [784]

Tribe CEUTORHYNCHINI Gistel, 1848

AMALORRHYNCHUS Reitter, 1913
   CEUTHORHYNCHIDIUS sensu Fowler, 1891
   partim non Jacquelin du Val, [1855]
melanarius (Stephens, 1831)

CALOSIRUS Thomson, C.G., 1859
   CEUTHORHYNCHUS sensu auctt. partim non
   Germar, 1824
terminatus (Herbst, 1795)

CEUTORHYNCHUS Germar, 1824 [785]
   CEUTHORHYNCHUS Schönherr, 1837 [40]
   CEUTHORRHYNCHUS Fowler, 1891 (misspelling)
   CEUTHORHYNCHIDIUS sensu Fowler, 1891
   partim non Jacquelin du Val, [1855]
alliariae Brisout de Barneville, H., 1860
   roberti sensu auctt. Brit. non Gyllenhal, 1837
assimilis (Paykull, 1792) [786]
   pleurostigma (Marsham, 1802) [787]
atomus Boheman, 1845
   setosus Boheman, 1845
cakilis (Hansen, 1917) [788]
chalybaeus Germar, 1824
   timidus Weise, 1883
   moguntiacus Schultze, 1895
cochleariae (Gyllenhal, 1813)
constrictus (Marsham, 1802)
contractus (Marsham, 1802) [789]
   minutus (Reich, 1797) non (Drury, 1773)
erysimi (Fabricius, 1787)

hepaticus Gyllenhal, 1837
hirtulus Germar, 1824
insularis Dieckmann, 1971
  contractus var. *pallipes* Crotch, 1865
obstrictus (Marsham, 1802)
  assimilis sensu auctt. non (Paykull, 1792) [787]
pallidactylus (Marsham, 1802)
  quadridens (Panzer, 1795) non (Fabricius, 1775)
parvulus Brisout de Barneville, Ch., 1869
pectoralis Weise, 1895
  chalybeus sensu auctt. Brit. ? Germar, 1824
pervicax Weise, 1883
  suturellus sensu auctt. non Gyllenhal, 1837
picitarsis Gyllenhal, 1837
pulvinatus Gyllenhal, 1837
pumilio (Gyllenhal, 1827)
  posthumus sensu auctt. Brit. non Germar, 1824
pyrrhorhynchus (Marsham, 1802)
querceti (Gyllenhal, 1813)
rapae Gyllenhal, 1837
resedae (Marsham, 1802)
sulcicollis (Paykull, 1800)
  cyanipennis Germar, 1824
syrites Germar, 1824
thomsoni Kolbe, 1900
  moguntiacus sensu auctt. non Schultze, 1895
turbatus Schultze, 1903
typhae (Herbst, 1795)
  floralis (Paykull, 1792) non (Olivier, 1790) [790]
  palustris (Edmonds, T., 1930)
unguicularis Thomson, C.G., 1871
  curvirostris Schultze, 1898

COELIODES Schönherr, 1837
rana (Fabricius, 1787)
  quercus sensu (Fabricius, 1787) non (Linnaeus, 1758)
  dryados (Gmelin in Linnaeus, 1790) [790]
ruber (Marsham, 1802)
transversealbofasciatus (Goeze, 1777)
  cinctus (Fourcroy, 1785) non (Drury, 1772)
  erythroleucos (Gmelin in Linnaeus, 1790) [790]
  erythroleucus Fowler, 1891 (misspelling)

COELIODINUS Dieckmann, 1972 [791]
  COELIODES sensu auctt. partim non Schönherr, 1837
nigritarsis (Hartmann, 1895)
rubicundus (Herbst, 1795)

DATONYCHUS Wagner, 1944
  CEUTORHYNCHUS sensu auctt. partim non Germar, 1824
angulosus (Boheman, 1845)
arquatus (Herbst, 1795) [792]
  arquata (Herbst, 1795) (incorrect original spelling)
melanostictus (Marsham, 1802)
urticae (Boheman, 1845)

DRUPENATUS Reitter, 1913
  DRUSENATUS Reitter, 1916
  POOPHAGUS sensu Fowler, 1891 partim non Schönherr, 1837
nasturtii (Germar, 1824)

ETHELCUS Reitter, 1916
  CEUTORHYNCHUS sensu auctt. partim non Germar, 1824
verrucatus (Gyllenhal, 1837)

GLOCIANUS Reitter, 1916
  CEUTORHYNCHUS sensu auctt. partim non Germar, 1824
distinctus (Brisout de Barneville, Ch., 1870)
  marginatus (Paykull, 1792) non (Fabricius, 1775)
  simillimus Edwards, J., 1911
moelleri (Thomson, C.G., 1868)
pilosellus (Gyllenhal, 1837)
punctiger (Gyllenhal, 1837)

HADROPLONTUS Thomson, C.G., 1859
  CEUTORHYNCHUS sensu auctt. partim non Germar, 1824
litura (Fabricius, 1775)
trimaculatus (Fabricius, 1775)

MICRELUS Thomson, C.G., 1859
  CEUTHORRHYNCHUS sensu Fowler, 1891 partim non Germar, 1824
ericae (Gyllenhal, 1813)

MICROPLONTUS Wagner, 1944
  CEUTORHYNCHUS sensu auctt. partim non Germar, 1824
campestris (Gyllenhal, 1837)
  chrysanthemi sensu auctt. Brit. non auctt. Europ. nec (Germar, 1824)
rugulosus (Herbst, 1795)
  cinereus (Marsham, 1802)
  melanostigma (Marsham, 1802)
  uniguttatus (Marsham, 1802)
triangulum (Boheman, 1845)

MOGULONES Reitter, 1916
  CEUTORHYNCHUS sensu auctt. partim non Germar, 1824
  BORAGINOBIUS Wagner, 1944
asperifoliarum (Gyllenhal, 1813)
euphorbiae (Brisout de Barneville, Ch., 1866)
geographicus (Goeze, 1777)

NEDYUS Schönherr, 1825
  CIDNORHINUS Thomson, C.G., 1859
  CIDNORRHINUS auctt. (misspelling)
  COELIODES sensu Fowler, 1891 partim non Schönherr, 1837
quadrimaculatus (Linnaeus, 1758)
  urticae (Marsham, 1802)

PARETHELCUS Dieckmann, 1972
  CEUTORHYNCHUS sensu auctt. partim non Germar, 1824
pollinarius (Forster, 1771)

POOPHAGUS Schönherr, 1837
sisymbrii (Fabricius, 1777)

SIROCALODES Voss, 1958
CEUTORHYNCHUS sensu auctt. partim non
Germar, 1824
depressicollis (Gyllenhal, 1813)
nigrinus (Marsham, 1802) non (Herbst, 1795)
mixtus (Mulsant & Rey, 1858)
quercicola (Paykull, 1792)

STENOCARUS Thomson, C.G., 1859
COELIODES sensu Fowler, 1891 partim non
Schönherr, 1837
ruficornis (Stephens, 1831)
cardui sensu auctt. Brit. non (Herbst, 1784)
fuliginosus (Marsham, 1802) non (Gmelin in
Linnaeus, 1790)
umbrinus (Gyllenhal, 1837) [793]

TAPEINOTUS Schönherr, 1826
TAPINOTUS Schönherr, 1826 [794]
sellatus (Fabricius, 1794)

THAMIOCOLUS Thomson, C.G., 1859
CEUTORHYNCHUS sensu auctt. partim non
Germar, 1824
viduatus (Gyllenhal, 1813)

TRICHOSIROCALUS Colonnelli, 1979
CEUTHORHYNCHIDIUS sensu auctt. non
Jacquelin du Val, [1855]
barnevillei (Grenier, 1866)
chevrolati (Crotch, 1865) [32]
dawsoni (Brisout de Barneville, Ch., 1869)
horridus (Panzer, 1801)
rufulus (Dufour, 1851)
thalhammeri (Schultze, 1906) [795]
troglodytes (Fabricius, 1787)

ZACLADUS Reitter, 1913
ALLODACTYLUS Weise, 1883 non Lataste &
Rochebrune, 1876
COELIODES sensu Fowler, 1891 partim non
Schönherr, 1837
exiguus (Olivier, 1807)
geranii (Paykull, 1800)
affinis (Paykull, 1792) non (Schrank, 1781)

Tribe AMALINI Wagner, 1936

AMALUS Schönherr, 1825
scortillum (Herbst, 1795)
haemorrhous (Herbst, 1795) non (Gmelin in
Linnaeus, 1790)

Tribe MONONYCHINI LeConte, 1876

MONONYCHUS Germar, 1824
punctumalbum (Herbst, 1784)
pseudacori (Fabricius, 1792) non (Rossi, 1790)

Tribe PHYTOBIINI Gistel, 1848

EUBRYCHIUS Thomson, C.G., 1859
velutus (Beck, 1817)

velatus (Germar, 1818) (misspelling)

NEOPHYTOBIUS Wagner, 1936
PHYTOBIUS sensu auctt. partim non Schönherr,
1833
muricatus (Brisout de Barneville, Ch., 1867)
quadrinodosus (Gyllenhal, 1813)
denticollis (Gyllenhal, 1837)

PELENOMUS Thomson, C.G., 1859
PHYTOBIUS sensu Dejean, 1835 non Schönherr,
1833
RHINONCUS sensu Fowler, 1891 partim non
Schönherr, 1825
canaliculatus (Fåhraeus, 1843)
comari (Herbst, 1795) [796]
olssoni (Israelson, 1972) [797]
quadricorniger (Colonnelli, 1986)
quadricornis (Gyllenhal, 1813) non (Paykull, 1792)
quadrituberculatus (Fabricius, 1787)
waltoni (Boheman, 1843)
zumpti (Wagner, 1939) [798]

PHYTOBIUS Schönherr, 1833
LITODACTYLUS Redtenbacher, 1849
leucogaster (Marsham, 1802)

RHINONCUS Schönherr, 1825
AMALORHINONCUS Wagner, 1936
albicinctus Gyllenhal, 1837
bruchoides (Herbst, 1784)
castor (Fabricius, 1792)
inconspectus (Herbst, 1795)
gramineus (Fabricius, 1792) non (Gmelin in
Linnaeus, 1790)
pericarpius (Linnaeus, 1758)
perpendicularis (Reich, 1797)

Tribe SCLEROPTERINI Schultze, 1902

RUTIDOSOMA Stephens, 1831
RHYTIDOSOMA auctt. (misspelling)
RHYTIDOSOMUS Schönherr, 1837 [40]
globulus (Herbst, 1795)

Subfamily COSSONINAE Schönherr, 1825

Tribe COSSONINI Schönherr, 1825

COSSONUS Clairville, 1798
linearis (Fabricius, 1775)
planatus Bedel, 1881
parallelepipedus (Herbst, 1795)
ferrugineus sensu Fowler, 1891 ? Clairville, 1798

RHOPALOMESITES Wollaston, 1873
MESITES sensu auctt. non Schönherr, 1838
tardyi (Curtis, 1825)
tardii auctt. (misspelling)

Tribe DRYOTRIBINI LeConte, 1876

COTASTER Motschulsky, 1851
uncipes (Boheman, 1838) [799]

Tribe ONYCHOLIPINI Wollaston, 1873

PSELACTUS Broun, 1886
CODIOSOMA Bedel, 1885 non Kirby, W.F., 1874
PHLOEOPHAGIA Aurivillius, 1924
spadix (Herbst, 1795)

PSEUDOPHLOEOPHAGUS Wollaston, 1873
CAULOTRUPIS sensu auctt. non Wollaston, 1854
CAULOTRYPIS Fowler, 1891 (misspelling)
aeneopiceus (Boheman, 1845)

STEREOCORYNES Wollaston, 1873
RHYNCOLUS sensu auctt. Brit. partim non Germar, 1817
truncorum (Germar, 1824)

Tribe PENTARTHRINI Lacordaire, 1865

ALLOPENTARTHRUM Kuschel in Wibmer & O'Brien, 1986
elumbe (Boheman, 1838) [800]

EUOPHRYUM Broun, 1909
confine (Broun, 1881)
rufum (Broun, 1880)

PENTARTHRUM Wollaston, 1854
huttoni Wollaston, 1854

Tribe PROECINI Voss, 1956

CONARTHRUS Wollaston, 1873
praeustus (Boheman in Schönherr, 1838) [801]

Tribe RHYNCOLINI Gistel, 1848

Subtribe RHYNCHOLINA Gistel, 1848

MACRORHYNCOLUS Wollaston, 1873
littoralis (Broun, 1880) [802]

RHYNCOLUS Germar, 1817
EREMOTES Wollaston, 1861
ater (Linnaeus, 1758) [803]
chloropus sensu auctt. non (Linnaeus, 1758)

Subtribe PHLOEOPHAGINA Voss, 1955

PHLOEOPHAGUS Schönherr, 1838 [804]
RHYNCOLUS sensu auctt. Brit. partim non Germar, 1817
lignarius (Marsham, 1802)

Subfamily CRYPTORHYNCHINAE Schönherr, 1825

Tribe CRYPTORHYNCHINI Schönherr, 1825

Subtribe CRYPTORHYNCHINA Schönherr, 1825

CRYPTORHYNCHUS Illiger, 1807
CRYPTORHYNCHIDIUS Pierce, 1919 non Champion, 1914
lapathi (Linnaeus, 1758)

Subtribe TYLODINA Lacordaire, 1865

ACALLES Schönherr, 1825 [805]
misellus Boheman, 1844
  ?*parvulus* Boheman, 1837
  turbatus sensu auctt. non Boheman, 1844
ptinoides (Marsham, 1802)

KYKLIOACALLES Stüben, 1999
ACALLES sensu auctt. partim non Schönherr, 1825
roboris (Curtis, 1835)

Subfamily CYCLOMINAE Schönherr, 1826

Tribe RHYTHIRRININI Lacordaire, 1863

Subtribe GRONOPINA Bedel, 1884

GRONOPS Schönherr, 1823
inaequalis Boheman, 1842 [806]
lunatus (Fabricius, 1775)

Subfamily ENTIMINAE Schönherr, 1823

Tribe ALOPHINI LeConte, 1874

GRAPTUS Schönherr, 1823
ALOPHUS Schönherr, 1826
triguttatus (Fabricius, 1775)

Tribe BRACHYDERINI Schönherr, 1826

BRACHYDERES Schönherr, 1826
incanus (Linnaeus, 1758) [807]
lusitanicus (Fabricius, 1781) [808]

NELIOCARUS Thomson, C.G., 1859
STROPHOSOMUS sensu auctt. partim non Schönherr, 1823
faber (Herbst, 1784)
nebulosus (Stephens, 1831)
  retusus (Marsham, 1802) non (Fabricius, 1781)
sus (Stephens, 1831)
  lateralis (Paykull, 1792) non (Panzer, 1789)

STROPHOSOMA Billberg, 1820
STROPHOSOMUS Schönherr, 1823
capitatum (De Geer, 1775)
  obesum (Marsham, 1802) non (Fabricius, 1775)

subrotundum (Marsham, 1802)
rufipes Stephens, 1831
fulvicorne (Walton, 1846)
curvipes Thomson, C.G., 1868 [809]
melanogrammum (Forster, 1771)
coryli (Fabricius, 1775)

Tribe CNEORHININI Lacordaire, 1863

ATTACTAGENUS Tournier, 1876
ATACTOGENUS Fowler, 1891 (misspelling)
CNEORHINUS sensu auctt. non Schönherr, 1823
CNEORRHINUS auctt. (misspelling)
plumbeus (Marsham, 1802)
exaratus (Marsham, 1802) non (Gmelin in Linnaeus, 1790)

PHILOPEDON Schönherr, 1826 [810]
plagiatum (Schaller, 1783)
geminatus (Fabricius, 1787)

Tribe GEONEMINI Gistel, 1848

BARYNOTUS Germar, 1817
moerens (Fabricius, 1792)
elevatus (Marsham, 1802)
obscurus (Fabricius, 1775)
squamosus Germar, 1824
schoenherri (Zetterstedt, 1838) non Dalman, 1823

Tribe OMIINI Shuckard, 1839

OMIAMIMA Silfverberg, 1977 [811]
OMIAS sensu auctt. non Germar, 1817
mollina (Boheman, 1834)

Tribe OTIORHYNCHINI Schönherr, 1826

OTIORHYNCHUS Germar, 1824 [812]
OTIORRHYNCHUS auctt. (misspelling)
BRACHYRHINUS Latreille, 1802 [813]
arcticus (Fabricius, O., 1780)
blandus Gyllenhal, 1834
armadillo (Rossi, 1792) [814]
atroapterus (De Geer, 1775)
aurifer Boheman, 1843 [815]
auropunctatus Gyllenhal, 1834
clavipes (Bonsdorff, 1785)
tenebricosus sensu (Olivier, 1790) non (Herbst, 1784)
fuscipes (Olivier, 1807)
lugdunensis Boheman, 1843
crataegi Germar, 1824 [816]
cribricollis Gyllenhal, 1834 [817]
desertus Rosenhauer, 1847
muscorum Brisout de Barneville, H., 1863
dieckmanni Magnano, 1979 [818]
setosulus sensu auctt. Brit. non Stierlin, 1861 [819]
ligneus (Olivier, 1807)
ligustici (Linnaeus, 1758)
morio (Fabricius, 1781)
ebeninus Gyllenhal, 1834
nodosus (Müller, O.F., 1764)
dubius (Strøm, 1783)

maurus (Gyllenhal, 1813) non (Marsham, 1802)
ovatus (Linnaeus, 1758)
porcatus (Herbst, 1795)
raucus (Fabricius, 1777)
rugifrons (Gyllenhal, 1813)
dillwyni Stephens, 1831
rugosostriatus (Goeze, 1777)
scabrosus (Marsham, 1802)
salicicola Heyden, 1908 [820]
scaber (Linnaeus, 1758)
septentrionis (Herbst, 1795)
singularis (Linnaeus, 1767)
picipes (Fabricius, 1777)
sulcatus (Fabricius, 1775)
uncinatus Germar, 1824

Tribe PERITELINI Lacordaire, 1863

PERITELUS Germar, 1824
sphaeroides Germar, 1824
griseus sensu (Olivier, 1807) non (Fabricius, 1775)

Tribe PHYLLOBIINI Schönherr, 1826

PHYLLOBIUS Germar, 1824

Subgenus PHYLLOBIUS Germar, 1824
pyri (Linnaeus, 1758)
vespertinus (Fabricius, 1792) [821]
artemisiae Desbrochers, 1873

Subgenus DIELETUS Reitter, 1916
argentatus (Linnaeus, 1758)

Subgenus METAPHYLLOBIUS Smirnov, 1913
glaucus (Scopoli, 1763)
calcaratus (Fabricius, 1792)
pomaceus Gyllenhal, 1834
urticae (De Geer, 1775) non (Scopoli, 1763)
alneti (Fabricius, 1792) non (Schrank, 1781)

Subgenus NEMOICUS Dillwyn, 1829
oblongus (Linnaeus, 1758)

Subgenus PARNEMOICUS Schilsky, 1911
roboretanus Gredler, 1882
virideaeris sensu auctt. Brit. non (Laicharting, 1781)
parvulus sensu (Olivier, 1807) non (Fabricius, 1792)
viridicollis (Fabricius, 1801)
pomonae (Olivier, 1807)

Subgenus PTERYGORRHYNCHUS Pesarini, 1969
maculicornis Germar, 1824

Subgenus SUBPHYLLOBIUS Schilsky, 1911
virideaeris (Laicharting, 1781)
viridiaeris auctt. (misspelling)

Tribe POLYDRUSINI Schönherr, 1823

LIOPHLOEUS Germar, 1817
tessulatus (Müller, O.F., 1776)
nubilus (Fabricius, 1777)
maurus (Marsham, 1802)

PACHYRHINUS Schönherr, 1823
lethierryi (Desbrochers, 1875) [822]
mustela (Herbst, 1797) [823]

POLYDRUSUS Germar, 1817 [824]
POLYDROSOS auctt. (misspelling)

Subgenus POLYDRUSUS Germar, 1817
tereticollis (De Geer, 1775)
undatus (Fabricius, 1781)

Subgenus CHRYSOPHIS des Gozis, 1882
THOMSONEONYMUS Desbrochers, 1902
formosus (Mayer, 1779) [825]
sericeus (Schaller, 1783) non (Goeze, 1777)
splendidus (Herbst, 1784)

Subgenus EUDIPNUS Thomson, C.G., 1859
mollis (Strøm, 1768)
micans (Fabricius, 1792)

Subgenus EURODRUSUS Korotyaev & Meleshko, 1997
confluens Stephens, 1831

Subgenus EUSTOLUS Thomson, C.G., 1859
TYLODRUSUS Stierlin, 1884
flavipes (De Geer, 1775)
pterygomalis Boheman, 1840

Subgenus METALLITES Germar, 1824
marginatus Stephens, 1831

Subgenus NEOEUSTOLUS Alonso-Zarazaga & Lyal, 1999
EUSTOLUS sensu auctt. non Thomson, C.G., 1859
cervinus (Linnaeus, 1758)
pilosus Gredler, 1866
pulchellus Stephens, 1831
chrysomela sensu auctt. Brit. non (Olivier, 1807) [826]
salsicola Fairmaire, 1852
insquamosus (Everts, 1921)

Tribe SCIAPHILINI Sharp, 1891 [827]

BARYPEITHES Jacquelin du Val, 1855
BARYPITHES Gemminger & Harold, 1871 [40]

Subgenus BARYPEITHES Jacquelin du Val, 1855
sulcifrons (Boheman, 1843)

Subgenus EXOMIAS Bedel, 1883
araneiformis (Schrank, 1781)
curvimanus (Jacquelin du Val, 1855)
pellucidus (Boheman, 1834)
duplicatus Keys, 1911
pyrenaeus Seidlitz, 1868

BRACHYSOMUS Schönherr, 1823
echinatus (Bonsdorff, 1785)
hirtus (Boheman, 1845)

SCIAPHILUS Schönherr, 1823
asperatus (Bonsdorff, 1785)
muricatus (Fabricius, 1792) non (Drury, 1773)

Tribe SITONINI Gistel, 1848 [828]

ANDRION Velázquez, 2007
SITONA sensu auctt. partim non Germar, 1817
regensteinense (Herbst, 1797)

CHARAGMUS Schönherr, 1826
SITONA sensu auctt. partim non Germar, 1817
griseus (Fabricius, 1775)

COELOSITONA González, 1971
SITONA sensu auctt. partim non Germar, 1817
cambricus (Stephens, 1831)
cinerascens (Fåhraeus, 1840) [829]
puberulus (Reitter, 1903)
cambricus sensu auctt. Brit. partim non (Stephens, 1831)
brevirostris (Solari, 1948)

SITONA Germar, 1817
ambiguus Gyllenhal, 1834
brevicollis sensu auctt. partim non Gyllenhal, 1834
lineellus sensu Lindberg, 1933, auctt. Brit. post 1930 non (Bonsdorff, 1785)
cylindricollis (Fåhraeus, 1840)
meliloti Walton, 1846
gemellatus Gyllenhal, 1834
gressorius (Fabricius, 1792) [830]
hispidulus (Fabricius, 1777)
humeralis Stephens, 1831
lepidus Gyllenhal, 1834
flavescens (Marsham, 1802) non (Fabricius, 1787)
lineatus (Linnaeus, 1758)
lineellus (Bonsdorff, 1785)
decipiens Lindberg, 1933
macularius (Marsham, 1802)
crinitus (Herbst, 1795) non (Gmelin in Linnaeus, 1790)
ononidis Sharp, 1866
puncticollis Stephens, 1831
striatellus Gyllenhal, 1834
tibialis (Herbst, 1795) non (Sparrman, 1785)
brevicollis Gyllenhal, 1834 partim
sulcifrons (Thunberg, 1798)
suturalis Stephens, 1831
waterhousei Walton, 1846

Tribe TANYMECINI Lacordaire, 1863

TANYMECUS Germar, 1817
palliatus (Fabricius, 1787)

Tribe TRACHYPHLOEINI Gistel, 1848 [831]

CAENOPSIS Bach, 1854
fissirostris (Walton, 1847)
waltoni (Boheman, 1843)

CATHORMIOCERUS Schönherr, [1842]
aristatus (Gyllenhal, 1827)
attaphilus Brisout de Barneville, Ch., 1880
myrmecophilus (Seidlitz, 1868)
britannicus Blair, 1934 [832]
socius Boheman, 1843
maritimus Rye, 1874
spinosus (Goeze, 1777)
asperatus (Boheman, 1843)
squamulatus sensu (Olivier, 1807) non (Herbst, 1795)
olivieri (Bedel, 1883)
alleni (Donisthorpe, 1948)

ROMUALDIUS Borovec, 2009
angustisetulus (Hansen, 1915) [833]
  bifoveolatus sensu auctt. Brit. partim non (Beck, 1817)
bifoveolatus (Beck, 1817)
  scaber sensu auctt. non (Linnaeus, 1758)

TRACHYPHLOEUS Germar, 1817
  TRACHYPHLAEUS Stephens, 1829 (misspelling)
alternans Gyllenhal, 1834
digitalis (Gyllenhal, 1827)
rectus Thomson, C.G., 1865
  laticollis sensu auctt. Brit. non Boheman, 1843 [834]
scabricul (Linnaeus, 1771)
  scabriculus auctt. (misspelling)
  scaber sensu Boheman, 1843 non (Linnaeus, 1758)
spinimanus Germar, 1824

Tribe TROPIPHORINI Marseul, 1863

TROPIPHORUS Schönherr, 1842
elevatus (Herbst, 1795)
  carinatus sensu (Müller, O.F., 1776) non (Linnaeus, 1767)
obtusus (Bonsdorff, 1785)
terricola (Newman, 1838)
  tomentosus (Marsham, 1802) non (Olivier, 1790)

Subfamily HYPERINAE Marseul, 1863 (1848)

HYPERA Germar, 1817
  PHYTONOMUS Schönherr, 1823 [835]

Subgenus HYPERA Germar, 1817
arator (Linnaeus, 1758)
  polygoni (Linnaeus, 1761)
fuscocinerea (Marsham, 1802)
  murina (Fabricius, 1792) non (Müller, O.F., 1764)
nigrirostris (Fabricius, 1775)
ononidis Chevrolat, 1863
plantaginis (De Geer, 1775)
postica (Gyllenhal, 1813)
  variabilis (Herbst, 1795) non (Fabricius, 1777)
suspiciosa (Herbst, 1795)
  pedestris (Paykull, 1792) non (Poda, 1761)
  miles (Paykull, 1792) [32]
venusta (Fabricius, 1781)
  trilineata (Marsham, 1802)

Subgenus ANTIDONUS Bedel, 1886
dauci (Olivier, 1807)
  fasciculata (Herbst, 1795) non (De Geer, 1775)
zoilus (Scopoli, 1763)
  punctata (Fabricius, 1775) non (Scopoli, 1763)
  ?*austriaca* (Schrank, 1781)

Subgenus BOREOHYPERA Korotyaev, 1999
diversipunctata (Schrank, 1798)
  elongata (Paykull, 1792) non (Fabricius, 1775)

Subgenus DAPALINUS Capiomont, 1868
meles (Fabricius, 1792)

Subgenus ERIRINOMORPHUS Capiomont, 1868
arundinis (Paykull, 1792)
pollux (Fabricius, 1801)
  adspersa (Fabricius, 1792) non (Fabricius, 1775)
  alternans (Stephens, 1831)

rumicis (Linnaeus, 1758)

Subgenus TIGRINELLUS Capiomont, 1868
pastinaceae (Rossi, 1790)
  tigrina Boheman, 1834

LIMOBIUS Schönherr, 1843
borealis (Paykull, 1792)
  dissimilis (Herbst, 1795)
mixtus (Boheman, 1834)

Subfamily LIXINAE Schönherr, 1823

Tribe LIXINI Schönherr, 1823

LARINUS Germar, 1824
planus (Fabricius, 1792)
  carlinae (Olivier, 1807)

LIXUS Fabricius, 1801 [836]

Subgenus LIXUS Fabricius, 1801
paraplecticus (Linnaeus, 1758)
  productus Stephens, 1831

Subgenus DILIXELLUS Reitter, 1916
angustatus (Fabricius, 1775)
  algirus sensu auctt. non (Linnaeus, 1758)
vilis (Rossi, 1790)
  bicolor (Olivier, 1807)
  lateralis Stephens, 1831

Subgenus EULIXUS Reitter, 1916
iridis Olivier, 1807
  paraplecticus sensu auctt. Brit. partim non (Linnaeus, 1758)
  turbatus Gyllenhal, 1824
  gemellatus Gyllenhal in Schönherr, 1836
scabricollis Boheman, 1843 [837]

Tribe CLEONINI Schönherr, 1826

BOTHYNODERES Schönherr, 1823
  CHROMODERUS Motschulsky, 1860
affinis (Schrank, 1781)
  fasciatus (Müller, O.F., 1776) non (Scopoli, 1763)
  albidus (Fabricius, 1787)

CLEONIS Dejean, 1821
  CLEONUS Schönherr, 1826
pigra (Scopoli, 1763)
  sulcirostris (Linnaeus, 1767)

CONIOCLEONUS Motschulsky, 1860
  CLEONUS sensu Fowler, 1891 partim non Schönherr, 1826
hollbergii (Fåhraeus, 1842)
  hollbergi auctt. (misspelling)
  glaucus (Fabricius, 1787) non (Scopoli, 1763)
nebulosus (Linnaeus, 1758)

Tribe RHINOCYLLINI Lacordaire, 1863

RHINOCYLLUS Germar, 1817
conicus (Frölich, 1792)

latirostris (Latreille, 1804)

Subfamily MESOPTILIINAE Lacordaire, 1863

Tribe MAGDALIDINI Pascoe, 1870

MAGDALIS Germar, 1817
Subgenus MAGDALIS Germar, 1817
duplicata Germar, 1818
memnonia (Gyllenhal in Faldermann, 1837)
phlegmatica (Herbst, 1797)

Subgenus EDO Germar, 1819
ruficornis (Linnaeus, 1758)
  pruni (Linnaeus, 1767) non (Scopoli, 1763)

Subgenus ODONTOMAGDALIS Barrios, 1984
  MAGDALINUS sensu auctt. non Germar, 1843
armigera (Fourcroy, 1785)
carbonaria (Linnaeus, 1758)

Subgenus PANUS Schönherr, 1826
barbicornis (Latreille, 1804)

Subgenus PORROTHUS Dejean, 1821
  NEOPANUS Reitter, 1916
cerasi (Linnaeus, 1758)

Subfamily MOLYTINAE Schönherr, 1823

Tribe MOLYTINI Schönherr, 1823

Subtribe MOLYTINA Schönherr, 1823

LIPARUS Olivier, 1807
  MOLYTES Schönherr, 1823
coronatus (Goeze, 1777)
  anglicanus (Marsham, 1802)
germanus (Linnaeus, 1758)

Subtribe LEIOSOMATINA Reitter, 1913

LEIOSOMA Stephens, 1829
  LEIOSOMUS Schönherr, 1842 [40]
  LIOSOMA Agassiz, 1847 non Brandt, 1835 [40]
deflexum (Panzer, 1795)
  ovatulum (Clairville, 1798)
oblongulum Boheman, 1842
troglodytes (Rye, 1873) [838]
  pyrenaeum sensu auctt. Brit. non Brisout de
    Barneville, Ch., 1866

Subtribe PLINTHINA Lacordaire, 1863

MITOPLINTHUS Reitter, 1897
  PLINTHUS Westwood, (1838) non Germar, 1817
  EPIPOLAEUS Weise, 1907
caliginosus (Fabricius, 1775)

Subtribe TYPODERINA Voss, 1965

ANCHONIDIUM Bedel, 1884
unguiculare (Aubé, 1850)

Tribe ANOPLINI Bedel, 1883

ANOPLUS Germar, 1820
plantaris (Naezen, 1794)
roboris Suffrian, 1840

Tribe HYLOBIINI Kirby, 1837

HYLOBIUS Germar, 1817
  CURCULIO sensu Bedel, 1888 non Linnaeus 1758

Subgenus CALLIRUS Dejean, 1821
abietis (Linnaeus, 1758)
transversovittatus (Goeze, 1777)
  fatuus (Rossi, 1790)

Tribe LEPYRINI Kirby, 1837

LEPYRUS Germar, 1817
capucinus (Schaller, 1783)
  binotatus (Fabricius, 1792)

Tribe PHRYNIXINI Kuschel, 1964

SYAGRIUS Pascoe, 1875
intrudens Waterhouse, C.O., 1903

Tribe PISSODINI Gistel, 1848

PISSODES Germar, 1817
castaneus (De Geer, 1775)
  notatus (Fabricius, 1787) non (Bonsdorff, 1785)
pini (Linnaeus, 1758)
validirostris (Sahlberg, C.R., 1834)

Tribe TRACHODINI Gistel, 1848

TRACHODES Germar, 1824
hispidus (Linnaeus, 1758)

Subfamily OROBITIDINAE Thomson, C.G., 1859

OROBITIS Germar, 1817
cyaneus (Linnaeus, 1758)

Subfamily SCOLYTINAE Latreille, 1804 [839]

Tribe SCOLYTINI Latreille, 1804 [840]

Subtribe SCOLYTINA Latreille, 1804

SCOLYTUS Geoffroy, 1762
  ECCOPTOGASTER Illiger, 1798
intricatus (Ratzeburg, 1837)
laevis Chapuis, 1873 [841]
mali (Bechstein & Scharfenberg, 1805)
  pruni (Ratzeburg, 1837)
multistriatus (Marsham, 1802)
pygmaeus (Fabricius, 1787) [842]
ratzeburgi Janson, 1856
  destructor sensu Ratzeburg, 1837 non Olivier, 1795

rugulosus (Müller, P.W.J., 1818)
scolytus (Fabricius, 1775)
    destructor Olivier, 1795

Subtribe CORTHYLINA LeConte, 1876

PITYOPHTHORUS Eichhoff, 1864
lichtensteinii (Ratzeburg, 1837) [843]
    lichtensteini auctt. (misspelling)
    scoticus Blandford, 1891
pubescens (Marsham, 1802)
    ramulorum (Perris, 1856)

Subtribe CRYPHALINA Lindemann, 1877

CRYPHALUS Erichson, 1836
asperatus (Gyllenhal, 1813)
    abietis (Ratzeburg, 1837)

ERNOPORICUS Berger, 1917
    ERNOPORUS sensu auctt. partim non Thomson,
    C.G., 1859
caucasicus (Lindemann, 1876)
fagi (Fabricius, 1798)

ERNOPORUS Thomson, C.G., 1859
tiliae (Panzer, 1793)

TRYPOPHLOEUS Fairmaire, 1864
    CRYPHALUS sensu Fowler, 1891 partim non
    Erichson, 1836
binodulus (Ratzeburg, 1837)
    asperatus sensu auctt. partim non (Gyllenhal, 1813)

Subtribe CRYPTURGINA LeConte, 1876

CRYPTURGUS Erichson, 1836
subcribrosus Eggers, 1933 [844]

Subtribe DRYOCOETINA Lindemann, 1877

DRYOCOETES Eichhoff, 1864
    DRYOCAETES auctt. (misspelling)
    DRYOCOETINUS Balachowsky, 1949
alni (Georg, 1856)
autographus (Ratzeburg, 1837)
villosus (Fabricius, 1792)

LYMANTOR Løvendal, 1889
    DRYOCOETES sensu Fowler, 1891 partim non
    Eichhoff, 1864
coryli (Perris, 1855)

TAPHRORYCHUS Eichhoff, 1878
bicolor (Herbst, 1794)
villifrons (Dufour, 1843) [845]

XYLOCLEPTES Ferrari, 1867
bispinus (Duftschmid, 1825)

Subtribe IPINA Bedel, 1888

IPS De Geer, 1775
    TOMICUS sensu Fowler, 1891 partim non Latreille,
    1802
acuminatus (Gyllenhal, 1827)
cembrae (Heer, 1836)
sexdentatus (Boerner, 1766)
typographus (Linnaeus, 1758)

ORTHOTOMICUS Ferrari, 1867
    ONTHOTOMICUS Ferrari, 1867
    TOMICUS sensu Fowler, 1891 partim non Latreille,
    1802
erosus (Wollaston, 1857)
laricis (Fabricius, 1792)
suturalis (Gyllenhal, 1827)

PITYOGENES Bedel, 1888
bidentatus (Herbst, 1784)
    ater (Fabricius, 1792)
chalcographus (Linnaeus, 1760)
quadridens (Hartig, 1834)
trepanatus (Nördlinger, 1848)

Subtribe XYLEBORINA LeConte, 1876

XYLEBORINUS Reitter, 1913
    XYLEBORUS sensu auctt. partim non Eichhoff,
    1864
saxesenii (Ratzeburg, 1837)
    saxeseni auctt. (misspelling)
    xylographus sensu auctt. Brit. non (Say, 1826)

XYLEBORUS Eichhoff, 1864
    ANISANDRUS Ferrari, 1867
dispar (Fabricius, 1792)
dryographus (Ratzeburg, 1837)
    villifrons sensu Donisthorpe, 1924 non (Dufour,
    1843)
    sampsoni Donisthorpe, 1940
    donisthorpei Schedl, 1951
monographus (Fabricius, 1792) [846]

Subtribe XYLOTERINA LeConte, 1876

TRYPODENDRON Stephens, 1830
    XYLOTERUS Erichson, 1836
domesticum (Linnaeus, 1758)
lineatum (Olivier, 1795)
signatum (Fabricius, 1792)

Tribe HYLESININI Erichson, 1836

Subtribe HYLESININA Erichson, 1836

HYLESINUS Fabricius, 1801
    LEPERISINUS Reitter, 1913
crenatus (Fabricius, 1787)
toranio (Danthoine in Bernard, 1788)
    oleiperda (Fabricius, 1792) [847]
    bicolor (Brullé, 1832)
varius (Fabricius, 1775)
    fraxini sensu auctt. Brit. non (Panzer, 1799) [848]

wachtli Reitter, 1887 [849]
ssp. ***orni*** Fuchs, 1906 [850]

HYLASTINUS Bedel, 1888
obscurus (Marsham, 1802)

KISSOPHAGUS Chapuis, 1869
    CISSOPHAGUS auctt. (misspelling)
vicinus (Comolli, 1837)
    hederae (Schmitt, 1843)

PTELEOBIUS Bedel, 1888
    ACRANTUS sensu auctt. Brit. non Broun, 1881
    HYLESINUS sensu Fowler, 1891 partim non
      Fabricius, 1801
vittatus (Fabricius, 1792)

Subtribe HYLASTINA LeConte, 1876

HYLASTES Erichson, 1836
angustatus (Herbst, 1794)
ater (Paykull, 1800) [851]
    pinicola Bedel, 1888 [852]
attenuatus Erichson, 1836
brunneus Erichson, 1836
cunicularius Erichson, 1836
opacus Erichson, 1836

HYLURGOPS LeConte, 1876
    HYLASTES sensu Fowler, 1891 partim non
      Erichson, 1836
palliatus (Gyllenhal, 1813)

Subtribe HYLURGINA Gistel, 1848

DENDROCTONUS Erichson, 1836
micans (Kugelann, 1794) [853]

TOMICUS Latreille, 1802
    BLASTOPHAGUS Eichhoff, 1864
    MYELOPHILUS Eichhoff, 1878
minor (Hartig, 1834)
piniperda (Linnaeus, 1758)

XYLECHINUS Chapuis, 1869
pilosus (Ratzeburg, 1837)

Subtribe PHLOEOTRIBINA Chapuis, 1869

PHLOEOSINUS Chapuis, 1869
aubei (Perris, 1855) [854]
    bicolor sensu auctt. non (Brullé, 1832) [855]
thujae (Perris, 1855)

PHLOEOTRIBUS Latreille, 1797
    PHLOEOPHTHORUS Wollaston, 1854
rhododactylus (Marsham, 1802)

Subtribe POLYGRAPHINA Chapuis, 1869

POLYGRAPHUS Erichson, 1836
grandiclava (Thomson, C.G., 1886) [856]

poligraphus (Linnaeus, 1758)
    pubescens (Fabricius, 1792)

103. Family PLATYPODIDAE Shuckard, 1839 [857]

Family author: M.G. Morris

PLATYPUS Herbst, 1793
cylindrus (Fabricius, 1792)

Families:	103
Genera:	1265
Species:	4072

Endnotes

1   Emended from original Sphaeriidae to remove homonymy with Sphaeriidae Deshayes, 1854 (1820) (Mollusca, Bivalvia), ICZN Op. 1957.
2   Name conserved, ICZN Op. 1957.
3   *Aulonogyrus striatus* (Fabricius), listed in Pope (1977), has been deleted from the British List.
4   Name conserved, ICZN Op. 1754.
5   Subgenera of *Haliplus* follow Holmen (1987).
6   Added (reinstated?) by Parry (1982a) and more fully by Parry (1983).
7   Synonymized by Lundmark, Drotz & Nilsson (2001).
8   This checklist is based on Nilsson (2003).
9   Use of the younger family name Hygrobiidae is in accordance with ICZN Article 40.2 and follows Bouchard *et al.* (2011).
10 Name rejected, ICZN Op. 280.
11 This checklist is based on Nilsson (2003[2006]).
12 Junior primary homonym of *biguttatus* Gmelin in Linnaeus, 1789.
13 Name conserved, ICZN Op. 1556.
14 Transferred from *Agabus* to *Ilybius* by Nilsson (2000).
15 Added by Owen *et al.* (1992a) and more fully by Owen *et al.* (1992b) [as *Agabus wasastjernae*].
16 Name rejected, ICZN Op. 289.
17 Name established by Nilsson & Holmen (1995).
18 Separated from *Copelatus* s.str. by Balke, Ribera & Vogler (2004).
19 Justified emendation of the original "Cybisterini".
20 *Hydaticus continentalis* Balfour-Browne, J., has been deleted from the British List (Foster, 2012).
21 *Hydroporus foveolatus* Heer, listed in Pope (1977), has been deleted from the British List.
22 Name conserved, ICZN Op. 2207.
23 Renamed by Fery (1999).
24 Name rejected, ICZN Op. 1556.
25 *N. canariensis* (Bedel), listed in Pope (1977) [in *Potamonectes*], has been deleted from the British List.
26 Added by Carr (1999) and more fully by Carr (2000).
27 Added by Foster & Spirit (1986).
28 Added by Foster & Friday (2011).
29 Added by Drake (2006).
30 Added by Carr (1984b).
31 *C. (Autocarabus) cancellatus* Illiger, listed by Pope (1977), is here treated as a non-established introduction.
32 Forgotten senior synonym.
33 Synonymized by Coulon (2004). See also Hammond (2002).
34 Added by Speight, Martinez & Luff (1986).
35 Added by Focarile (1964); see also Speight, Martinez & Luff (1986).
36 Junior primary homonym of *varium* Müller, O.F., 1776.
37 Added by Telfer (2001a).
38 Added by Levey & Pavett (1999).
39 Junior primary homonym of *minimum* Müller, O.F. 1776.
40 Unjustified emendation.

41 Junior primary homonym of *littoralis* Olivier, 1795.
42 Junior primary homonym of *aethiops* Herbst, 1784.
43 Junior primary homonym of *longicollis* Lichtenstein, 1796.
44 Junior primary homonym of *niger* Linnaeus, 1764.
45 Added by J.A. Owen in Jones (1989) and more fully by Luff (1990).
46 Junior primary homonym of *vernalis* Müller, O.F., 1776.
47 *Abax parallelus* (Duftschmid) has been moved to the list of Non-established introductions.
48 Added by Anderson & Luff (1994).
49 Junior primary homonym of *mollis* Strøm, 1768. ICZN Op. 1723 rejects Strøm's name for Principle of Priority but not for Principle of Homonymy.
50 Junior primary homonym of *rotundatus* Gmelin in Linnaeus, 1789.
51 Added by Anderson (1985).
52 Junior primary homonym of *ovata* Paykull, 1790.
53 Junior primary homonym of *infima* Gravenhorst, 1807.
54 Junior primary homonym of *fulva* De Geer, 1774.
55 Confirmed as British by Owen (1996).
56 Junior primary homonym of *cordatus* Scopoli, 1763.
57 Added by Telfer (2001b).
58 *Stenolophus comma* (Fabricius) has been moved to the list of Non-established introductions.
59 Added by Telfer (2003).
60 British records reviewed by Denton (2007), who showed that the species has been temporarily established in Britain in the past.
61 Added by Hammond (1982a).
62 Junior primary homonym of *agilis* Müller, O.F., 1776.
63 Added by Eversham & Collier (1997).
64 Hansen (1991) elevated the Helophorinae (Hydrophilidae) to the level of a family within the Hydrophiloidea. Classification follows Angus (1999).
65 Name rejected, ICZN Op. 1724.
66 Synonymized by Angus (1982).
67 Name conserved, ICZN Op. 1629.
68 Hansen (1991) elevated the Georissinae (Hydrophilidae) to the level of a family within the Hydrophiloidea.
69 Name rejected, ICZN Op. 1891.
70 Hansen (1991) elevated the Hydrochinae (Hydrophilidae) to the level of a family within the Hydrophiloidea.
71 Synonymized by Hansen (1999).
72 Added by van Berge Henegouwen (1988).
73 Hansen (1991) elevated the Spercheinae (Hydrophilidae) to the level of a family within the Hydrophiloidea.
74 Subfamilies and sequence of genera follows Löbl & Smetana (2004), after Foster (2005).
75 Status revised by van Berge Henegouwen (1986).
76 Name conserved, ICZN Op. 1577.
77 Added by Foster (2005). See also Levey (2005).
78 Added by Sharp (1914) but since synonymized by authors including Pope (1977); status revised by Hansen (1987).
79 Added by Newbery (1914) but since synonymized by authors including Pope (1977); status revised by Hansen (1987).

[80] Added by Foster (1984) [as *Enochrus isotae*].

[81] Synonymized by Schödl (1997).

[82] Added by Hansen (1982).

[83] Junior primary homonym of *niger* Fabricius, 1775.

[84] British records of *L. obscuratus* Rottenberg require confirmation. See Gentili & Chiesa (1975) for a putative record.

[85] Added by Gentili & Chiesa (1975) [as "*L. simulator*"].

[86] Synonymized by Gentili & Ribera (1998).

[87] Added by Allen (2004).

[88] Arrangement of subgenera follows Hansen (1987).

[89] Added by Owen & Mendel (1990).

[90] Synonymized by Allen (1969).

[91] Synonymized by Hansen (1999).

[92] Status revised by van Berge Henegouwen (1989).

[93] Nomenclature and arrangement of this family follow Mazur (1984).

[94] Added by Lackner (2005).

[95] Added by Nash (1982).

[96] Added by Allen & Hance (2009).

[97] Synonymized by Hansen (1998).

[98] Added by Carr (1984a).

[99] Raised from level of subgenus of *Ochthebius* by Hansen (1998).

[100] Three further species are in need of confirmation as British. Two specimens of *O. (Hymenodes) difficilis* Mulsant, labelled from "Britannia" (M. von Pfaundler coll.), are in the Zoologische Staatssammlung, Munich (Hebauer, 1985); one male specimen of *O. (Hymenodes) pilosus* Waltl, labelled from "Anglia", is in the Deutsches Entomologisches Institut, Eberswalde (Jäch, 1989); and a specimen of *O. (Hymenodes) quadrifoveolatus* Wollaston, labelled from "Anglia" (leg. Jekel), is in the Universitets Zoologiska Museum, Helsingfors (Jäch, 1989). In the absence of specimens with more accurate collection data the evidence that these three species have occurred in Britain must remain inconclusive.

[101] Added by Foster (2008) and more fully by O'Callaghan *et al.* (2009).

[102] For a review of the *O. viridis* species group, see Foster (2009).

[103] Name conserved, ICZN Op. 1277.

[104] Synonymized by Johnson (2001a).

[105] Added by Johnson (2001b).

[106] Synonymized by Biström & Silfverberg (1979); see also Johnson (1987a).

[107] Not Erichson, 1845; name conserved, ICZN Op. 1277.

[108] Reinstated by Johnson (1987a).

[109] Now a separate genus (Dybas & Dybas, 1990).

[110] Transferred from Ptinellini to Ptiliini by Darby (2012).

[111] Added by Johnson (1987a).

[112] Transferred from Ptiliini to Ptinellini by Darby (2012).

[113] Synonymized by Johnson (1982b); see also Johnson (1987a).

[114] Name rejected, ICZN Op. 1307.

[115] Checklist is based on Johnson (1987b). This paper showed that *A. championi* and *A. sarae*, listed as '*species incertae sedis*' in Pope (1977), are foreign species which have no valid place on the British Isles List.

[116] Listed as a '*species incertae sedis*' in Pope (1977); synonymized by Johnson (1987b).

[117] See Johnson (2003).

[118] *thomsoni* Erichson is an invalid varietal name lacking description, so has no standing in nomenclature (Johnson, 1987b).

[119] Added by Johnson (1992b).

[120] See Sörensson (2003).

[121] Added by Johnson (1967), but shown to be a misidentification of *A. rugulosa* by Johnson (1975a).

[122] Added by Johnson (1987a).

[123] The checklist for this subfamily is based on the checklist in Cooter (1996b). *Aglyptinus agathidioides* Blair has been moved to the list of Non-established introductions.

[124] Reidentified by Cooter (1992).

[125] See Cooter (1996a).

[126] Reidentified by Cooter (2004).

[127] Name rejected, ICZN Op. 1810.

[128] Added by Collier (1986).

[129] Added by Schilthuizen (1990) [as *Choleva septentrionis septentrionis*] and confirmed by Růžička & Vávra (2003) [as *C. lederiana*].

[130] Added by Schilthuizen (2010).

[131] Junior primary homonym of *tristis* Rossi, 1790.

[132] Allocation of genera to tribes follows Herman (2001b).

[133] Zanetti (1987) used the name *Lathrimaeum* for this genus, because he considered that the type species of *Anthobium* is a species of *Eusphalerum*. However, current usage of *Anthobium* has been conserved by redesignation of the type species (Herman, 2001a).

[134] Synonymized by Assing & Schülke (2012).

[135] Name conserved, ICZN Op. 2077.

[136] Current usage of this name has been conserved by redesignation of its type species (Herman, 2001a).

[137] Name conserved, ICZN Op. 2086.

[138] Regarded as a subspecies by Zanetti (1987) and Smetana (2004a), but merely as a synonym by Hansen (1996).

[139] Added by Lott (1989).

[140] Name conserved, ICZN Op. 1722.

[141] Tottenham (1954) used this name for *A. minuta*, which had been recently found for the first time in Britain (Steel, 1957), although Joy (1932) used the name for the more widespread *A. sulcula*.

[142] Raised to generic rank by Steel (1970), though not followed by most subsequent authors until generic status reaffirmed by Thayer (2003).

[143] Synonymy appears in Zanetti (1987).

[144] Synonymized in Hammond (1996).

[145] Not listed by Smetana (2004a), but specific status and presence in Britain confirmed by Jászay & Hlavac (2006).

[146] Added by Jászay & Hlavac (2006).

[147] Name conserved, ICZN Op. 2053.

[148] Added by Hammond (2007b).

[149] Raised to generic rank by Steel (1970) and most contemporary authors, though still combined with *Phloeonomus* by Poole & Gentili (1996).

[150] Raised to generic rank by Steel (1970) and most contemporary authors, though still combined with *Hapalaraea* by Poole & Gentili (1996).

[151] Name conserved, ICZN Op. 2053.

[152] The type series of *brunnipennis* is conspecific with *depressus* (Hammond, pers. comm.), so this name

[153] cannot be used to replace *concinnus* as suggested by Hansen (1996).

[153] Raised to generic rank by Steel (1970).

[154] Added by Allen & Booth (2005).

[155] Junior homonym replaced in Cuccodoro & Löbl (1997).

[156] This name was previously used as a misidentification of what is now known as *M. prosseni* (Cuccodoro & Löbl, 1997). Consequently, great care needs to be taken in interpreting the name *depressus* in the literature and it is recommended that when using this name in future, it should be quoted as '*depressus* (=*sinuatocollis*)'.

[157] Synonymized by Cuccodoro & Löbl (1997).

[158] Synonymized by Zerche (1998).

[159] Name protected by Herman (2001a).

[160] First replaced with the name *serrifer* (Muona, 1977), which was later synonymized by Gusarov (1992).

[161] Synonymized by Assing & Schülke (2012).

[162] Formerly regarded as a separate family, the Pselaphinae were placed within the Staphylinidae by Newton & Thayer (1995). Supertribes, tribes and subtribes follow Löbl & Besuchet (2004).

[163] The species in this genus have been confused in the British literature up until very recently. Names in old lists should be interpreted with caution.

[164] Added by Bowestead & Eccles (1987).

[165] Subspecific status reaffirmed by Besuchet (1999).

[166] This name has been frequently misapplied to *E. bescidicus* by British authors including Pearce (1957). See also Allen (1994b).

[167] Junior homonym replaced in Löbl & Besuchet (2004).

[168] Until recently the name, *E. punctatus*, was applied to two British species formerly considered to be of subspecific status.

[169] Raised from subspecific rank by Besuchet (1989).

[170] Added by Johnson & Eccles (1983).

[171] The species in this genus have been confused in the British literature in the past. Names in old lists should be interpreted with caution.

[172] Added by Johnson (2012c).

[173] Formerly considered to be a subspecies of *haematica* (Reichenbach), but raised to specific rank by Besuchet (1999) and recorded from Britain by Sabella *et al.* (2004).

[174] Löbl (1998) considered this name to be a junior synonym of *hipponensis* (Saulcy) because he mistakenly believed the date of publication to be 1882. In fact Rye (1869) had published the name earlier to replace *simplex* (Waterhouse) which was preoccupied.

[175] Usage follows Löbl & Besuchet (2004).

[176] Considered to be an intraspecific male form of *longicornis* by Besuchet (1989), but regarded as a separate species by Odegaard (2001).

[177] Lectotype designated by Schülke (1999), establishing synonymy with *Quedius scitus* (Gravenhorst, 1806). All uses of the name *Bolitobius analis* (Fabricius) are therefore misidentifications.

[178] Raised to generic rank by Campbell (1993).

[179] Synonymized by Schülke (2012).

[180] Formerly treated as a subgenus of *Mycetoporus*, but raised to full generic rank by Campbell (1991).

[181] Added by Schülke (2011b).

[182] Transferred from *Bolitobius* by Assing & Schülke (2001).

[183] Prior to Hammond (1973), this name was also used for three other closely related species, which were then unrecognised.

[184] Species groups have been recognised by Ullrich (1975).

[185] Formerly used as a subgenus, but synonymized by Campbell (1973).

[186] Synonymy appears in Lohse (1989). The Scottish paratype is also conspecific with *pallipes* (Hammond, pers. comm.).

[187] Usage and synonymy reinstated by Schülke (2005).

[188] Added by Booth (1988).

[189] The subgenus *Palporus* was created for this species by Campbell (1979).

[190] The list of tribes follows Newton & Thayer (1992) and is the same as that used by Smetana (2004d). Subtribes used by Smetana (2004d) are also included.

[191] This tribe is included within the Homalotini by Ashe (1998).

[192] Most subgenera are assigned to full generic rank by Seevers (1978), but this practice has not been followed by recent authors (e.g. Maus *et al.*, 2000). Subgenera and species names follow Welch (1997), except where indicated.

[193] Added by Welch (1969) [as *Aleochara verna*]. This species was misidentified as *Aleochara pauxilla* by Lohse (1974).

[194] Reidentified by Welch (1990b).

[195] Added by Owen (1990c) [as *Aleochara pauxilla*].

[196] Synonymized by Maus (2001).

[197] This name has been previously misapplied to *A. punctatella* (Lohse, 1985b).

[198] Only known from the unique type specimen taken in Dorset in 1937; listed as a synonym of *A. obscurella* by Smetana (2004d), but regarded as a good species by Welch (1997).

[199] Combined with *Emplenota* in Welch (1997), but its distinctness was emphasised by Maus *et al.* (2000).

[200] Synonymized by Welch (1997).

[201] Replacement due to Likovský (1984).

[202] Previously placed in the tribe Hoplandriini, but transferred to the Aleocharini by Hanley (2002).

[203] The generic limits in this subtribe are in need of review and are likely to be fluid for the foreseeable future (see also notes in family introduction). There are inconsistencies between the checklists published in Central Europe and Scandinavia (see e.g. Lohse 1985a, 1987; Muona 1987). The list of genera in the Palaearctic catalogue (Smetana 2004d) largely follows central European usage while incorporating as subgenera several generic names recognised in Scandinavian checklists. While this approach outlines divisions in the large amorphous genus, *Atheta*, it leaves several fairly distinctive groups within *Atheta* unrecognised in binomial names. Consequently, this checklist continues the expedient approach of splitting *Atheta* into manageable groups and recognising them as genera. A note is given wherever appropriate to explain how taxa are treated differently by Smetana (2004d).

[204] Species groups have been identified by Brundin (1952).

[205] Synonymized by Allen (1994a).

[206] Added by Lott (2009b).

[207] Added by A.C. Galsworthy & R.G. Booth in Hodge (2005).

[208] Synonymized by Nikitsky et al. (1998).

[209] Synonymized by Lohse (1988).

[210] Synonymized by Gusarov (2003).

[211] Listed by Smetana (2004d) as a subgenus of Atheta, but widely ranked as a genus by previous authors. Several species listed in Lohse's Atheta 'Mischgruppe II' are included here.

[212] Added by P.J. Hodge in Hodge (2003) [as Atheta linderi].

[213] Listed by Smetana (2004d) in Oreostiba.

[214] Listed by Smetana (2004d) in Traumoecia.

[215] British records of Alevonota spp. require review, following publication of Assing & Wunderle (2008).

[216] Synonymized by Assing & Wunderle (2008).

[217] Added by Last (1980).

[218] Reidentified by Lott (1993b).

[219] Added by Hammond (1981).

[220] Raised to generic rank by Muona (1979a) and Köhler & Klausnitzer (1998), though not by Assing & Schülke (2007).

[221] Before Last (1952), this name was misapplied to A. longicollis.

[222] Before Last (1952), this name was misapplied to A. coulsoni.

[223] Synonymized by Muona (1990).

[224] Synonymized by Williams (1979a).

[225] Listed by Smetana (2004d) as a subgenus of Atheta, but ranked as a genus by Seevers (1978) and Silfverberg (2004).

[226] Synonymized by Benick (1970); formerly placed in Microdota.

[227] Added by Allen (1994a) [as Atheta verulamii].

[228] Synonymized by Vogel (2004).

[229] Kevan (1963) placed this species in Atheta subgenus Amidobia. Muona (1979b) found that the species had been misinterpreted by Brundin (1948), who, according to Allen (1994a) and Assing et al. (1998), confused it with A. puberula, a species assigned to Anopleta by Benick (1970).

[230] Formerly placed in Microdota but transferred to Anopleta by Benick (1970).

[231] Used as a subgenus of Atheta by Likovský in Boháč (1993) for A. liturata and A. nigritula and retained by Smetana (2004d).

[232] Added by Lyszkowski (1992).

[233] Listed by Pope (1977) but first formally recorded by Allen (1994a).

[234] The list below contains species assigned to Atheta s. str. by Smetana (2004d) that should be treated provisionally as belonging to the genus Atheta. The list is made up of various species that were placed in Lohse's 'Mischgruppen I and III/IV' in Atheta sensu lato (Freude et al., 1974). It contains a number of species groups that are not necessarily closely related, and which await a proper revision of generic relationships within the Athetini (see discussion by Gusarov (2003)).

[235] Listed by Pope (1977) but first formally recorded by Allen (1994a).

[236] Replacement due to Muona (1979b).

[237] Included in Lohse's 'Mischgruppe I', but assigned to Dimetrota by Brundin (1963) and Smetana (2004d).

[238] Placed in the genus Plataraea by Muona (1979a), but not by Lohse (1985a).

[239] Nomenclature as in Smetana (2004d).

[240] Listed by Smetana (2004d) as a subgenus of Atheta, and not ranked as a genus by recent authors. Treated as separate from Dimetrota by Muona (1979a) and Baranowski (1982), though not by Lohse (1989). Synonymized with Atheta rather than Dimetrota by Poole & Gentili (1996).

[241] Listed by Smetana (2004d) as a subgenus of Atheta, but ranked as a genus by Seevers (1978).

[242] Regarded as a good species by Lohse (1989).

[243] Formerly included in Atheta subgenus Actophylla; name protected by Vogel (2004).

[244] Listed as a synonym of Dimetrota by Smetana (2004d).

[245] Added by Owen (1983) [as Atheta hansseni].

[246] Synonymized by Assing (2001).

[247] Listed by Smetana (2004d) as a subgenus of Atheta, and not ranked as a genus by recent authors.

[248] Synonymy by Benick (1974) confirmed by Muona (1979b).

[249] Listed by Smetana (2004d) as a subgenus of Atheta, and not ranked as a genus by recent authors.

[250] Listed by Smetana (2004d) as a subgenus of Atheta, and not ranked as a genus by recent authors, but regarded as a subgenus of Acrotona by Seevers (1978).

[251] Generic assignment established by Gusarov (2003) following discussion by Muona (1979b).

[252] Listed by Smetana (2004d) as a subgenus of Atheta, but ranked as a genus by Seevers (1978).

[253] In Joy (1932) and Strand & Vik (1964), this name was misapplied to D. dadopora.

[254] Listed on its own in Atheta subgenus Datostiba by Likovský in Boháč (1993).

[255] In Joy (1932), this name was misapplied to D. nigra.

[256] Listed by Smetana (2004d) as a subgenus of Atheta, but ranked as a genus by Seevers (1978) and Silfverberg (2004).

[257] This is the type species of Dralica, which was retained as a separate subgenus of Atheta by Likovský in Boháč (1993) and Vogel (2004).

[258] Listed by Smetana (2004d) as a subgenus of Atheta, but ranked as a genus by Seevers (1978) and Lohse et al. (1990). The traditional concept of Dimetrota has been threatened by Blackwelder's designation of Homalota tristicula (=Cadaverota cadaverina) as the type species (Muona, 1979b), but Lohse & Smetana (1985) argued that Fenyes' previous designation of D. marcida is valid. Smetana (2004d) includes both Cadaverota cadaverina and traditional Dimetrota species within the genus.

[259] The true identity of this species was established by Lohse & Smetana (1985).

[260] Synonymized by Lott (2002).

[261] Name conserved, ICZN Op. 2098.

[262] Listed incorrectly as a junior synonym of H. puncticeps by Smetana (2004d).

[263] Listed as British by Hammond (2000a) on the basis of an as yet unpublished revision of the British species of Halobrecta. Diagnostic characters are

[264] given by Gusarov (2004). Listed incorrectly as a synonym of *H. flavipes* by Smetana (2004d).

[265] Not listed by Smetana (2004d).

[266] Synonymized by Assing & Schülke (2001).

[267] Added by Allen & Eccles (1988) [as *Hydrosmectina delicatissima*].

[268] Allen (1992a), following Lohse (1988), replaced *H. subtilissima* in the British list with this name, but it is synonymized in Smetana (2004d).

[269] Synonymy appears in Muona (1979a) and is followed by subsequent checklists.

[270] Listed as a subspecies by Muona (1979a) but Assing & Schülke (2007) argued that it should be regarded as a synonym.

[271] Placed in *Atheta* by Poole & Gentili (1996), though assigned to *Liogluta* by all recent European authors and Seevers (1978).

[272] Synonymy confirmed by Assing & Schülke (1999).

[273] Listed by Smetana (2004d) as a subgenus of *Atheta*, but ranked as a genus by Seevers (1978). Species groups have been identified by Brundin (1948).

[274] Added (reinstated) by Allen (1994a). In Joy (1932) this name was misapplied to *M. glabricula*.

[275] In Joy (1932) this name was misapplied to *Anopleta soedermani*.

[276] Listed as a synonym of *Acrotona* by Smetana (2004d). Carter & Owen (1987) listed *M. gilvicollis* Scheerpeltz as a British species, although there are no published British records for this species, whose taxonomic status in any case requires further scrutiny, preferably as part of a wider study of the *M. fungi* species group.(Assing & Schülke, 2001).

[277] Hodge & Jones (1995) reported that some British records of *M. clientula* were referable to *M. negligens* (Mulsant & Rey), but provided no supporting evidence.

[278] Listed by Smetana (2004d) as a subgenus of *Atheta*, and not ranked as a genus by recent authors.

[279] Pace (2006) placed this species in *Atheta* subgenus *Oxypodera* Bernhauer, 1915.

[280] Replaced by Zerche (1991).

[281] Listed by Smetana (2004d) as a subgenus of *Atheta*, but ranked as a genus by Brundin (1940) and Likovský in Boháč (1993).

[282] Listed by Smetana (2004d) as a subgenus of *Atheta*, but ranked as a genus by Silfverberg (2004).

[283] In Joy (1932) this name was misapplied to *Microdota atricolor*.

[284] Listed by Smetana (2004d) as a subgenus of *Atheta*, but ranked as a genus by Seevers (1978).

[285] Listed by Smetana (2004d) as a subgenus of *Atheta*, even though the promotion of this taxon to full generic status was proposed by Muona (1995) and it was used as a genus by Seevers (1978) and Lohse *et al.* (1990). Species groups have been recognised by Brundin (1943).

[286] In Strand & Vik (1964), this name was misapplied to *P. malleus*.

[287] Synonymized by Muona (1995).

[288] Listed by Smetana (2004d) as a subgenus of *Atheta*, and not ranked as a genus by recent authors.

[289] Added by Owen (1990b) [as *Schistoglossa benicki*].

[290] Synonymized by Mahler & Vagtholm-Jensen (2002).

[291] Listed by Smetana (2004d) as a subgenus of *Atheta*, and not ranked as a genus by recent authors.

[291] Added by Heal (1993).

[292] Listed by Smetana (2004d) as a subgenus of *Atheta*, but ranked as a genus by Seevers (1978). The American concept of this genus is much wider than that used here. It includes a number of species whose relationships are unclear and which in Europe would be placed in *Atheta* sensu lato.

[293] Synonymized by Good (1998).

[294] Correct name established by Good (1998).

[295] Listed as *Myrmecoporus* sp. by Hammond (2000a); see note below.

[296] Added by Assing (1997a), but considered to be the true *brevipes* by Hammond (2000a) and listed as such in his list of British coastal Staphylinidae.

[297] Synonymized by Assing (1999c).

[298] Usage conserved by designation of *Leptusa pulchella* as type species (ICZN Op. 2113).

[299] Added by Owen (1991) [as *Pseudomicrodota jelineki*].

[300] Synonymized by Vogel (2004).

[301] Placed in *Hygroecia* (= *Philhygra*) by Joy (1932).

[302] Taxonomic status doubtful (Hammond, 1996); only known from the unique type specimen taken in Hertfordshire in 1930.

[303] In Joy (1932) this name is misapplied to *Gyrophaena congrua*.

[304] A junior homonym according to Nikitsky *et al.* (1998).

[305] Added by Welch (2000).

[306] Added by R.G. Booth and A.C. Galsworthy in Booth (2008).

[307] Added by Welch (1998).

[308] British records of *Cypha ovulum* (Heer) require confirmation. See Hyman (1994).

[309] This name was used for *Cypha pulicaria* by Joy (1932).

[310] Synonymized by Dauphin (2003, 2004).

[311] Added by Welch (1995).

[312] Placed in the genus *Oligota* by Williams (1979c), although this has been overlooked by subsequent authors who allocate it to the genus *Holobus*.

[313] Treated as a genus by Maruyama (2006) following Kistner (1971).

[314] Species groups have been identified by Muona (1991).

[315] Usage follows Nikitsky *et al.* (1998).

[316] Muona (1979b) considered this name to be a *nomen dubium*.

[317] Added by Owen (1982) [as *Meotica lohsei*]. Before 1976, this name was misapplied to *Meotica anglica* by British coleopterists.

[318] Species groups have been identified by Assing (1996a).

[319] Synonymized by Assing (2003).

[320] Raised to generic rank by Assing & Schülke (2001).

[321] Added by Williams (1979b) [as *Ocyusa nitidiventris*], but central European and British records are referred to *Cousya nigrata* by Assing & Schülke (2007).

[322] See discussion by Smetana (2004d).

[323] Synonymized by Assing (1999b).

[324] Added by Lott (1993a).

[325] Added by Owen (1994b) and more fully by Owen (1994d).

[326] Listed as a synonym of *Ocyusa* by Poole & Gentili (1996), even though its generic status has long been

established in Europe and even though Seevers (1978) suggested that it was distinct.

327 Synonymized by Assing & Schülke (2001).

328 This name is now used for a species in the Athetine genus Nehemitropia (Zerche, 1991).

329 This synonymy appears in an update of the Central European list (Lohse, 1989) and is followed in other checklists.

330 The use of this name follows Zerche (1994). Not recorded since before 1970 (Hyman, 1994).

331 Added by Williams (1990) from a single female found at Dungeness, Kent.

332 Regarded as a long-winged form of Oxypoda brachyptera by Pope (1977), but raised to specific rank in most European works.

333 Synonymized by Lohse (1984).

334 Added by R.G. Booth in Hodge (1999).

335 Synonymized by Dauphin (2001).

336 Reidentified by Whitehead (1994).

337 Nomenclatural change due to Assing & Schülke (2007).

338 Listed by Smetana (2004d) under the tribe Athetini, but transferred to the Oxypodini by Lohse (1989) and Hansen (1996).

339 The name as first used by Scheerpeltz (1968) is unavailable, and the correct authorship was established by Assing & Schülke (2001).

340 Raised to generic rank in Assing et al. (1998).

341 Included in Tachyusa by Assing et al. (1998) but assigned to Ischnopoda by Paśnik (2006b).

342 The type species of Tachyusa and Ischnopoda were set by ICZN Op. 600.

343 Status revised by Paśnik (2006a). British records of Tachyusa coarctata (Erichson) apparently all relate to T. concinna Heer (R.G. Booth, pers. comm.).

344 Added by R.G. Booth and P.J. Hodge in Hodge (2003). British records summarised by Telfer (2007b).

345 Added by R.G. Booth in Hodge (1999).

346 Formerly regarded as a separate family, but incorporated into the Staphylinidae as a subfamily on grounds of larval morphology by Kasule (1966). This approach was supported by Lawrence & Newton (1982) and adopted by most subsequent authors except for Hansen (1997). Tribal classification and arrangement follow Löbl (1997).

347 Tribal classification follows Herman (2001b). Hansen (1996) gives an alternative classification. Gildenkov (2003) erected a new tribe, Mandini, which includes the genera Manda and Planeustomus.

348 Name conserved, ICZN Op. 1722.

349 Described as new to science by Schülke (2009). For further British records see Lott (2011).

350 Added by Owen (1997b).

351 Herman (1986) used species groups in his world revision rather than existing subgenera. However, subgenera are widely referred to in the literature and are included here as listed by Smetana (2004e).

352 Status follows Assing & Schülke (2012).

353 The distinctness of this species from B. pallipes is not widely recognised outside Britain and is in need of review. Several authors including Anderson et al. (1997) have followed Pope (1977) in using this name in a wider sense to include B. pallipes s. str. when referring to the British fauna, as did Herman (1986),

while continuing to use the name pallipes in listing the continental fauna.

354 This name was misapplied in part by Horion (1963) to B. pallipes.

355 All British records were referred to B. annae in Pope (1977), but B. pallipes was listed as a separate species by Herman (1986) and this name is now applied to one of the more frequently encountered species of the subgenus in Western Europe.

356 Status revised by Schülke (2010).

357 Name replacement due to Herman (1986).

358 Name conserved, ICZN Op. 2053.

359 British populations were referred to this subspecies by Hammond (2000a).

360 Name conserved, ICZN Op. 2053.

361 Synonymized by Schülke (2011a).

362 Described as new to science by Schülke (2011a).

363 Added by Owen (1979).

364 Added by Lott (2009a).

365 Recognised for some time (P.M. Hammond, unpublished) but first formally listed in Boyce (1991) and Fowles & Boyce (1992).

366 Previous British records based on misidentifications, but listed by Hammond (2000a) as British on the basis of a single museum specimen collected at Gretna in 1934. The species was not keyed in the monograph by Lott (2009a).

367 Status revised by Lott (2009a).

368 Lott (2009a) briefly mentions a single male found in Leicestershire in 1990, although the species was not included in the keys because there is no evidence of an established breeding population in Britain. C. alutaceus is included in the British list here on the basis that the species is putatively a natural vagrant (cf. Oxypoda praecox).

369 Relegated to a subspecies by Gildenkov (2001).

370 Species list based on Lott (2008).

371 A reassessment of the identity of this taxon is needed following work on the types of related species by Makranczy & Schülke (2001).

372 Synonymized by Schülke (1998).

373 Name conserved, ICZN Op. 2129.

374 Regarded as a subspecies by Muona (1979a), but merely treated as a synonym by Schülke in Assing & Schülke (1999).

375 Added by Lott (1993c).

376 Position as a subfamily within the Staphylinidae, and as a sister group to Steninae + Euaesthetinae, follows Grebennikov & Newton (2009). Higher classification for the Scydmaeninae follows Newton & Franz (1998). This work uses an alphabetic listing of genera within family groups, and subgenera within genera.

377 Added by R.G. Booth in Booth (2001).

378 Originally Nevraphes, treated as an incorrect original spelling by Newton & Franz (1998).

379 Treated as a valid subgenus by Newton & Franz (1998), but according to Franz & Besuchet (1971) Cyrtoscydmus includes exilis (Erichson) which is the type species of Stenichnus s.str., so Cyrtoscydmus has been synonymized until this problem can be resolved.

380 Puthz (1967) discussed the lack of compatibility between the subgenera listed here and a natural phylogenetic classification. However, these

subgenera are well established and continue to be widely used as a matter of convenience to break up this species-rich genus into manageable groups.

381 This name was formerly applied to the subgenus *Metastenus*.

382 Synonymized by Ádám (1987) and followed by Herman (2001b).

383 Originally proposed as a generic name (Ádám, 1987), but downgraded by Puthz (1999). Species groups have been recognised by Puthz (1972) [as *Hemistenus*].

384 Considered to be a short-winged form of *picipes* by Puthz (1972) and listed as a subspecies by Smetana (2004f), but Kevan & Allen (1961) argued that it was distinct.

385 Junior primary homonym of *boops* Gravenhorst, 1806.

386 Added by Shirt (1987).

387 Added by R.G. Booth in Hodge (2003).

388 Added by Lott & Foster (1990).

389 Synonymized by Puthz (1974).

390 Synonymized by Puthz (1993).

391 Species groups have been recognised by Puthz (1968).

392 Added by Lott & Anderson (2011).

393 Tribes and subtribes are arranged according to Lott & Anderson (2011).

394 Synonymized by Assing (2008).

395 Synonymized by Gusarov (1991).

396 Added by Lott *et al.* (2007).

397 Marked as a junior primary homonym in Pope (1977), but without further details.

398 The subgenera listed here have been accorded generic rank by Coiffait (1982a), Assing & Schülke (2001, 2007), Herman (2003) and Smetana (2004g), but are retained as subgenera here pending a world revision of paederine genera.

399 Junior primary homonym of *fulvipenne* Turton in Linnaeus, 1802.

400 Name conserved, ICZN Op. 2126.

401 Synonymized by Gusarov (1992).

402 Raised to generic rank by Coiffait (1982a).

403 Raised to generic rank by Assing in Assing & Schülke (2012).

404 Synonymized by Gusarov (1994).

405 Duff (1995) and Herman (2003) have pointed out that *Hypomedon* and *Chloecharis* share the same type species, because Blackwelder (1939) designated *Lithocharis debilicornis* as the type species of *Hypomedon*. Most authors follow Bordoni (1975a) in separating *debilicornis* from species now placed in *Sunius*, so *Hypomedon* should be removed from synonymy with *Sunius*.

406 Added by Drane (1994) [as *Chloecharis debilicornis*].

407 Drugmand (1994) regarded *ochracea* as a junior synonym of *tricolor* (Fabricius), although the latter name is traditionally applied to a species of *Xantholinus*.

408 Species groups have been recognised by Coiffait (1984).

409 Treated as a genus by Coiffait (1984), but still combined with *Lithocharis* by Poole & Gentili (1996).

410 Raised to generic rank by Coiffait (1982a) and Frank (1988), but still combined with *Paederus* by Poole & Gentili (1996).

411 Species groups have been allocated to subgenera by Coiffait (1984).

412 Reidentified by Allen (1995b).

413 Junior primary homonym of *fragilis* Latreille, 1804. Synonymized by Herman (2003).

414 Added by Skidmore (1988).

415 Synonymized by Assing (1997b).

416 Genus re-established by Smetana (1995), who identified species groups.

417 Name conserved, ICZN Op. 2053.

418 Added by Owen (1993b).

419 Transferred from *Gabrius* by Smetana (1995).

420 Species groups have been recognised by Coiffait (1974) and amended by Smetana (1995).

421 According to Silfverberg (1991) and Herman (2001a), *subnigritulus* Reitter, 1909 is unavailable and *subnigritulus* Joy is a different species.

422 Identified as a junior synonym of *coxalus* by Gusarov (1991), which in turn was synonymized by Schillhammer (1997).

423 Synonymized by Bordoni (1975b).

424 Species groups have been recognised by Coiffait (1974) and amended by Smetana (1995).

425 Synonymized by Smetana (1995).

426 Allen (1970) drew attention to an old British record of *P. albipes* var. *alpinus*. However, Allen (*op. cit.*) never examined the specimen in question and *P. alpinus* was not listed as British by Pope (1977). Regarded as British by Hyman (1994) and Hodge & Jones (1995) on the basis of reports of a more recent unpublished record from Kent. However, it cannot be included in a British Isles checklist on currently published evidence.

427 The name, *P. carbonarius*, has also been widely applied to *P. tenuicornis* Mulsant & Rey until very recently. Therefore, great care is needed when interpreting the use of this name.

428 Synonymized in Coiffait (1974).

429 Introduced to the British list as a distinct species by Smetana (1966), but considered to be a variety by Coiffait (1972).

430 The familiar use of the name *immundus* for this species is based on an early misidentification by Gyllenhal (Smetana & Herman, 1999).

431 Added by Allen & Owen (1997).

432 Lott & Anderson (2011) consider that some specimens standing as *varians* may be referable to *P. pseudovarians* Strand, A., 1941.

433 Raised to generic rank by Coiffait (1974).

434 Status as a separate genus questioned by Smetana (1995) and included in *Cafius* by Hammond (2000a), but retained by Herman (2001b) and Smetana (2004h).

435 Added by P.J. Hodge in Booth (2006) and more fully by Hance (2007).

436 Junior primary homonym of *binotatus* Thunberg, 1796.

437 Added by J.A. Owen in Jones (1989).

438 Whitehead (2002), Smetana (2004h) and Solodovnikov in Assing & Schülke (2012) followed Israelson (1979) in regarding this taxon as a subspecies or junior synonym of *H. praevius*.

[439] Usage conserved by designation of *Q. obscuripennis* as type species (ICZN Op. 2115).

[440] Described as new to science by Lott (2010).

[441] Herman (2001b) has shown that Gravenhorst's (1802) usage of this name is a *nomen nudum*.

[442] Usage of *simplicifrons* follows Pope (1977) and Herman (2001b), although Bordoni (1976) regarded *hispanicus* as the correct name for the species that occurs in Britain.

[443] Synonymized by Assing (1999a).

[444] Listed as a separate species by Smetana (2004h).

[445] Appears as a synonym in Pope (1977) following Korge (1963). Synonymy reaffirmed by Assing (1999a).

[446] Placed in the genus *Platydracus* Thomson by Pope (1977), though not by other authors.

[447] Pope (1977) listed all *Ocypus* species under *Staphylinus,* unlike other contemporary authors.

[448] Status follows Smetana & Davies (2000).

[449] Synonymized by Smetana & Davies (2000).

[450] Raised to generic rank by Coiffait (1974) but retained as subgenus by Smetana & Davies (2000).

[451] Raised to generic rank by Coiffait (1974) and by Smetana & Davies (2000).

[452] Pope (1977) listed all *Tasgius* species under *Staphylinus,* unlike other contemporary authors.

[453] Raised to generic rank by Coiffait (1974) as *Alapsodus* Tottenham, but retained as subgenus by Smetana & Davies (2000).

[454] Junior primary homonym of *compressus* Fourcroy, 1785. Synonymized by Steel (1948).

[455] Regarded as a synonym by Pope (1977) and others until re-established as a good species by Lohse (1989). Again synonymized by Assing (1996b).

[456] Name conserved, ICZN Op. 2173.

[457] Raised to generic rank by Coiffait (1972) as *Phalacrolinus,* which was later synonymized by Smetana (1982).

[458] Raised to generic rank by Coiffait (1972).

[459] Added by Hammond (1982b) [as *Phacophallus tricolor*]. The use of *pallidipennis* as the name for this species follows Bordoni (2002) and subsequent authors.

[460] Replacement due to Drugmand (1994).

[461] Drugmand's (1994) interpretation of the identity of Fabricius' *Staphylinus tricolor,* as the species traditionally known as *Lithocharis ochraceus,* was not accepted by Herman (2001a).

[462] The generic placement of the species follows that of Zunino (1984) and Löbl & Smetana (2006). The use of *Anoplotrupes* and *Trypocopris* is accepted throughout the European literature (e.g. Baraud, 1992; Král, 1993; Krell, 1998; Martín-Piera & López-Colón, 2000) and the Palaearctic Catalogue (Löbl & Smetana, 2006).

[463] Junior primary homonym of *spiniger* Gmelin in Linnaeus, 1790. The name *puncticollis* (Malinowsky, 1811) has been applied to this species (e.g. Martín-Piera & López-Colón).

[464] The synonmy of *Scarabaeus fovetaus* Marsham, 1802 under *G. stercorarius* (Linnaeus, 1758) has been found to be incorrect and it should be placed under *G. spiniger* (Marsham, 1802) (D.J. Mann, unpublished).

[465] The nomenclature of the Trogidae follows that of Löbl & Smetana (2006).

[466] The nomenclature of the Lucanidae follows that of Löbl & Smetana (2006).

[467] The nomenclature follows that of Löbl & Smetana (2006).

[468] The taxonomy of the European members of *Psammoporus* was reviewed by Pittino (2006). *P. insularis* is at present only known from Great Britain.

[469] Name conserved, ICZN Op. 1890.

[470] The placement of the species follows Dellacasa, Bordat & Dellacasa (2001) and Löbl & Smetana (2006).

[471] Authorship of *Aphodius* follows the recommendation of Alonso-Zarazaga & Krell (2011).

[472] Dellacasa, Bordat & Dellacasa (2001) elevate all of the current subgenera under *Aphodius* to generic level. The subgenera under *Aphodius* are placed in alphabetical order, as the taxonomic placement of most subgenera is uncertain.

[473] *A. (Coprimorphus) scrutator* (Herbst, 1789) is listed in Löbl & Smetana (2006) as occurring in Britain, but this is believed to be in error.

[474] Name conserved, ICZN Op. 1890.

[475] Name conserved, ICZN Op. 1890.

[476] Added by Wilson (2001). A species currently confused under *A. fimetarius* (Linnaeus, 1758). Dellacasa & Dellacasa (2007) still consider *pedellus* to be synonymous with *A. fimetarius* (Linnaeus). The status of *pedellus* as a valid species is considered valid by Löbl & Smetana (2006) and here.

[477] Name conserved, ICZN Op. 2099.

[478] Added by D.J. Mann in Hodge (2000). A species currently confused under *A. sphacelatus* (Panzer, 1798).

[479] Name conserved, ICZN Op. 2150.

[480] Name suppressed, ICZN Op. 2150.

[481] Name conserved, ICZN Op. 2150.

[482] The name *sticticus* (Panzer) has long been used in the continental literature (e.g. Dellacasa, 1988; Baraud, 1992) and is followed here.

[483] Added by Angus *et al.* (2003).

[484] The generic placement of this species is discussed by Mann & Booth (2000).

[485] Added by D.J. Mann in Hodge (1998).

[486] The nomenclature follows Zunino (1979) and Löbl & Smetana (2006).

[487] *O. (Palaeonthophagus) ovatus* (Linnaeus, 1758) is listed by Löbl & Smetana (2006) as occurring in 'GB', but this is believed to be in error as no specimens have been found in collections.

[488] Confirmed as British by Johnson (1991).

[489] Status revised by Rössner, Schönfeld & Ahrens (2010). The true *O. vacca* is not currently known from the British Isles.

[490] Junior primary homonym of *verticicornis* Fabricius, 1775, but see Branco (2002).

[491] The name *farinosa* (Linnaeus) is used in the continental literature (e.g. Baraud, 1992). However, the use of the name *philanthus* (Füessly) was re-established by Krell (1991).

[492] Name conserved, ICZN Op. 1754.

[493] Re-identification due to Krell & Rößner (2009).

[494] The nomenclature of the Cetoniinae follows Krajčík (1998).

[495] Name conserved, ICZN Op. 2186.

[496] Name rejected because it was published in a work that was not consistently binominal (Krell, 2012).

[497] Added by Johnson (1997).

[498] Added by Johnson (1992b).

[499] Members of the *marginata*-group of *Elodes* were transferred to *Odeles* by Klausnitzer (2004).

[500] Added by Skidmore (1985b).

[501] Synonymized by Klausnitzer (1998).

[502] The higher classification of the Buprestidae is currently undergoing a period of change. The most recent summary of the higher classification is Bellamy (2003).

[503] Added by Cooter (1992b).

[504] This enigmatic species was described from specimens collected in Dumfriesshire. It is given as a synonym of *viridis* Linnaeus in G.R. Waterhouse's 1858 *Catalogue of British Coleoptera*. It is not certain if Waterhouse was the originator of this synonymy, or if he was following an earlier opinion. The true identity of *littlei* will only be resolved if the type material is found.

[505] Name conserved, ICZN Op. 2221.

[506] Junior primary homonym of *biguttata* Scopoli, 1763, but this name is used by all European authors.

[507] Added by P.J. Hodge in Booth (2009) and more fully by Hodge (2010).

[508] Added by James (1994).

[509] Added by Levey (2012).

[510] Name conserved, ICZN Op. 1323.

[511] Name conserved, ICZN Op. 1754.

[512] Name rejected, ICZN Op. 1396.

[513] Added by Johnson (1978).

[514] Added by Parry (1980).

[515] Synonymized by Nelson (1990).

[516] Added by Johnson (1992b).

[517] Split from *Heterocerus* Fabricius, 1792, by Pacheco (1964).

[518] Added by Mann (2006).

[519] Added by Mendel, Jeffery & Pledger (2011).

[520] Added by Speight (1986).

[521] Junior primary homonym of *vittatus* Gmelin in Linnaeus, 1789.

[522] Added by Owen *et al.* (1985) [as *Panspoeus guttatus*].

[523] Status revised by Mendel (2004).

[524] Added by Mendel (2002).

[525] Arrangement of the subfamily follows Kazantsev (2004).

[526] Added by Allen (1989).

[527] Synonymized by Alexander (2003).

[528] Added by Barclay & Kopetz (2003).

[529] Added by Hammond & Barham (1982) [as *Laricobius erichsoni*].

[530] The checklist for this family is taken from Peacock (1993), slightly simplified.

[531] Added by Peacock (1993).

[532] Added by Peacock (1979).

[533] Added by Shaw (1999).

[534] Added by Thompson (1978).

[535] Added by Adams (1978).

[536] Added by Hinton (1945) but not listed in Pope (1977).

[537] Added by Adams (1988).

[538] Name conserved, ICZN Op. 1754.

[539] Justified emendation of the original *oequinoctiale*. Added by Bellès & Halstead (1985).

[540] British records of *P. pilosus* Müller, P.W.J. require confirmation.

[541] Name rejected, ICZN Op. 1809.

[542] Added by Hodge & Parry (1981).

[543] Added by Johnson (1992b).

[544] Added by Mendel & Hatton (2012).

[545] Added by Mendel (1982) [as *Hemicoelus nitidus*].

[546] Added by M.V.L. Barclay in Booth (2002a) and more fully by Barclay (2005).

[547] Name conserved, ICZN Op. 1754.

[548] Added by M.V.L. Barclay in Booth (2009).

[549] Rejected spelling, ICZN Op. 1810.

[550] Added by Mendel & Owen (1991).

[551] Described as new to science by Bercedo & Arnáiz (2010).

[552] Long recognised as a stored-product pest (e.g. Mound, 1989) but not included in Pope (1977).

[553] Arrangement and nomenclature of the Cleridae generally follow Gerstmeier (1998). The Thaneroclerinae were raised to family rank by Kolibáč (1992), followed by Gerstmeier (1998), but this checklist retains it as a subfamily of Cleridae following Lawrence & Newton (1995).

[554] Justified emendation of the original *buquet*.

[555] Separation of Dasytidae from Melyridae follows Löbl & Smetana (2007).

[556] Added by Johnson (1975a) [as *Dasytes caeruleus*].

[557] Separation of Malachiidae from Melyridae follows Löbl & Smetana (2007).

[558] Added by Welch (2008).

[559] Added by Key (1983).

[560] Added by Allen (1984b).

[561] Rejected spelling, ICZN Op. 1862.

[562] Name conserved, ICZN Op. 1916.

[563] Synonymized by Biström (1977).

[564] Synonymized by Audisio (1993).

[565] Added by P.M. Hammond & R.G. Booth in Booth (2008).

[566] Synonymized by Kurochkin & Kirejtshuk (2006).

[567] Synonymized by Audisio (1993).

[568] Species groups are recognised by Kirk-Spriggs (1996).

[569] Synonymized by Bacchus & Kirk-Spriggs (1991).

[570] Cited incorrectly as a synonym of *M. lugubris* in Pope (1977); reinstated by Kirk-Spriggs (1988).

[571] Added by Parry (1990); see also Kirk-Spriggs (1992).

[572] Added by R.G. Booth in Hodge (in prep.).

[573] New sense from 1993 onwards, cf. *persicus* (Faldermann).

[574] Synonymized by Silfverberg (1979) .

[575] Omitted from Pope (1977) in error; reinstated by Booth (1986).

[576] Added by James (2011).

[577] Added by Prance (2001).

578 Nomenclature for this family largely follows Löbl & Smetana (2007) except that the treatment of *Cyanostolus* follows Peacock (1977).

579 Rejected spelling, ICZN Op. 1810.

580 Added by King & Fielding (1989).

581 Added by R.G. Booth in Booth (2001).

582 Name rejected, ICZN Op. 1771.

583 Nomenclature for this family follows Löbl & Smetana (2007).

584 Rejected spelling, ICZN Op. 1810.

585 Added by Johnson (1988).

586 Synonymized by Johnson (2012a).

587 Added by Johnson (1992b).

588 Synonymized by Johnson (1992a).

589 Added by B. Levey & P.M. Pavett in Hodge (1999). See also Johnson (2002).

590 Added by R.G. Booth & A.C. Galsworthy in Booth (2008),

591 Arrangement of subfamilies follows Leschen (2003).

592 Added by Hammond (2007c).

593 Name conserved, ICZN Op. 1754.

594 Added by Eccles & Bowestead (1987).

595 Names and arrangement of subfamilies follows Tomaszewska (2000).

596 Added by Hawkins (2001).

597 Added by D.A. Coleman per R.G. Booth in Hodge (2000).

598 Added by R.G. Booth & A. Salisbury in Booth (2004).

599 Junior primary homonym of *arcuatus* Fabricius, 1787.

600 Added by Denton (2004).

601 Added by R.M. Lyszkowski in Hodge & Jones (1995).

602 Added by D.S. Hackett in Booth (2002a).

603 Possibly a valid species.

604 Added by Constantine & Majerus (1994).

605 Added by Majerus et al. (1997). See also Muggleton (1999).

606 Added by Majerus & Roy (2005).

607 Added by I.S. Menzies in Hodge (1999) and more fully by Menzies & Spooner (2000). The generic placement of *argus* is probably not correct.

608 This checklist is based upon that in Bowestead (1999).

609 Added by A.C. Galsworthy & R.G. Booth in Hodge (2007). For identification characters see Bowestead (2002).

610 Status revised by Bowestead (1999). Name resurrected by Bowestead for the species long known as *Orthoperus mundus*.

611 Synonymized by Bowestead (1999).

612 Nomenclature for this family follows Löbl & Smetana (2007).

613 Added by Levey (1997).

614 Added by Johnson (2007).

615 Reidentified by Krell et al. (2005).

616 See Johnson (2012b).

617 Added by Lyszkowski, Owen & Taylor (1992) [as *Corticaria abietorum*].

618 Added by Johnson (1992b).

619 Added by Johnson (1986).

620 Added by A.C. Galsworthy & R.G. Booth in Hodge (2007) [as *Migneauxia orientalis*].

621 Name rejected, ICZN Op. 1754.

622 Added by Dorning, Halstead & Hammond (2012).

623 Added by Harrison (1996).

624 Added by Allen & Booth (2008).

625 Hallomeninae was transferred from Melandryidae to Tetratomidae by Nikitsky (1998).

626 Junior primary homonym of *binotatus* Gmelin in Linnaeus, 1789.

627 Justified emendation of the original misspelling *Desmaretsii*.

628 Junior primary homonym of *quercinus* Gmelin in Linnaeus, 1789.

629 Reidentified by Batten (1986).

630 Added by Batten (1986).

631 Added by Allen (1986) [as *Mordellistena parvuloides*]; see also Owen (1999).

632 Synonymized by Horák (1996).

633 Described as new to science by Allen (1999). Synonymized by Levey (2002).

634 Added by Batten (1986). The status of *M. pseudopumila* needs further study.

635 Added by Cooter (1991) [as *Mordellistena pygmeola*]. The status of *M. pygmaeola* needs further study.

636 Added by Levey (1999).

637 Added by Allen (1986).

638 Described as new to science by Allen (1995c). Synonymized by Levey (2002).

639 Added by Allen (1986).

640 Added by Mendel & Owen (1987) [as *Cicones undata*].

641 Name rejected, ICZN Op. 1495.

642 Confirmed as a British species by Hammond (2007a).

643 Geoffroy's name is suppressed for purposes of the Principle of Priority but not for those of the Principle of Homonymy (ICZN Op. 1754).

644 Added by Skidmore & Hunter (1980).

645 Added by Allen (1988b).

646 Added by Whitehead (1992).

647 Junior primary homonym of *rugosus* Thunberg, 1781.

648 Name conserved, ICZN Op. 1754.

649 Junior primary homonym of *ater* De Geer, 1774.

650 Junior primary homonym of *planirostris* Piller & Mitterpacher, 1783.

651 Genera of Anthicidae follow Chandler, Nardi & Telnov (2004). See also Telnov (2010).

652 Synonymized by Chandler et al. (2004).

653 Junior primary homonym of *constrictus* Say, 1826.

654 Name conserved, ICZN Op. 1754.

655 An application is being prepared to give *maculata* (Geoffroy in Fourcroy, 1785) precedence over *melanopa* Forster, 1771.

656 Levey (2002) suggested that *A. thoracica* and *A. septentrionalis* belong to a single variable species.

657 Levey (1996) synonymized *A. schilskyana* with *septentrionalis*.

658 Levey (2003) revised the European species of this subgenus.

659 Csiki (1915) gives *fuscescens* as a junior synonym of *A. thoracica* (Linnaeus, 1758), however all the specimens standing over this name in the Stephens

collection in the Natural History Museum, London are *costai* Emery. For reasons of stability it is not proposed to invoke the principle of priority and it is likely that the prevailing usage of the name *costai* would be validated under article 23.9 of the Fourth Edition of the International Code of Zoological Nomenclature.

[660] This checklist is based on that of Rejzek (2004), which contains details of the higher classification used.

[661] Synonymized by Uhthoff-Kaufmann (1989).

[662] Name rejected, ICZN Op. 1473.

[663] Added by Mendel & Barclay (2008).

[664] Name conserved, ICZN Op. 1525.

[665] Established introduction in the Reading, Berkshire, area in the late 19th and early 20th Centuries (Barclay, 2003b).

[666] Described as new to science by Wallin, Nylander & Kvamme (2009).

[667] Added by Harrison (1992a, 1992b).

[668] Synonymized by Petitpierre (2000).

[669] Name conserved, ICZN Op. 1809.

[670] Name rejected, ICZN Op. 1809.

[671] Added by Hammond & Harvey (2011).

[672] Added by D. Hance & P.J. Hodge in Cox (2001).

[673] Added by Cox (2001).

[674] Added (reinstated) by Aldridge & Pope (1986).

[675] Added by Hodge (1997).

[676] *Oulema melanopus* sensu stricto added by Cox (1995a).

[677] Added by Booth (1994b) [as *Oulema duftschmidi*].

[678] Synonymized by Monrós (1956), who showed that the beetle known to European coleopterists as *Lamprosoma concolor* Sturm is not congeneric with *L. bicolor* Kirby, the type species of *Lamprosoma* Kirby, and that the valid generic name for the European species is *Oomorphus* Curtis, 1831.

[679] Added (reinstated?) by Kendall (1981).

[680] Subgenera are recognised by European authors. Species-level nomenclature for this genus follows Bieńkowski (2001).

[681] First recorded indoors by Johnson (1963), then as a possibly established introduction by Halstead (1996).

[682] Justified emendation of the original misspelling *bankii*.

[683] A breeding population was documented by Salisbury, Malumphy & Halstead (2012).

[684] See also Cox (2000).

[685] Status revised by Cox (2000).

[686] Synonymized by Shute in Cox (2000).

[687] Added as an accidental introduction by Smith (1990), then recorded as probably wild by Buckland & Skidmore (1999).

[688] Cox (2007) describes morphological variation within British and Irish *G. tanaceti* and discusses the possibility that another *Galeruca* species may be present.

[689] Added by Ostojá-Starzewski (2005).

[690] Added by Johnson & Booth (2004).

[691] See Cox (2000) for a discussion of the identity of *A. atrovirens* of British authors.

[692] Added by Sinclair & Hutchins (2009).

[693] Junior primary homonym of *dorsalis* Fabricius, 1777.

[694] Synonymized by Döberl (1987); see also Allen (1993).

[695] Added by Booth (1994a).

[696] Synonymized by Booth (1998).

[697] Added by Harrison (2010).

[698] Name conserved, ICZN Op. 1754.

[699] Added by P.M. Hammond in Cox (2000).

[700] Replacement name due to Warchałowski (1995).

[701] Synonymy follows Cox (2007).

[702] Synonymized by Arnold (1990).

[703] Junior primary homonym of *aurata* Fabricius, 1775.

[704] *C. aridula* (Gyllenhal) and *C. conducta* (Motschulsky) have been deleted from the list, as all records are considered to be due to misidentification or mislabelling. See Cox (2000).

[705] Added by Booth & Owen (1997).

[706] Synonymized by Booth & Owen (1997).

[707] Added by Cox (1995b).

[708] Synonymized by Döberl (1995).

[709] The checklists for the families of Curculionoidea follow Morris (2003), with more recent amendments referred to in relevant endnotes.

[710] Following Thompson (1992) and others.

[711] See Alonso-Zaraga & Lyal (1999) for a discussion of this family and the nomenclature of its constituents.

[712] Subfamily Nemonychinae contains only *Nemonyx* (not British).

[713] *Rhinomacer*, and family-group names based thereon, have been effectively suppressed (ICZN Op. 1754).

[714] Given precedence over Choragidae (ICZN Op. 1756). Nomenclature and arrangement follow M. Trýzna & B.D. Valentine in Löbl & Smetana (2011).

[715] See Alonso-Zaraga & Lyal (1999).

[716] Generic assignment per Löbl & Smetana (2011).

[717] Placed as a subgenus of *Tropideres* in Pope (1977).

[718] Given family rank by Crowson (1984) and so treated by Morris (1990, 1995). Though currently regarded as a subfamily of Anthribidae, as here, this may be controversial.

[719] (Re)discovered in Britain by Hyman (1987b) [as *Bruchella rufipes*]; see also Morris (1990).

[720] British records of *L. sericeus* (Herbst) require confirmation; see Morris & Johnson (2005).

[721] See Morris (1990).

[722] Generic status follows Löbl & Smetana (2011).

[723] Not now regarded as a subfamily of Brentidae (cf. Gønget, 1997). Nomenclature and arrangement follow M.A. Alonso-Zarazaga in Löbl & Smetana (2011).

[724] Synonymized by Thompson & Alonso-Zarazaga (1988).

[725] Added by Fowles & Morris (1994).

[726] Added by P.J. Hodge in Booth (2009) and more fully in Hodge (2011).

[727] Added by Morris & Péricart (1988).

[728] Taxonomy of *Ceratapion* follows Alonso-Zarazaga (1991) and Wanat (1995); see also Morris (1993a).

[729] Synonymized by Wanat (1995); see also Morris & Booth (1997).

[730] Incorrect original spelling, established by Wanat (1995); see also Morris & Booth (1997).

[731] Junior primary homonym of *laevigatum* Fabricius, 1792. Species-group name unchanged from Pope

(1977), restored from Morris (1990); see also Morris (1993a).

[732] Added by Foster, Morris & Whitehead (2001).

[733] Added by Jones (2006).

[734] Taxonomy of group clarified by Dieckmann (1976).

[735] Wrongly placed in *Apion (Eutrichapion)* in Pope (1977).

[736] Added by Parry (1982b) [as *Apion intermedium*].

[737] Status as a good species doubtful, but to be investigated.

[738] Name rejected, ICZN Op. 1526.

[739] Included in Cossoninae by Pope (1977).

[740] Name rejected, ICZN Op. 1286.

[741] Name rejected, ICZN Op. 1287.

[742] Nomenclature and arrangement follow R. Caldara in Löbl & Smetana (2011).

[743] This synonymy (Alonso-Zarazaga & Lyal, 1999) was overlooked by Morris (2002), who incorrectly separated *aethiops* from the other species. It is *bimaculatus* which should be placed in a separate genus, following recent research which is continuing. For a radically different treatment of *Notaris* sensu lato see for example Abbazzi & Osella (1992).

[744] Junior primary homonym of *scirpi* Rossi, 1790.

[745] The taxonomy and nomenclature of the species in this genus were revised by Thompson (2006).

[746] Added by Booth (1993). The synonymy was established by Booth (2002b).

[747] Treated as a subfamily of Curculionidae (Gonatoceri) by many authors, including Osella (1977) and Lawrence & Newton (1995).

[748] Added as a casual introduction by Williams (1968), then recorded as an established introduction by Thompson (1995) and Owen (1995) [as *Raymondionymus marqueti*]. See also Owen (1997a).

[749] *A. britannus* Desbrochers is here treated as a non-established introduction.

[750] Often treated as a full genus.

[751] Added by Read (1981) [as *Furcipus rectirostris*].

[752] Added by R.G. Booth in Hodge (in prep.).

[753] British records of *E. scanicus* (Paykull) require confirmation.

[754] Subgenera, used by Alonso-Zarazaga & Lyal (1999) and widely in the Continental literature, are not employed here because both Dieckmann (1986) (the German fauna) and O'Brien (1970) (the Nearctic species) found them to be unsustainable. They have not been used to any extent in the British literature.

[755] Junior primary homonym of *affinis* (Schrank, 1781). Some Continental literature already uses *edoughensis* for this species.

[756] Synonymized by Dieckmann (1979); see also Morris & Booth (1997).

[757] The treatment of this tribe follows the phylogenetic analysis of Caldara (2001).

[758] Owen (1993a) is followed in believing that this taxon, at least in Britain, is to be referred to *graminis*. This was the opinion of the late Dr L. Dieckmann, who examined some British examples of putative *degorsi* (M.G. Morris *in litt.*); see also Owen (1988). It is uncertain whether a good species, *degorsi*, is part of the European fauna, but it seems doubtful.

[759] *G. lloydi* Donisthorpe is now treated as a synonym of *Sibinia pyrrhodactyla* (Marsham).

[760] Transferred from *Gymnetron* by Caldara (2001).

[761] Treatment of *Rhynchaenus* and *Orchestes* as different genera follows the detailed study by Kojima & Morimoto (1996).

[762] Added (reinstated?) by Parry (1981) [as *Rhynchaenus populi*].

[763] Added by Allen (1988a) [as *Rhynchaenus calceatus*]. See also Mendel (1994), Morris & Booth (1997), and Morris & Owen (1997).

[764] Synonymized by Morris & Booth (1997).

[765] See the conclusions of Thompson (1994).

[766] Added by Hammond (2000b).

[767] Junior primary homonym of *lonicerae* de Razoumovsky, 1789.

[768] Added by Tempère (1982) [as *Rhynchaenus pseudostigma*]. See also Morris (1987, 1993b).

[769] *S. pellucens* (Scopoli) has been omitted from the current list. Its history as a 'British' species is in Fowler (1891), but its status in Britain is thought to be dubious. See Morris & Johnson (2005) for details.

[770] Correctly '*primitus*', but treated as adjectival in form by later revisers and in recent European checklists; see Morris & Booth (1997).

[771] Synonymized by Caldara (1990a); see also Morris & Booth (1997).

[772] Synonymized by Morris (1989).

[773] Not regarded as a valid genus or subgenus by Caldara (1990b).

[774] Added by R.G. Booth in Hodge (1999).

[775] Rediscovered as a resident species by Morris (2004a, 2004b).

[776] Synonymized by Caldara (1983, 1990a); see also Allen (1984a) and Morris & Booth (1997).

[777] The subgenera of *Bagous* used here have no phylogenetic significance and are unsustainable on a world basis (Caldara & O'Brien, 1998; Alonso-Zarazaga & Lyal, 1999); they are retained here because of their familiarity and use in the identification of species (Dieckmann, 1964; Morris, 2002).

[778] Synonymized by Caldara & O'Brien (1998).

[779] Synonymized by Allen (1992b); see also Morris & Booth (1997).

[780] Synonymized by Caldara & O'Brien (1994); see also Morris & Booth (1997).

[781] Nomenclature and arrangement follow J. Prena in Löbl & Smetana (2011).

[782] *B. chlorizans* Germar is noted by Fowler (1891) but has not been confirmed as British and is omitted from the current list. See Morris & Johnson (2005) for details.

[783] Taxonomy and nomenclature of the European taxa clarified by Dieckmann (1991) and briefly discussed by Morris & Booth (1997).

[784] The checklist for this subfamily follows Morris (1991a) with the necessary changes for an alphabetical arrangement and incorporating important nomenclatural amendments by Colonnelli (1998, 2004).

[785] Subgenera of *Ceutorhynchus* are not used by Alonso-Zarazaga & Lyal (1999), probably do not express phylogenetic relationships, and are mostly unfamiliar to British workers. Consequently, they are not included here.

[786] Junior primary homonym of *assimilis* Fabricius, 1775.

[787] Synonymized by Colonnelli (1998). See also Morris & Booth (1997).

[788] Added to the British list by Owen (1990a), and to the Irish list by Morris (1992).

[789] Junior primary homonym of *contractus* Fourcroy, 1785.

[790] Synonymized by Colonnelli (1998).

[791] Treated as a full genus in Alonso-Zarazaga & Lyal (1999).

[792] Rediscovered as a resident species by Fowles & Morris (1999).

[793] Synonymized by Colonnelli (1990).

[794] Rejected alternative original spelling.

[795] First recognised in Britain by J. Parry. First published record appears in Atty (1983). See Owen (1994d).

[796] Retained over *commari* (Panzer, 1795) (Colonnelli, 1998); Alonso-Zarazaga & Lyal (2002) have shown that Colonnelli's interpretation is based on an error of dating.

[797] Added by Johnson (1982a) [as *Phytobius olssoni*].

[798] Added by Morris (1991b) [as *Phytobius zumpti*].

[799] Added by Hammond (2008).

[800] Added by Turner (2011) [as *Pentarthrum elumbe*].

[801] Added by Allen & Turner (2012).

[802] Added by Welch (1990a).

[803] Name conserved, ICZN Op. 1655; see also Morris & Booth (1997).

[804] British records of *P. gracilis* (Rosenhauer) require confirmation.

[805] Revised by Stüben *et al.* (2003).

[806] Added by Clemons (1983).

[807] Added by Denton (2005).

[808] Added by M.V.L. Barclay in Hodge (2007).

[809] Synonymized by Dieckmann (1980).

[810] The genus *Philopedon* must be treated as neuter; see Morris & Booth (1997).

[811] Replacement name; see Morris (1997).

[812] British records of *O. coecus* Germar, 1824 (= *niger* (Fabricius, 1775) non (Drury, 1773)) require confirmation.

[813] Name rejected, ICZN Op. 1999.

[814] Added by M.V.L. Barclay in Hodge (2000) and more fully by Barclay (2003a).

[815] Added by Hyman (1987a).

[816] The species was first detected by MAFF in the mid-1980s but was formally added by Morris (1997).

[817] Added by Harrison (2008).

[818] Re-identification due to Barclay (2009).

[819] Added by N.F. Heal; see Morris (1997).

[820] Added by S.G. Cole in Booth (2002a) and more fully by Barclay (2003a).

[821] Generally considered to be a good species distinct from *P. pyri* but not accepted as such by Pesarini (1981).

[822] Added by Plant, Morris & Heal (2006).

[823] Added by Denton (2005).

[824] *P. prasinus* (Olivier), included by Pope (1977), is omitted as being very doubtfuly British (Morris, 1997). See also Morris & Johnson (2005).

[825] cf. Morris (1997).

[826] Synonymized by Roudier (1963), who established that the true *Polydrusus chrysomela* is confined to Portugal; see also Morris (1996).

[827] *Eusomus ovulum* (Germar) is omitted; see Morris (1997) and Morris & Johnson (2005).

[828] Classification of the Sitonini follows Velázquez *et al.* (2007).

[829] Recently confirmed as an established and probably native species (P.M. Hammond, pers. comm.).

[830] Added by Cunningham (in prep.).

[831] Arrangement of this tribe follows Morris (2011).

[832] Detailed work by Piper *et al.* (2001) has failed to confirm this taxon as a good species.

[833] Shown to be distinct from *T. bifoveolatus* by Jermiin *et al.* (1991). Added to the British list by Harrison (1993).

[834] Synonymized by Borovec (1991); see also Morris & Booth (1997).

[835] Unjustified replacement name.

[836] The single British record of *L. (Lixochelus) filiformis* (Fabricius) [listed as *L. elongatus* (Goeze) in Pope (1977)] is of dubious provenance and the species has been omitted from the current list. See Mann, Hancock & Morris (2005) and Morris & Johnson (2005) for details.

[837] Added by Heal (1992).

[838] Tempère (1979) established that *L. troglodytes* is a good species and not a subspecies of *L. pyrenaeum*; see also Morris & Booth (1997).

[839] The genera in this group are not included in Alonso-Zarazaga & Lyal (1999), but it is regarded by them as a subfamily of Curculionidae. Nomenclature and arrangement follow M. Knížek in Löbl & Smetana (2011).

[840] Only two tribes of Scolytinae are recognised here (cf. Silfverberg, 1992). Tribes and subtribes follow Wood (1982), with adjustments for that author's view that Scolytinae have family (not subfamily) rank.

[841] Added by Atkins, O'Callaghan & Kirby (1981).

[842] Added by Heal (2003).

[843] Confirmed as a British breeding species by Lyszkowski (1993). See also Owen (1994a), who showed that some earlier records were inaccurate.

[844] Added by Winter (1990); see also Owen (1993a).

[845] Added by Heal (2006).

[846] Added by Telfer (2007a).

[847] Synonymized by Wood & Bright (1992).

[848] Synonymy follows Pope (1977) and Owen (1993a), but see Silfverberg (1992).

[849] Status of *Hylesinus orni* Fuchs, 1906 follows Löbl & Smetana (2011).

[850] Synonymized with *Leperisinus varius* in error by Pope (1977).

[851] Junior primary homonym of *ater* Rossi, 1792.

[852] Possibly a misidentification of *ater* Fabricius, 1792. Replacement name for *ater* Paykull, 1800 non Rossi, 1792.

[853] Added by Cooter (1982, 1983). See also King & Fielding (1989).

[854] Added as a casual introduction by Winter (1991) [as *P. bicolor*], then recorded as apparently established by Winter (1998).

[855] Synonymy follows Schedl (1981).

[856] Added by Chuter (2010).

[857] Genera in this group are not included in Alonso-
Zarazaga & Lyal (1999), who regard it as a family
separate from Curculionidae. The treatment in the
current list follows Pope (1977).

Species found as fossils in Quaternary sediments

P.I. Buckland & P.C. Buckland

For over 40 years the study of fossil insects has provided a significant contribution to our understanding of environmental and climate change. Beetles, by way of their frequency as fossils in Quaternary terrestrial and freshwater sediments, relative ease of identification as fossils, niche specificity and apparent morphological and physiological stability over the Quaternary and beyond[1], have been the most studied group. Their study has been particularly instrumental in elucidating climate change over the past 250,000 years and in defining the rapidity of climatic transitions during the last glaciation and into the present interglacial, the Holocene. Although a relatively small group, palaeoentomologists continue to provide an important contribution to Quaternary science and archaeology, as well as to the broader issues of biodiversity, landscape change and human impact. What follows is a list of those species which have been found in Quaternary sediments in the British Isles, but which have not been recorded as native in historic times. This includes both species that have Holocene records and are now extinct in Britain and species that have only pre-Holocene records, and thus no Red Data Book status. The status of a number of species is inevitably uncertain with some previously known only as fossils being discovered either as residents or introductions[2], and others having doubtfully extant breeding populations.

All data are extracted from the Bugs Coleopteran Ecology Package (BugsCEP), a database of Coleopteran ecology, distribution and fossil records, which is available for free download from http://www.bugscep.com. BugsCEP is currently being developed and extended by Phil Buckland (Umeå University, Sweden, e-mail phil.buckland @arke.umu.se) and Paul Buckland (Sheffield, UK, e-mail paul.buckland @bugscep.com), and provides a variety of tools to aid the entomologist and palaeoentomologist in teaching and research. These include, among other things, complex habitat queries, graphical ecology summaries, and palaeotemperature reconstruction. Please see the website for details, and also access to a full bibliography (Qbib) of relevant papers. A mapping interface, and the ability to cross-query with other proxy data such as plant macrofossils and geochemistry, is being developed as part of the Strategic Environmental Archaeology Database (SEAD, http://www.sead.se).

Key
Q = (prior to Marine Isotope Stage 5e / >~130kaBP).
I = last interglacial (=Ipswichian / Eemian / Marine Isotope Stage 5e / ~130-115kaBP)
G = last glacial (=Devensian / Weichselian / Marine Isotope Stages 5d-2 / ~115-15kaBP)
LG = late glacial (= Pollen Zones I-III / ~13.5-10kaBP)
H = Holocene
* = species with historical records but probably only as non-established introductions

Note: records described as **G** or **LG** do not necessarily equate with cold (stadial) phases and may indicate conditions at least as warm as the present day, but in an open (=interstadial) landscape.

When publishing data from this list, or the BugsCEP computer package, please use the full citation given for Buckland & Buckland (2006) in "Bibliography" on page 136. Further information on using BugsCEP and examples can be found in Buckland (2007) and on the project website.

---

[1] Most recently on this see Fikáček, Prokin & Angus (2011).
[2] See for example Mendel, Jeffrey & Pledger (2011).

Family GYRINIDAE Latreille, 1810
Gyrinus colymbus Erichson **H**

Family DYTISCIDAE Leach, 1815
Agabus serricornis (Paykull) **G**
Bidessus grossepunctatus Vorbringer **H**
Colymbetes dolabratus (Paykull) **Q;G;LG**
Colymbetes paykulli Erichson **G;LG**
Colymbetes striatus (Linnaeus) **LG;H**
Hydroporus lapponum (Gyllenhal) **LG**
Hydroporus nigellus Mannerheim **LG**
Hydroporus notabilis LeConte **LG**
Hygrotus (Coelambus) unguicularis Crotch **LG**
Ilybius angustior (Gyllenhal) **LG**
Ilybius vittiger (Gyllenhal) **G**

Family RHYSODIDAE Laporte, 1840
Rhysodes sulcatus (Fabricius) **Q;I;H**

Family CARABIDAE Latreille, 1802
*Abax parallelus (Duftschmid) **Q**
Agonum consimile (Gyllenhal) **G;LG**
Agonum exaratum (Mannerheim) **Q;G**
Amara erratica (Duftschmid) **G**
Amara interstitialis Dejean **G**
Amara municipalis (Duftschmid) **Q;G**
Amara torrida (Panzer) **G;LG**
Asaphidion cyanicorne (Pandellé) **LG**
*Bembidion callosum Küster **LG**
Bembidion dauricum (Motschulsky) **G;LG**
Bembidion difficile (Motschulsky) **Q;G;LG**
Bembidion elongatum Dejean **I;Q**
Bembidion fellmanni (Mannerheim) **Q;G;LG**
Bembidion grisvardi Dewailly / *ibericum* Piochard de la Brulerie **LG**
Bembidion hastii Sahlberg, C.R. **Q;G;LG**
Bembidion hyperboraeum Munster **G;LG**
Bembidion lapponicum Zetterstedt **G;LG**
Bembidion mckinleyi Fall **G;LG**
Bembidion obscurellum (Motschulsky) **G;LG**
Bembidion petrosum Gebler **G;LG**
Bembidion transparens (Gebler) **Q;G;LG**
Bembidion velox (Linnaeus) **Q;LG**
*Carabus (Autocarabus) cancellatus Illiger **G**
Carabus (Carabus) menetriesi Hummel **G**
Carabus (Homoeocarabus) maeander Fischer von Waldheim **G**
Carabus (Oreocarabus) hortensis Linnaeus **G**
*Carabus (Tomocarabus) convexus Fabricius **LG**
Chlaenius costulatus Motschulsky **Q;G;LG**
Chlaenius festivus (Panzer) **I**
Chlaenius quadrisulcatus (Paykull) **I**
Chlaenius sulcicollis (Paykull) **Q;I;H**
Chlaenius variegatus (Geoffroy in Fourcroy) **I**
Cymindis angularis Gyllenhal **G;LG**
Cymindis humeralis (Geoffroy in Fourcroy) **LG**
Diacheila arctica (Gyllenhal) **G;LG**
Diacheila polita (Faldermann) **Q;G;LG**
Dyschirius nigricornis Motschulsky
   = septentrionis Munster **Q;G;LG**
Oodes gracilis Villa & Villa **Q;I;H**
Pterostichus blandulus Miller **G;LG**
Pterostichus brevicornis (Kirby) **Q**
Pterostichus kokeilii Miller **G**
Pterostichus magus Mannerheim **LG**

Pterostichus middendorffi (Sahlberg, J.R.) **Q**
Syntomus parallelus (Ball) **LG**
Trechus amplicollis Fairmaire **Q**

Family HYDROCHIDAE Thomson, C.G., 1859
Hydrochus flavipennis Küster **G**

Family HELOPHORIDAE Leach, 1815
Helophorus aquaticus (Linnaeus) *s. str.* **G**
Helophorus arcticus Brown **Q**
Helophorus aspericollis Angus **G**
Helophorus discrepans Rey **G**
Helophorus glacialis Villa **G;LG**
Helophorus lapponicus Thomson, C.G. **G;LG**
Helophorus mongoliensis Angus **G**
Helophorus oblongus LeConte **Q;LG**
Helophorus obscurellus Poppius **Q;G;LG**
Helophorus orientalis Motschulsky **G**
Helophorus pallidus Gebler **G**
Helophorus praenanus (Łomnicki) **G;LG**
Helophorus sibiricus (Motschulsky) **G;LG**
Helophorus splendidus Sahlberg, J. **G;LG**

Family HYDROPHILIDAE Latreille, 1802
Hydrobius arcticus Kuwert **G**
Laccobius decorus (Gyllenhal) **G**

Family HISTERIDAE Gyllenhal, 1808
Chetabraeus globulus (Creutzer) **Q;G**
Hister funestus Erichson **G**
Margarinotus terricola Germar **LG**

Family HYDRAENIDAE Mulsant, 1844
Hydraena coopei Angus **Q**
Hydraena cf. *latebricola* Jäch **G**
Ochthebius foveolatus Germar **LG**
Ochthebius kaninensis Poppius **G**
Ochthebius pedicularius Kuwert **LG**

Family SILPHIDAE Latreille, 1806
Ptomascopus zhangla Háva, Schneider & Růžička **G**
Thanatophilus lapponicus (Herbst) **LG**

Family STAPHYLINIDAE Latreille, 1802
Acidota quadrata (Zetterstedt) **Q;G;LG;H**
Anotylus gibbulus (Eppelsheim) **Q;G**
Anotylus mendus Herman **Q**
Anotylus tetratoma Czwalina **Q**
Anthophagus bicornis (Block) **LG**
Aploderus caesus (Erichson) **Q;G**
Arpedium quadrum (Gravenhorst) **Q;LG;H**
Batrisus formicarius Aubé **H**
Bledius cribricollis Heer, 1839
Bledius littoralis Heer **Q;G**
Boreaphilus henningianus Sahlberg, C.R. **Q;G;LG**
Eucnecosum cf. *tenue* LeConte **LG**
Eudectus giraudi Redtenbacher **LG**
Geodromicus kunzei Heer **Q;LG**
Holoboreaphilus nordenskioldi (Mäklin) **Q;G;LG**
Leptobium gracile (Gravenhorst) **Q**
Micropeplus hoogendorni Matthews
   =dokuchaevi (Ryabukhin) **Q**
Ocypus picipennis (Fabricius) **LG**
Olophrum boreale (Paykull) **Q;G;LG**

Olophrum rotundicolle (Sahlberg, C.R.) **G;LG**
Philonthus linkei Solsky **G**
Pycnoglypta lurida (Gyllenhal) **Q;G;LG**
Stenus plicipennis (Casey) **LG**
Tachinus arcticus (Motschulsky) **G**
Tachinus caelatus Ullrich **Q;G**
Tachinus jacuticus Poppius **Q;G**
Tachinus marginatus Gyllenhal **G**

Family SCARABAEIDAE Latreille, 1802
Aphodius (Acrossus) carpetanus Gräells **Q;I**
Aphodius (Alocoderus) holdereri Reitter **G**
Aphodius (Amidorus) obscurus (Fabricius) **LG**
Aphodius (Calaphodius) bonvouloiri Harold **G**
Aphodius (Chilothorax) jacobsoni Koshantschikov **G**
Aphodius (Esymus) quadriguttatus (Herbst) **H**
Aphodius (Neagolius) montivagus Erichson **LG**
Aphodius (Nialus) varians Duftschmid **H**
Aphodius (Phaeaphodius) costalis Gebler **G**
Aphodius (Teuchestes) brachysomus Solsky **G**
Ataenius horticola Harold **Q**
Caccobius histeroides (Ménétries) **I**
Caccobius schreberi (Linnaeus) **Q;I;H**
Drepanocerus sp. **I**
Euoniticellus fulvus (Goeze) **I**
Onthophagus furcatus (Fabricius) **I**
Onthophagus gibbulus (Pallas) **I;G**
Onthophagus massai Baraud **I**
Onthophagus opacicollis Reitter **I**
Osmoderma eremita (Scopoli) **Q**
Rhyssemus algiricus Lucas **Q**
Triodonta sp. **Q**
Valgus hemipterus (Linnaeus) **Q;I;H**

Family BUPRESTIDAE Leach, 1815
Buprestis rustica Linnaeus **H**

Family BYRRHIDAE Latreille, 1804
Curimopsis cyclolepidia (Munster) **G;LG**
Simplocaria metallica (Sturm) **Q;G;LG**

Family ELMIDAE Curtis, 1830
Dupophilus brevis Mulsant & Rey **Q**
Esolus pygmaeus (Müller, P.W.J.) **Q;H**
Limnius opacus Müller, P.W.J. **Q**
Stenelmis consobrina Dufour **I;H**

Family LIMNICHIDAE Erichson, 1846
Pelochares versicolor (Waltl) **Q;I**

Family HETEROCERIDAE MacLeay, 1825
Heterocerus intermedius Kiesenwetter **G;LG**

Family EUCNEMIDAE Eschscholtz, 1829
Dromaeolus barnabita (Villa) **H**

Family ELATERIDAE Leach, 1815
Hypnoidus rivularius (Gyllenhal) **G;LG**
Idolus picipennis (Bach) **Q**
Porthmidius austriacus (Schrank) **H**

Family DERMESTIDAE Latreille, 1804
Dermestes laniarius Illiger **I;H**

Family PTINIDAE Latreille, 1802
Stagetus borealis Israelson **H**

Family TROGOSSITIDAE Latreille, 1802
Peltis grossa (Linnaeus) **Q;H**
Temnochila caerulea (Olivier) **H**
Tenebroides fuscus (Goeze) **H**

Family CLERIDAE Latreille, 1802
Opetiopalpus scutellaris (Panzer) **Q;I;G**

Family KATERETIDAE Kirby, 1837
Heterhelus scutellaris (Heer) **Q**

Family SILVANIDAE Kirby, 1837
Airaphilus elongatus (Gyllenhal) **Q;LG;H**

Family LAEMOPHLOEIDAE Ganglbauer, 1899
Cryptolestes corticinus (Erichson) **H**
Notolaemus castaneus (Erichson) **H**

Family EROTYLIDAE Latreille, 1802
Leucohimatium sp. **H**

Family BOTHRIDERIDAE Erichson, 1845
Bothrideres bipunctatus (Gmelin) **I;H**

Family ENDOMYCHIDAE Leach, 1815
Mycetina cruciata (Schaller) **H**

Family COCCINELLIDAE Latreille, 1807
Anisosticta strigata (Thunberg) **G**
Ceratomegilla ulkei Crotch **G;LG**
Hippodamia arctica (Schneider, D.H.) **Q;G;LG**
Hippodamia baratovskii Semenov & Dobzhansky **G**
Hippodamia septemmaculata (De Geer) **Q**
Nephus bipunctatus (Kugelann) **G;LG**

Family CIIDAE Leach, 1819
Ropalodontus baudueri Abeille de Perrin **H**

Family COLYDIIDAE Billberg, 1820
Pycnomerus terebrans (Olivier) **Q;I;H**

Family TENEBRIONIDAE Latreille, 1802
Diaclina fagi (Panzer) **Q**

Family PROSTOMIDAE Thomson, C.G., 1859
Prostomis mandibularis (Fabricius) **H**

Family ANTHICIDAE Latreille, 1819
Anthicus ater (Panzer) **G;LG**
Cordicomus gracilis (Panzer) **H**
Cordicomus sellatus (Panzer) **I**

Family CHRYSOMELIDAE Latreille, 1802
Chaetocnema obesa (Boieldieu) **G**
Chrysolina limbata (Fabricius) **Q;G**
Chrysomela collaris Linnaeus **G;LG**
Chrysomela septentrionalis (Ménétries) **G;LG**
Donacia polita Kunze **Q**
Entomoscelis adonidis (Pallas) **G;LG**

Neocrepidodera interpunctata (Motschulsky) **G**
Pachnephorus tesselatus (Duftschmid) **Q**

Family RHYNCHITIDAE Gistel, 1848
Rhynchites pubescens (Fabricius) **G**

Family CURCULIONIDAE Latreille, 1802
Aphanommata filum (Mulsant & Rey) **I**
Aulacobaris violaceomicans (Solari) **LG**
Cathormiocerus curviscapus Seidlitz **G**
Cathormiocerus validiscapus Rouget **Q;G;H**
Cyphocleonus trisulcatus (Herbst) **H**
Hypera obovata (Csiki) **G;LG**
Isochnus flagellum (Ericson) **LG**
Otiorhynchus fuscipes (Olivier) **G;LG**
Otiorhynchus mandibularis Redtenbacher **G**

Otiorhynchus politus Gyllenhal **G**
Otiorhynchus proximus Stierlin **I**
Phthorophloeus spinulosus Rey **Q**
Pissodes gyllenhali (Sahlberg, C.R.) **H**
Pseudocleonus cinereus (Schrank) **G**
Rhyncolus elongatus (Gyllenhal) **Q;G;H**
Rhyncolus punctulatus Boheman **Q;H**
Rhyncolus sculpturatus Waltl **I;H**
Rhyncolus strangulatus Perris **Q;H**
Scolytus carpini (Ratzeburg) **I**
Scolytus koenigi Schevyrev **Q;I**
Stenoscelis submuricatus (Schönherr) **Q;I**
Stomodes gyrosicollis (Boheman) **Q**
Tachyerges rufitarsis (Germar) **Q**

Family PLATYPODIDAE Shuckard, 1839
Platypus oxyurus Dufour **Q**

Non-established introductions

A. G. Duff

All published records of the following species appear to have been of non-breeding introductions, or at least no definite evidence has yet been forthcoming that the specimens found are natural vagrants or represent established populations, even if they have occurred apparently 'in the wild'. Non-established introductions are not considered to be an integral part of our fauna and so have been omitted from the main checklist. This checklist has mainly been compiled from a limited range of sources (Aitken, 1975; Coquempot, 2006; Hincks, 1953; Hinton, 1945; Hodge & Jones, 1995; Ostojá-Starzewski & Hammon, 2008; Winter, 1991) and is likely to be incomplete. Note that the taxonomy and nomenclature of some species may have changed since the original record(s).

Family DYTISCIDAE Leach, 1815
Megadytes costalis (Fabricius)

Family CARABIDAE Latreille, 1802
Abax parallelus (Duftschmid)
Anchista binotata (Dejean)
Bembidion (Nepha) callosum Küster
Carabus (Autocarabus) cancellatus Illiger
Carabus (Tomocarabus) convexus Fabricius
Casnoidea indica (Thunberg)
Celaenephes parallelus Schmidt-Göbel
Coleolissus iris Andrewes
Colliuris suturalis Péringuey
Cymindis setifensis Lucas
Elaphropus quadrisignatus (Duftschmid)
Mochtherus tetraspilotus (MacLeay)
Plochionus pallens (Fabricius)
Porotachys bisulcatus (Nicolai)
Stenolophus comma (Fabricius)
   = *plagiatus* Gorham
Syntomus fuscomaculatus (Motschulsky)
Tachyta guineensis Alluaud

Family HYDROPHILIDAE Latreille, 1802
Cercyon variegatus Sharp

Family HISTERIDAE Gyllenhal, 1808
Carcinops troglodytes (Paykull)
Macrolister major (Linnaeus)
Platysoma elongatum (Thunberg)
   = *oblongum* (Fabricius)
Saprinus chalcites (Illiger)
Teretrius punctulatus Fåhraeus

Family LEIODIDAE Fleming, 1821
Aglyptinus agathidioides Blair

Family STAPHYLINIDAE Latreille, 1802
Coenonica puncticollis (Kraatz)
Oligota chrysopyga Kraatz
Oligota pseudoparva Williams
Reichenbachia coronatus Westwood

Family TROGIDAE MacLeay, 1819
Trox suberosus (Fabricius)

Family SCARABAEIDAE Latreille, 1802
Anisoplia agricola (Poda)
Aphodius (Nialus) varians Duftschmid
Aphodius (Subrinus) sturmi Harold
Aphodius (Trichonotulus) scrofa (Fabricius)
Archophileurus santafeanus Arrow
Bothynus dasypleurus (Germar)
Catharsius molossus (Linnaeus)
Catharsius satyrus Kolbe
Heteronychus licas (Klug)
Megasoma gygas Herbst
Onitis philemon Fabricius
Onthophagus flavocinctus Klug
Onthophagus furcatus (Fabricius)
Oxythyrea funesta (Poda)
Phyllotocus rufipennis (Boisduval)
Smaragdesthes africana (Drury)
Strategus simson (Linnaeus)
Trochalus fraterculus Kolbe
Valgus hemipterus (Linnaeus)

Family BUPRESTIDAE Leach, 1815
Anthaxia dimidiata Thunberg
Buprestis aurulenta Linnaeus
Buprestis novemmaculata Linnaeus
Buprestis rusticorum Kirby
Buprestis splendens Fabricius
Chrysobothris chrysostigma (Linnaeus)
Chrysobothris femorata (Olivier)
Eurythrea austrica Linnaeus
Stenocera sternicornis (Linnaeus)

Family ELATERIDAE Leach, 1815
Drasterius bimaculatus (Rossi)
Heteroderes rufangulus (Gyllenhal)

Family DERMESTIDAE Latreille, 1804
Dermestes laniarius Illiger
Phradonoma cercyonoides (Reitter)
Phradonoma nobile (Reitter)
Thorictodes dartevellei John
Thorictus grandicollis Germar
Thorictus indicus Grouvelle
Trogoderma anthrenoides (Sharp)
Trogoderma sternale Jayne

Family BOSTRICHIDAE Latreille, 1802
Amphicerus cornutus (Pallas)
Apate monachus (Fabricius)
Bostrychoplites cornutus (Olivier)
Bostrychus jesuita Fabricius
Dinoderus bifoveolatus (Wollaston)
Dinoderus brevis Horn
Dinoderus distinctus Lesne
Dinoderus minutus (Fabricius)
Dinoderus ocellaris Stephens
Heterobostrychus aequalis (Waterhouse)
Heterobostrychus brunneus (Murray)
Heterobostrychus hamatipennis (Lesne)
Lyctoxylon dentata Pascoe
Lyctoxylon japonicum Reitter
Lyctus africanus Lesne
Lyctus cavicollis LeConte
Lyctus discedens Blackburn
Lyctus planicollis LeConte
Lyctus simplex Reitter
Lyctus sinensis Lesne
Micrapate xyloperthoides (Jacquelin du Val)
Minthea rugicollis Walker
Octodesmus parvulus Lesne
Phonopate stridula Lesne
Scobicia declivis LeConte
Scobicia pustulata (Fabricius)
Sinoxylon anale Lesne
Sinoxylon conigerum Gerstaecker
Sinoxylon crassum Lesne
Sinoxylon doliolum Lesne
Sinoxylon pugnax Lesne
Sinoxylon ruficorne Fåhraeus
Tetrapriocera longicornis (Olivier)
Tristaria grouvellei Reitter
Trogoxylon impressum (Comolli)
Trogoxylon parallelopipedum (Melsheimer)
Xylion adusta Fåhraeus
Xylion collaris (Erichson)
Xylion cylindricus (MacLeay)
Xylion securifer Lesne
Xylobiops basilaris (Say)
Xylodeleis obsipa (Germar)
Xyloperthella crinitarsis (Imhoff)
Xyloperthella picea (Olivier)
Xylothrips flavipes (Illiger)

Family PTINIDAE Latreille, 1802
Mezium americanum Laporte
Nicobium castaneum (Olivier)
Ptinus incisicollis Pic
Ptinus japonicus Reitter
Stegatus palliatus (Chevrolat)
Tricorynus herbarium (Gorham)
Trypopitys carpini Herbst
Xyletinus ater (Creutzer)

Family TROGOSSITIDAE Latreille, 1802
Tenebroides oblongus Sharp

Family CLERIDAE Latreille, 1802
Cylidroctenus chalybaeum (Westwood)
Cylidrus fasciatus Laporte
Cymatodera cylindricollis Chevrolat
Exkorynetes analis (Klug)
Opilo domesticus Sturm
Tillus notatus Klug

Family MELYRIDAE Leach, 1815
Melyris oblonga Fabricius

Family MALACHIIDAE Fleming, 1821
Anthocomus bipunctatus Harrer
Ebaeus bicaudatus Thomson

Family NITIDULIDAE Latreille, 1802
Brachypeplus depressus Erichson
Brachypeplus niger Murray
Brachypeplus pilosellus (Murray)
Brachypeplus rubidus Murray
Carpophilus fumatus Boheman
Carpophilus lugubris Murray
Colopterus morio Erichson
Colopterus truncatus Randall
Glischrochilus fasciatus (Olivier)
Glischrochilus vittatus (Say)
Lasiodactylus tibialis Boheman
Stelidota ferruginea Reitter
Stelidota strigosa (Gyllenhal)

Family SILVANIDAE Kirby, 1837
Cathartosilvanus vulgaris (Grouvelle)
Monanus concinnulus (Walker)
Oryzaephilus acuminatus Halstead
Oryzaephilus gibbosus Aitken
Parasilvanus fairmairei (Grouvelle)
Parasilvanus ocellatus (Grouvelle)
Protosilvanus lateritius (Reitter)
Silvanoprus cephalotes (Reitter)
Silvanoprus scuticollis (Walker)
Silvanus castaneus MacLeay
Silvanus difficilis Halstead
Silvanus inarmatus Wollaston
Silvanus lateritius (Broun)
Silvanus lewisi Reitter
Silvanus productus Halstead
Silvanus proximus Grouvelle

Family PASSANDRIDAE Blanchard, 1845
Laemotmetus rhizophagoides (Walker)

Family LAEMOPHLOEIDAE Ganglbauer, 1899
Placonotus politissimus (Wollaston)

Family CRYPTOPHAGIDAE Kirby, 1826
Micrambe punctata Grouvelle
 = *aubrooki* Donisthorpe
Salebius tarsalis Casey

Family EROTYLIDAE Latreille, 1802
Leucohimatium arundinaceum (Forskål)
Pharaxonotha kirschi Reitter

Family ENDOMYCHIDAE Leach, 1815
Trochoideus desjardinsi Guérin-Méneville

Family COCCINELLIDAE Latreille, 1807
Calvia decemguttata (Linnaeus)
Chilocorus distigma Klug
?*Cynegetis impunctata* (Linnaeus) [3]
Eriopis connexa Germar

---

[3]   Added by Roy *et al.* (2011) but the identification has
    since been challenged.

Exochomus nigromaculatus (Goeze)
Hippodamia convergens (Guérin-Méneville)
Hyperaspis festiva Mulsant
Procula douei Mulsant
Scymnus impexus Mulsant

Family CORYLOPHIDAE LeConte, 1852
Arthrolips croceus Matthews

Family LATRIDIIDAE Erichson, 1842
Cartodere sp. cf. *constricta* (Gyllenhal)
Dienerella costulata (Reitter)
Dienerella lurida Rücker

Family CIIDAE Leach, 1819
Cis sp. cf. *delagoensis* Reitter

Family MELANDRYIDAE Leach, 1815
Serropalpus barbatus (Schaller)

Family COLYDIIDAE Billberg, 1820
Aprostoma planifrons Westwood
Bitoma siccana (Pascoe)
Colobicus parilis Pascoe
Colobicus specialis Grouvelle
Microprius confusus Grouvelle
Microprius raffragi Grouvelle
Tyrtaeus dobsoni Hinton

Family TENEBRIONIDAE Latreille, 1802
Ammophorus peruvianus Guérin-Méneville
Ammophorus rubripes Solier
Anemia sardoa Géné
Anthrenopsis scriptipennis (Fairmaire)
Blaps binominata Escalera
Blapstinus punctulatus Solier
Coelopalorus carinatus (Blair)
Coelopalorus foveicollis (Blair)
Cossyphus depressus (Fabricius)
Cynaeus angustus (LeConte)
Gonocephalum bilineatum (Walker)
Gonocephalum prolixum (Erichson)
Gonocephalum rusticum (Olivier)
Gonocephalum simplex (Fabricius)
Helenophorus collaris Linnaeus
Himatismus villosus (Haag-Rutenberg)
Leiochrodes varicolor Westwood
Lyphia depressa Hinton
Lyphia orientalis Blair
Lyprops curticollis Fairmaire
Meneristes australis (Blessinger)
Mesomorphus villiger (Blanchard)
Mesostena puncticollis Solier
Mesostenopa arabica Gestro
Mesostenopa rathjensi Gebien
Morica planata (Fabricius)
Ocnera hispida (Forskål)
Opatroides judaicus Baudi
Opatroides punctulatus Brullé
Opatrum verrucosum Germar
Pachychila salzmanni Solier
Palorinus humeralis (Gebien)
Palorus beesoni Blair
Palorus cerylonoides (Pascoe)
Palorus ficicola (Wollaston)
Palorus genalis Blair
Pimelia rugosa (Fabricius)

Platydema convexifrons Chevrolat
Polycoelogastridion tenuipes Kaszab
Rhytinota laticollis Schaufuss
Rhytinota praelonga Reiche
Scotobius crispatus Germar
Sitophagus hololeptoides (Laporte)
Tenebrio guineensis Imhoff
Tentyria cypria Kraatz
Tribolium anaphe Hinton
Tribolium audax Halstead
Tribolium madens (Charpentier)

Family MELOIDAE Gyllenhal, 1810
Meloe decorus (Brandt & Erichson)

Family ANTHICIDAE Latreille, 1819
Anthicus crinitus Laferté-Sénectère
Anthicus hesperi King
Anthicus troilus Hinton
Floydwernerius australis (King)
Formicomus pedestris Rossi
Notoxus monodon (Fabricius)
Notoxus numidicus (Lucas)

Family CERAMBYCIDAE Latreille, 1802
Acanthocinus griseus (Fabricius)
Ambona electus Gahan
Ancylonotus tribulus (Fabricius)
Anoplophora chinensis (Forster)
Anoplophora glabripennis (Motschulsky)
Aphoplistis pilosellus (Chevrolat)
Apriona germari (Hope)
Arhopalus productus LeConte
Asemum moestum Haldeman
Batocera rufomaculata (De Geer)
Chlorophorus annularis (Fabricius)
Chlorophorus annulatus Hope
Chlorophorus figuratus (Scopoli)
Chlorophorus pilosus (Forster)
Chlorophorus varius (Müller, O.F.)
Chlorophorus viticis Gressitt & Rondonare
Clytus rhamni Germar
Compsocerus euqestris (Guérin-Méneville)
Coptocerus rubripes Boisduval
Coptops aedificator (Fabricius)
Cordylomera spinicornis (Fabricius)
Cordylomera suturalis Chevrolat
Cyllene caryae Gahan
Cyrtophorus verrucosus (Olivier)
Deilus fugax (Olivier)
Diaxenes dendrobii Gahan
Eburia quadrigeminata (Say)
Elaphidion nanum (Fabricius)
Enaphalodes rufulus (Haldeman)
Ergates faber (Linnaeus)
Ergates spiculatus LeConte
Euderces pini (Olivier)
Geloneatha hirta (Fairmaire)
Hesperophanes gayi Plavilstshikov
Leptura obliterata (Haldeman)
Lagocheirus undulatus (Voet)
Mallodon downesi (Hope)
Megacyllene guttata (Chevrolat)
Monochamus galloprovincialis (Olivier) ssp. *pistor*
(Germar)
Monochamus rosenmuelleri (Cederhjelm)
Monochamus sartor (Fabricius)
Monochamus scutellatus (Say)

Monochamus sutor (Linnaeus)
Monochamus titillator (Fabricius)
Morimus asper (Sulzer)
Neoclytus acuminatus (Fabricius)
Neoclytus caprea (Say)
Neoclytus scutellaris (Olivier)
Oemona hirta (Fabricius)
Oxypleurus nodieri Mulsant
Pachydissus hector Kolbe
Paramoeocerus barbicornis (Fabricius)
Phoracantha recurva Newman
Phoracantha semipunctata (Fabricius)
Phryneta leprosa (Fabricius)
Plagionotus detritus (Linnaeus)
Plocaederus basalis Gahan
Plocaederus viridipennis (Hope)
Rhagiomorpha lepturoides (Boisduval)
Romaleum rufulum Haldeman
Rosalia alpina (Linnaeus)
Smodicum cucujiforme (Say)
Sophronica cinerascens Breuning
Stenygrinum quadrinotatum Bates
Stromatium barbatum (Fabricius)
Tetropium cinnamopterum Kirby, W.
Tetropium fuscum (Fabricius)
Trichoferus campestris (Faldermann)
Trichoferus griseus (Fabricius)
Xylotrechus colonus (Fabricius)

Family CHRYSOMELIDAE Latreille, 1802
Altica carduorum (Guérin-Méneville)
Bruchidius albopubens (Pic)
Bruchidius atrolineatus (Pic)
Bruchidius lividimanus (Gyllenhal)
Bruchidius pauper (Boheman)
Bruchidius quinqueguttatus (Olivier)
Bruchus dentipes (Baudi)
Bruchus emarginatus Allard
Bruchus lentis Frölich
Bruchus natalensis (Pic)
Bruchus signaticornis Gyllenhal
Bruchus tristis Boheman
Callosobruchus analis (Fabricius)
Callosobruchus phaseoli (Gyllenhal)
Callosobruchus rhodesianus (Pic)
Calopepla leayana Latreille
Caryedon longipennis Pic
Caryedon serratus (Olivier)
Caryedon sudanensis Southgate
Caryopemon cruciger Stephens
Cerotoma trifurcata (Forster)
Charidotis circulus Boheman
Chelymorpha variabilis Boheman
Chrysolina viridana (Küster)
Chrysomela cuprea Baly
Chrysomela superba Thunberg
Coptocycla laqueata Spaeth
Crioceris duodecimpunctata (Linnaeus)
Deloyala cruciata (Linnaeus)
Euspermophagus sericeus (Geoffroy in Fourcroy)
Hilarocassis albida (Germar)
Hispa atra Linnaeus
Leptinotarsa decemlineata (Say)
Mesomphalia sextillata Boheman
Pachymerus cardo (Fåhraeus)
Pachymerus nigriventris Pic
Pachynephorus cylindricus Lucas
Plagiometriona tenella (Klug)

Poecilaspis angulata (Germar)
Pseudomesophila indigacea (Boheman)
Smaragdina salicina (Scopoli)
Specularius erythraeus (Pic)
Specularius impressithorax (Pic)
Spermophagus elizabethae Pic
Spermophagus natalensis Fåhraeus
Stolas cyanea (Linnaeus)
Stolas impluviata (Boheman)
Stolas lacordairei var. *subrugosa* (Boheman)
Stolas redtenbacheri imperialis (Boheman)
Zabrotes subfasciatus (Boheman)

Family ANTHRIBIDAE Billberg, 1820
Anaemerus fuscus (Olivier)
Trigonorhinus secundus (Wolfrum)

Family APIONIDAE Schönherr, 1823
Piezotrachelus sp. cf. *gerstaeckeri* (Faust)

Family DRYOPHTHORIDAE Schönherr, 1825
Cosmopolites sordidus (Germar)
Diocalandra impressicollis (Quedenfeldt)
Sipalinus gigas (Fabricius)
Sitophilus linearis (Herbst)
Trochorhopalus strangulatus (Gyllenhal)

Family CURCULIONIDAE Latreille, 1802
Achopera alternata Lea
Anthonomus britannus Desbrochers
 = *pubescens* sensu auctt. non (Fabricius)
Antliarhinus verdcourti Marshall, G.
Balanogastris kolae (Desbrochers)
Baris janthina (Boheman)
Blosyrus asellus (Olivier)
Brachycerus albidentatus Gyllenhal
Brachytemnus porcatus (Germar)
Caulophilus oryzae (Gyllenhal)
Coccotrypes congonus Eggers
Coccotrypes dactyliperda (Fabricius)
Conarthrus cylindricus Wollaston
Conorhynchus brevirostris (Gyllenhal)
Conotrachelus histrio Boheman
Cryphaloides donisthorpei Formanek
Cryphalus buscki Hopkins
Cryphalus pallidus Eichhoff
Cryphalus piceae (Ratzeburg)
Crypturgus borealis Swaine
Crypturgus cinereus (Herbst)
Crypturgus pusillus (Gyllenhal)
Cycloderes espanoli Roudier
Dendroctonus brevicornis LeConte
Dendroctonus pseudotsugae Hopkins
Dendroctonus rufipennis (Kirby)
Dryocoetes affaber (Mannerheim)
Episus cyathiformis Gyllenhal
Gnathotrichus sulcatus (LeConte)
Hylastes ruber Swaine
Hylurgus ligniperda (Fabricius)
Hylurgops porosus (LeConte)
Hylurgops rugipennis (Mannerheim)
Hyperodes bonariensis Kuschel
Hypothenemus eruditus Westwood
Hypothenemus hampei (Ferrari)
Hypothenemus moschatae (Schauffer)
Ips amitinus (Eichhoff)
Ips curvidens Germar

Ips duplicatus (Sahlberg)
Ips latidens (LeConte)
Ips pini (Say)
Larinus turbinatus Gyllenhal
Leurostenus filum Marshall
Listroderes aequivoca Kuschel
Listroderes costirostris ssp. *obliquus* (Klug)
Lixus ochraceus Boheman
Lixus terminalis LeConte
Mesites pallidipennis Boheman
Orchidophilus aterrimus (Waterhouse)
Orthorhinus cylindrirostris (Fabricius)
Orthotomicus caelatus (Eichhoff)
Orthotomicus proximus (Eichhoff)
Otiorhynchus corruptor (Host)
Otiorhynchus parvicollis Gyllenhal
Phaenomerus strigicollis Faust
Phlyctinus callosus Boheman
Pityogenes plagiatus (LeConte)
Poecilips myristicae (Roepke)
Polydrusus (Metallites) impar des Gozis
Polygraphus rufipennis (Kirby)
Polygraphus subopacus (Thomson)
Pseudohylesinus tsugae Swaine
Pseudostenotrupis marshalli Zimmermann
Pseudostoreus placitus Lea
  = *harrisoni* (Pool)
Rhyncolus elongatus (Gyllenhal)
Rhytideres plicatus (Olivier)

Scierus annectens LeConte
Sclerorhinus bubalus (Olivier)
Scolytus dimidiatus Chapuis
Scolytus laricis Blackman
Scolytus unispinosus LeConte
Sipalinus gigas (Fabricius)
Sipalinus squalidus Kolbe
Stephanoderes cassiae Eichhoff
Stephanoderes coffeae Hagedorn
Sternochetus mangiferae (Fabricius)
Trypodendron retusum (LeConte)
Trypophloeus granulatus (Ratzeburg)
Tychius pumilus Brisout
Xyleborus eurygraphus (Ratzeburg)
Xyleborus ferrugineus (Fabricius)
Xyleborus mascarensis Eichhoff
Xylosandrus morigerus (Blandford)
Zeugenia rosacea Heller

Family PLATYPODIDAE Shuckard, 1839
Doliopygus bidentatus ssp. *kenyaensis* Schedl
Doliopygus crinitus (Chapuis)
Doliopygus serratus (Strohmeyer)
Platypus curtus Chapuis
Platypus hintzi Schaufuss
Platypus linearis Stephens
Platypus parallelus (Fabricius)
Platypus penetralis Sampson

Bibliography

ABBAZZI, P. & OSELLA, G. 1992. Elenco sistematico-faunistico degli Anthribidae, Rhinomaceridae, Attelabidae, Apionidae, Brentidae, Curculionidae Italiani (Insecta, Coleoptera, Curculionoidea). *Redia* **75**: 267-414.

ÁDÁM, L. 1987. Staphylinidae of the Kiskunság National Park (Coleoptera). In: Mahunka, S. (ed.) *The Fauna of Kiskunság National Park II.* Budapest: Akademiai Kiado, pp. 126-148.

ADAMS, R.G. 1978. The first British infestation of *Reesa vespula* (Milliron) (Coleoptera: Dermestidae). *Entomologist's Gazette* **29**: 73-75.

—— 1988. *Anthrenus olgae* Kalik new to Britain (Coleoptera: Dermestidae) with notes on its separation from *A. caucasicus* Reitter. *Entomologist's Gazette* **39**: 207-210.

AHN, K.-J. & ASHE, J.S. 1995. Systematic position of the intertidal genus *Bryobota* Casey and a revised phylogeny of the falagriine genera of America north of Mexico (Coleoptera: Staphylinidae: Aleocharinae). *Annals of the Entomological Society of America* **88**: 143-154.

AITKEN, A.D. 1975. *Insect Travellers. Vol. I Coleoptera.* Ministry of Agriculture, Fisheries and Food Agricultural Development and Advisory Service Pest Infestation Control Laboratory Technical Bulletin 31. London: HMSO. xvi+191 pp.

ALDRIDGE, R.J.W. & POPE, R.D. 1986. The British species of *Bruchidius* Schilsky (Coleoptera: Bruchidae). *Entomologist's Gazette* **37**: 181-193.

ALEXANDER, K.N.A. 2003. Is *Malthodes brevicollis* (Paykull) (Cantharidae) a British beetle? *The Coleopterist* **12**: 35-39.

ALLEN, A.A. 1967. A review of the status of certain Scarabaeoidea (Col.) in the British fauna; with the addition to our list of *Onthophagus similis* Scriba. *Entomologist's Record & Journal of Variation* **79**: 201-206, 220-224, 257-262, 284-290.

—— 1969. *Cercyon laminatus* Sharp (Col. Hydrophilidae) new to Britain; with corrections to our list of species, and further notes. *Entomologist's Record & Journal of Variation* **81**: 211-216.

—— 1970. Notes on various little-known, doubtful, or misidentified British Staphylinidae (Col.). *Entomologist's Monthly Magazine* **105**: 193-196.

—— 1984a. Note on the restoration of a name in the genus *Tychius* (Col., Curculionidae). *Entomologist's Monthly Magazine* **120**: 143.

—— 1984b. A genus and species of Malachiinae (Col.: Melyridae) new to Britain. *Entomologist's Record & Journal of Variation* **96**: 243-244.

—— 1986. On the British species of *Mordellistena* Costa (Col. Mordellidae) resembling *parvula* Gyll. *Entomologist's Record & Journal of Variation* **98**: 47-50.

—— 1988a. *Rhynchaenus calceatus* Germ. (Col., Curculionidae), an addition to the British list. *Entomologist's Monthly Magazine* **124**: 147-148.

—— 1988b. A fourth species of *Ischnomera* Steph. (Col.: Oedemeridae) in Britain. *Entomologist's Record & Journal of Variation* **100**: 199-202.

—— 1989. *Lamprohiza splendidula* (L.) (Col., Lampyridae) taken in Kent in 1884. *Entomologist's Monthly Magazine* **125**: 182.

—— 1992a. *Hydrosmecta septentrionum* Benick, not *H. subtilissima* (Kr.) (Col., Staphylinidae), a British species. *Entomologist's Monthly Magazine* **128**: 42.

—— 1992b. On *Bagous arduus* Sharp and *B. rudis* Sharp (Col.: Curculionidae). *Entomologist's Record & Journal of Variation* **104**: 199-201.

—— 1993. *Longitarsus longiseta* Weise (Col.: Chrysomelidae) a British species, and a new synonymy. *Entomologist's Record & Journal of Variation* **105**: 175-176.

—— 1994a. Notes on some British Staphylinidae (Col.) - 5. The genus *Atheta* Thoms.: three additions to the fauna, a reinstatement and two synonymies. *Entomologist's Monthly Magazine* **130**: 165-71.

—— 1994b. *Euplectus brunneus* Grimm. (Col.: Pselaphidae) and its status in Britain. *Entomologist's Record & Journal of Variation* **106**: 171-172.

—— 1995a. *Hylesinus orni* Fuchs (Scolytidae) not a synonym of *H. varius* (Fabricius). *The Coleopterist* **4**: 19.

—— 1995b. The name *Scopaeus ryei* Wollaston, 1872 (Col., Staphylinidae) reinstated. *Entomologist's Monthly Magazine* **131**: 45.

—— 1995c. An apparently new species of *Mordellistena* (Col.: Mordellidae) in Britain. *Entomologist's Record & Journal of Variation* **107**: 25-26.

—— 1999. Another new species of *Mordellistena* Costa (Col. Mordellidae) in Britain. *Entomologist's Record & Journal of Variation* **111**: 205-206.

ALLEN, A.A. & ECCLES, T.M. 1988. *Hydrosmectina delicatissima* Bernhauer (Col., Staphylinidae), new to Britain. *Entomologist's Monthly Magazine* **124**: 215-220.

ALLEN, A.J. 2004. *Dactylosternum abdominale* (Fabricius) (Hydrophilidae) in Dorset - new to Britain. *The Coleopterist* **13**: 1-3.

ALLEN, A.J. & BOOTH, R.G. 2005. *Xylostiba bosnica* (Bernhauer) (Staphylinidae) in Britain. *The Coleopterist* **14**: 93-96.

—— 2008. *Berginus tamarisci* Wollaston, 1854 (Mycetophagidae) in Surrey - new to Britain. *The Coleopterist* **17**: 205-206.

ALLEN, A.J. & HANCE, D. 2009. *Hololepta plana* (Sulzer, 1776) (Histeridae) in Norfolk - new to Britain. *The Coleopterist* **18**: 153-154.

ALLEN, A.J. & OWEN, J.A. 1997. *Philonthus spinipes* Sharp (Staphylinidae) in Dorset - new to Britain. *The Coleopterist* **6**: 81-83.

ALLEN, A.J. & TURNER, C.R. 2012. *Conarthrus praeustus* (Boheman in Schönherr, 1838)

(Curculionidae: Cossoninae) in Britain. *The Coleopterist* **21**: 21-23.

ALONSO-ZARAZAGA, M.A. 1989. Revision of the supraspecific taxa in the Palaearctic Apionidae Schönherr, 1823. 1. Introduction and subfamily Nanophyinae Seidlitz, 1891 (Coleoptera, Curculionoidea). *Fragmenta Entomologica* **21**: 205-262.

ALONSO-ZARAZAGA, M.A. & KRELL, F.-T. 2011. Change of authorship of *Aphodius* and *Oryctes* to Hellwig, 1798 (Insecta: Coleoptera: Scarabaeidae). *Zootaxa* **3060**: 67-68.

ALONSO-ZARAZAGA, M.A. & LYAL, C.H.C. 1999. *A World Catalogue of Families and Genera of Curculionoidea (Insecta: Coleoptera (Excepting Scolytidae and Platypodidae))*. Barcelona: Entomopraxis.

—— 2002. Addenda and corrigenda to 'A world catalogue of families and genera of Curculionoidea (Insecta: Coleoptera)'. *Zootaxa* **63**: 1-37.

ANDERSON, R. 1985. *Agonum lugens* (Duftschmid) new to the British Isles (Col., Carabidae). *Entomologist's Monthly Magazine* **121**: 133-135.

ANDERSON, R. & LUFF, M.L. 1994. *Calathus cinctus* Motschulsky, a species of the *Calathus melanocephalus / mollis* complex (Col., Carabidae) in the British Isles. *Entomologist's Monthly Magazine* **130**: 131-135.

ANDERSON, R., MCFERRAN, D. & CAMERON, A. 2000. *The ground beetles of Northern Ireland (Coleoptera-Carabidae)*. Ulster Museum, Belfast.

ANDERSON, R., NASH, R. & O'CONNOR, J.P. 1997. Irish Coleoptera. A revised and annotated list. *Irish Naturalists' Journal, Special Entomological Supplement*: 1-81.

ANGUS, R.B. 1982. Separation of two species standing as *Helophorus aquaticus* (L.) (Coleoptera, Hydrophilidae) by banded chromosome analysis. *Systematic Entomology* **7**: 265-281.

—— 1999. Insecta: Coleoptera: Hydrophilidae: Helophorinae. *Süßwasserfauna von Mitteleuropa*, 20/10-2. Gustav Fischer Verlag, 144 pp.

—— 2010. *Boreonectes* gen. n., a new genus for the *Stictotarsus griseostriatus* (De Geer) group of sibling species (Coleoptera: Dytiscidae), with additional karyosystematic data on the group. *Comparative Cytogenetics* **4**: 123-131.

ANGUS, R.B., WILSON, C.J., MATÉ, J.F., HAMMOND, P.M. & MANN, D.J. 2003. *Saprosites mendax* (Blackburn) and *S. natalensis* (Peringuey) (Scarabaeoidea, Aphodiidae), two species introduced into Britain. *Proceedings of the Second Pan-European Conference on Saproxylic Beetles, 2003*. People's Trust for Endangered Species, pp. 72-76.

ARNDT, E. 1998. Phylogenetic investigation of Carabidae (Coleoptera) using larval characters. In: Ball, G.E., Casale, A. & Vigna Taglianti, A. (eds.) *Phylogeny and Classification of Caraboidea (Coleoptera: Adephaga)*. Torino: Museo Regionale di Scienze Naturali, pp 171-190.

ARNOLD, U. 1990. Zur Kenntnis der Gattung *Altica* 1(8)(Coleoptera, Chrysomelidae, Alticinae). *Entomologische Nachrichten, Berlin* **34**: 167-170.

ASHE, J.S. 1991. The systematic positions of *Placusa* Erichson and *Euvira* Sharp: the tribe Placusini described (Coleoptera: Staphylinidae: Aleocharinae). *Systematic Entomology* **16**: 383-400.

—— 1998 (version 22 September 1998). Aleocharinae In: The Tree of Life web site (http://phylogeny.arizona.edu/tree/phylogeny.html).

ASSING, V. 1996a. A revision of the European species of *Calodera* Mannerheim (Coleoptera, Staphyliniae, Aleocharinae). *Beiträge zur Entomologie* **46**: 3-24.

—— 1996b. Über *Gyrohypnus angustatus* Stephens, 1833, und *G. scoticus* (Joy, 1913) (Col.: Staphylinidae). *Entomologische Blätter für Biologie und Systematik der Käfer* **92**: 78-84.

—— 1997a. A revision of the Western Palaearctic species of *Myrmecopora* Saulcy, 1864, *sensu lato* and *Eccoptoglossa* Luze, 1904. (Coleoptera: Staphylinidae, Aleocharinae, Falagriini). *Beiträge zur Entomologie* **47**: 69-151.

—— 1997b. A revision of *Othius* Stephens, 1829 III. The species of the Western Palaearctic Region exclusive of the Atlantic Islands (Coleoptera, Staphylinidae: Xantholininae). *Beiträge zur Entomologie, Nova Supplementa Entomologica, Berlin* **10**: 3-130.

—— 1999a. Zur Kenntnis und Synonymie einiger mitteleuropäischer Arten der Gattung *Quedius* Stephens (Coleoptera, Staphylinidae). *Entomologische Blätter für Biologie und Systematik der Käfer* **95**: 35-46.

—— 1999b. A revision of *Ilyobates* Kraatz, 1856 (Coleoptera: Staphylinidae, Aleocharinae, Oxypodini). *Beiträge zur Entomologie* **49**: 295-342.

—— 1999c. A revision of the Palaearctic species of *Myrmecopora* Saulcy, 1864 and *Eccoptoglossa* Luze, 1904 (Coleoptera: Staphylinidae, Aleocharinae, Falagriini). Supplement I. *Entomologische Zeitung* **109**: 499-504.

—— 2001. A revision of *Callicerus* Gravenhorst, 1802, *Pseudosemiris* Machulka, 1935, and *Saphocallus* Sharp, 1888 (Coleoptera: Staphylinidae, Aleocharinae, Athetini). *Beiträge zur Entomologie* **51**: 247-334.

—— 2003. A revision of *Calodera* Mannerheim. II. A new species, new synonyms and additional records (Coleoptera: Staphylinidae, Aleocharinae). *Beiträge zur Entomologie* **53**: 217-230.

—— 2008. On the taxonomy and zoogeography of some Palaearctic Paederinae and Xantholinini (Coleoptera: Staphylinidae). *Linzer Biologische Beiträge* **40**: 1237-1294.

ASSING, V., FRISCH, J., KAHLEN, M., LÖBL, I., LOHSE, G.A., PUTHZ, V., SCHÜLKE, M., TERLUTTER, H., UHLIG, M., VOGEL, J., WUNDERLE, P. & ZERCHE, L. 1998. Familie Staphyliniden. In: Lucht, W.H. & Klausnitzer, B. (eds.) *Die Käfer Mitteleuropas*. Band 15, 4 Supplementband, pp. 119-197. Krefeld: Goecke & Evers.

ASSING, V. & SCHÜLKE, M. 1999. Supplemente zur mitteleuropäischen Staphylinidenfauna (Coleoptera, Staphylinidae). *Entomologische Blätter für Biologie und Systematik der Käfer* **95**: 1-31.

—— 2001. Supplemente zur mitteleuropäischen Staphylinidenfauna (Coleoptera, Staphylinidae). II.

Entomologische Blätter für Biologie und Systematik der Käfer **97**: 121-176.

—— 2007. Supplemente zur mitteleuropäischen Staphylinidenfauna (Coleoptera, Staphylinidae). III. *Entomologische Blätter für Biologie und Systematik der Käfer* **102**: 1-78.

—— 2012. *Freude-Harde-Lohse-Klausnitzer—Die Käfer Mitteleuropas.* Band 4: Staphylinidae I. (2nd revised edn.). Spektrum Akademischer Verlag, pp.xii+560.

ASSING, V. & WUNDERLE, P. 2008. On the *Alevonota* species of the western Palaearctic region (Coleoptera: Staphylinidae: Aleocharinae: Athetini). *Beiträge zur Entomologie, Keltern* **58**: 145-189.

ATKINS, P.M., O'CALLAGHAN, D.P. & KIRBY, S.G. 1981. *Scolytus laevis* (Chapuis) (Coleoptera: Scolytidae) new to Britain. *Entomologist's Gazette* **32**: 280.

ATTY, D.B. 1983. *Coleoptera of Gloucestershire.* Cheltenham: privately published.

AUDISIO, P.A. 1993. Coleoptera Nitidulidae - Kateretidae. *Fauna d'Italia* 33. Bologna: Edizioni Calderini, 972 pp.

BACCHUS, M.E. & KIRK-SPRIGGS, A.H. 1991. The *Meligethes* (Col., Nitidulidae) described by T. Marsham & J.F. Stephens - changes in nomenclature and type designation. *Entomologist's Monthly Magazine* **127**: 209-214.

BALKE, M., RIBERA, I. & VOGLER, A.P. 2004. MtDNA phylogeny and biogeography of Copelatinae, a highly diverse group of tropical diving beetles (Dytiscidae). *Molecular Phylogenetics & Evolution* **32**: 866-880.

BALL, G.E., CASALE, A. & VIGNA TAGLIANTI, A. 1998. Introduction. In: Ball, G.E., Casale, A. & Vigna Taglianti, A. (eds.) *Phylogeny and Classification of Caraboidea. (Coleoptera: Adephaga).* Torino: Museo Regionale di Scienze Naturali, pp 15-52.

BARANOWSKI, R. 1982. *Atheta (Badura) ehnstremi* n. sp. (Coleoptera: Staphylinidae), with short review of other known species of *Badura. Entomologica Scandinavica* **13**: 33-40.

BARAUD, J. 1992. *Coléoptères Scarabaeoidea d'Europe.* Faune de France **78**: i-ix, 1-856.

BARCLAY, M.V.L. 2003a. *Otiorhynchus (s. str.) armadillo* (Rossi, 1792) and *Otiorhynchus (s. str.) salicicola* Heyden, 1908 (Curculionidae: Entiminae: Otiorhynchini) - two European vine weevils established in Britain. *The Coleopterist* **12**: 41-56.

—— 2003b. On some British examples of *Poecilium lividum* (Rossi) (= *Phymatodes lividus*) (Cerambycidae) from the New Forest: are they authentic? *The Coleopterist* **12**: 114-118.

—— 2005. *Priobium carpini* (Herbst, 1793) (Coleoptera: Anobiidae), a European woodworm established in Britain. *Entomologist's Monthly Magazine* **141**: 43-48.

—— 2009. British *Otiorhynchus 'setosulus'* re-identified as *Otiorhynchus dieckmanni* Magnano, with notes on its distribution and identification. *The Coleopterist* **18**: 59-65.

BARCLAY, M.V.L. & KOPETZ, A. 2003. *Malthodes lobatus* (Kiesenwetter) (Cantharidae) new to Britain. *The Coleopterist* **12**: 97-100.

BATTEN, R. 1986. A review of the British Mordellidae (Coleoptera). *Entomologist's Gazette* **37**: 25-235.

BELLAMY, C.L. 2003. *An Illustrated Summary of the Higher Classification of the Superfamily Buprestoidea (Coleoptera).* Folia Heyrovskyana Supplementum 10. Zlín, Czech Republic: V. Kabourek Ltd.

BÈLLES, X. & HALSTEAD, D. 1985. Identification & geographical distribution of *Gibbium aequinoctiale* Boieldieu & G. *psylliodes* (Czenpinski) (Coleoptera - Ptinidae). *Journal of Stored Products Research* **21**: 151-155.

BENICK, G. 1970. Revision der Untergattung *Anopleta* Mulsant Rey (Genus *Atheta* Staphyl.). *Entomologische Blätter für Biologie und Systematik der Käfer* **66**: 83-110.

—— 1974. Neue Atheten (Col. Staphyl.) aus Deutschland und den zunächst gelegenen Ländern. *Stuttgarter Beiträge zur Naturkunde A* **273**: 1-23.

BERCEDO, P. & ARNÁIZ, L. 2010. A new species of the genus *Mirosternomorphus* Español, 1977 from England (United Kingdom) (Coleoptera: Ptinidae: Dorcatominae). *The Coleopterist* **19**: 105-110.

BERGE HENEGOUWEN, A.L. van 1986. Revision of the European species of *Anacaena* Thomson (Coleoptera: Hydrophilidae). *Entomologica Scandinavica* **17**: 393-407.

—— 1988. *Hydrochus megaphallus,* a new and widespread European water beetle described from the Netherlands (Coleoptera: Hydrophilidae). *Balfour-Browne Club Newsletter* **42**: 8-22.

—— 1989. *Sphaeridium marginatum* reinstated as a species distinct from S. *bipustulatum* (Coleoptera: Hydrophilidae). *Entomologische Berichten, Amsterdam* **49**: 168-170.

BESUCHET, C. 1989. Familie: Pselaphidae. In: Lohse, G.A. & Lucht, W.H. (eds.) *Die Käfer Mitteleuropas.* Band 12, Supplementband mit Katalogteil, pp. 240-243. Krefeld: Goecke & Evers.

—— 1999. Pselaphides paléarctiques. Note taxonomiques et faunistiques (Coleoptera Staphylinidae Pselaphinae). *Revue Suisse de Zoologie* **106**: 45-67.

BIEŃKOWSKI, A.O. 2001. A study on the genus *Chrysolina* Motschulsky, 1860, with a checklist of all the described subgenera, species, subspecies, and synonyms (Coleoptera: Chrysomelidae: Chrysomelinae). *Genus* **12**: 105-235.

BISTRÖM, O. 1977. Nomenclatoric notes on Coleoptera. *Notulae Entomologicae* **57**: 17-18.

BISTRÖM, O. & SILFVERBERG, H. 1979. The type species of the European genera of Ptiliidae (Coleoptera). *Annales Entomologici Fennici* **45**: 12-15.

BLACKWELDER, R.E. 1939. A generic revision of the staphylinid beetles of the tribe Paederini. *Proceedings of the U.S. National Museum* **87**: 93-125.

—— 1952. The generic names of the beetle family Staphylinidae: with an essay on genotypy. *U.S. National Museum Bulletin* **200**, pp. iv+483.

BOHÁČ, J. 1993. Staphylinidae. In: Jelinek, J. (ed.) *Checklist of Czechoslovak Insects IV (Coleoptera)*, Folia Heyrovskyana. Supplementum 1, pp. 39-63.

BOOTH, R.G. 1986. Some interesting Coleoptera from the Yorkshire Wolds. *Entomologist's Monthly Magazine* **12**: 78.

—— 1988. The identity of *Tachyporus chrysomelinus* (Linnaeus) and the separation of *T. dispar* (Paykull) (Coleoptera; Staphylinidae). *Entomologist* **107**: 127-133.

—— 1993. *Thryogenes fiorii* Zumpt, 1928 (Curculionidae) new to Britain. *The Coleopterist* **2**: 19-20.

—— 1994a. *Longitarsus longiseta* Weise rediscovered and *Longitarsus obliteratoides* Gruev (Chrysomelidae) new to Britain. *The Coleopterist* **3**: 4-5.

—— 1994b. *Oulema 'melanopus'* (Chrysomelidae) in Britain. *The Coleopterist* **3**: 4-5.

—— 1998. The taxonomic status of *Longitarsus bearei* Kevan (Col., Chrysomelidae). *Entomologist's Monthly Magazine* **134**: 335-338.

—— 2001. 2000 Annual Exhibition, Imperial College, London SW7 - 11 November 2000. Coleoptera. *British Journal of Entomology and Natural History* **14**: 160-168.

—— 2002a. 2001 Annual Exhibition. Imperial College, London SW7 - 10 November 2001. Coleoptera. *British Journal of Entomology and Natural History* **15**: 173-180.

—— 2002b. The specific characters of *Thryogenes fiorii* and *T. atrirostris* (Coleoptera, Curculionoidea, Erirhinidae). *Mitteilungen aus dem Museum für Naturkunde Berlin, Deutsch Entomologische Zeitchrift* **49**: 273-274.

—— 2004. 2003 Annual Exhibition. Imperial College, London SW7 - 8 November 2003. Coleoptera. *British Journal of Entomology and Natural History* **17**: 173-180.

—— 2006. 2005 Annual Exhibition. Imperial College, London SW7 - 12 November 2005. Coleoptera. *British Journal of Entomology and Natural History* **19**: 185-189.

—— 2008. 2007 Annual Exhibition. Imperial College, London SW7 - 10 November 2007. Coleoptera. *British Journal of Entomology and Natural History* **21**: 176-184.

—— 2009. 2008 Annual Exhibition. Imperial College, London SW7 - 8 November 2008. Coleoptera. *British Journal of Entomology and Natural History* **22**: 178-184.

BOOTH, R.G. & OWEN, J.A. 1997. *Chaetocnema picipes* Stephens (Chrysomelidae: Alticinae) in Britain. *The Coleopterist* **6**: 85-89.

BORDONI, A. 1975a. Morfologia cefalica e addominale della sottotribu Medonina nov. e del genere *Medon*. *Redia* **56**: 417-446.

—— 1975b. *Neobisnius cerrutii* Gridelli, 1943 è sinonimo di *lathrobioides* (Baudi, 1848) (Coleoptera, Staphylinidae). *Bollettino del Museo di Zoologia dell'Università di Torino* **1975**: 81-84.

—— 1976. Studi sulla sistematica e la geonemia del genere *Quedius* Steph. – IV – I *Quedius* s.str. della fauna italiana. *Redia* **59**: 85-107.

—— 2002. *Xantholinini della Regione Orientale (Coleoptera: Staphylinidae). Classificazione, Filogenesi e Revisione Tassonomica.* Monografie di Museo Regionale di Scienze Naturali, Torino 33, pp. 998.

BOROVEC, R. 1991. Revision der *Trachyphloeus laticollis*-Gruppe (Insecta, Coleoptera, Curculionidae: Otiorhynchinae). *Entomologische Abhandlungen, Staatliches Museum für Tierkunde, Dresden* **54**: 47-40.

BOUCHARD, P., BOUSQUET, Y., DAVIES, A.E., ALONSO-ZARAZAGA, M.A., LAWRENCE, J.F., LYAL, C.H.C., NEWTON, A.F., REID, C.A.M., SCHMITT, M., ŚLIPIŃSKI, S.A. & SMITH, A.B.T. 2011. Family-group names in Coleoptera (Insecta). *ZooKeys* **88**: 1-972.

BOUSKELL, F. 1903. *Aphodius sturmi* Harold: a British insect. *Entomologist's Record & Journal of Variation* **15**: 92.

BOWESTEAD, S., 1999. *A revision of the Corylophidae (Coleoptera) of the West Palaearctic region.* Geneva: Museum d'Histoire Naturelle. pp. 203.

—— 2002. *Sericoderus brevicornis* Matthews (Coleoptera: Corylophidae) redescribed and new to New Zealand. *New Zealand Entomologist* **25**: 65-67.

BOWESTEAD, S. & ECCLES, T.M. 1987. *Euplectus bonvouloiri rosae* Raffray (Col., Pselaphidae) new to Britain. *Entomologist's Monthly Magazine* **123**: 109-111.

BOYCE, D.C. 1991. Coleoptera recording in Ceredigion during 1990. *Dyfed Invertebrate Group Newsletter* **20**: 15-20.

BRANCO, T. 2002. *Scarabaeus verticicornis* Laicharting, 1781 (currently *Onthophagus* (*Palaeonthophagus*) *verticicornis*), a valid name (Coleoptera, Scarabaeidae). *Boletín de la Sociedad Entomológica Aragonesa* **31**: 125-127.

BRUNDIN, L. 1940. Studien über die Untergattung *Oreostiba* Ganglb. (Col: Staphylinidae). *Entomologisk Tidskrift* **61**: 56-130.

—— 1943. Monographie der palaearktischen Arten der *Atheta*-Untergattung *Hygroecia* (Coleoptera, Staphylinidae). *Analen der Naturhistorischen Museums in Wien* **53**: 129-301.

—— 1948. *Microdota*-Studien. *Entomologisk Tidskrift* **69**: 8-65.

—— 1952. *Acrotona*-Studien. *Entomologisk Tidskrift* **73**: 93-128.

—— 1963. Die Palaearktischen Arten der *Atheta*-Untergattung *Dimetrota* Muls. et Rey (Col., Staphylinidae). *Arkiv för Zoologi* **5**: 369-434.

BUCKLAND, P. 2007. *The Development and Implementation of Software for Palaeoenvironmental and Palaeoclimatological Research: The Bugs Coleopteran Ecology Package (BugsCEP).* Ph.D. thesis, Environmental Archaeology Lab., Department of Archaeology & Sámi Studies. University of Umeå, Sweden. Archaeology and Environment 23, 236 pp + CD. Available online: http://www.diva-portal.org/umu/abstract.xsql?dbid=1105.

BUCKLAND, P.C. & SKIDMORE, P. 1999. *Xanthogaleruca luteola* (Müller) (Chrysomelidae) in Britain. *The Coleopterist* **8**: 97-99.

BUCKLAND, P.I. & BUCKLAND, P.C. 2006. *BugsCEP Coleopteran Ecology Package.* IGBP PAGES/World

Data Center for Paleoclimatology Data Contribution Series # 2006-116. NOAA/NCDC Paleoclimatology Program, Boulder CO, USA.

BULLOCK, E. 1930. Some new records of Coleoptera from Ireland. *Entomologist's Monthly Magazine* **66**: 140-141.

CALDARA, R. 1983. Studio dei sintipi di *Tychius* custoditi nelle collezione Banks,Marsham, Stephens, Kirby, Walton e Waterhouse del British Museum (Natural History). *Bollettino della Società Entomologica Italiana* **115**: 86-90.

—— 1990a. Su alcuni problemi nomenclaturiali e sinonimici in specie della tribu' Tychiini (Coleoptera Curculionidae). *Bollettino della Società Entomologica Italiana* **122**: 21-26.

—— 1990b. Revisione tassonomica delle specie paleartiche del genere *Tychius* Germar (Coleoptera Curculionidae). *Memorie della Società Italiana di Scienze Naturali e del Museo Civico di Storia Naturale de Milano* **25**: 53-218.

—— 2001. Phylogenetic analysis and higher classification of the trbe Mecinini (Coleoptera: Curculionidae, Curculioninae). *Koleopterologische Rundschau* **71**: 171-203.

CALDARA, R. & O'BRIEN, C.W. 1994. On the systematic position and nomenclature of some species of the genus *Bagous* Germar, 1817 (Coleoptera: Curculionidae). *Giornale Italiano di Entomologia* **7**: 1-4.

—— 1998. Systematics and evolution of weevils of the genus *Bagous*. VI. Taxonomic treatment of the species of the Western Palaearctic Region (Coleoptera Curculionidae). *Memorie della Società Entomologica Italiana* **76**[1997]: 31-347.

CAMPBELL, J.M. 1973. A revision of the genus *Tachinus* (Coleoptera: Staphylinidae) of North and Central America. *Memoirs of the Entomological Society of Canada* **90**: 1-137.

—— 1979. A revision of the genus *Tachyporus* Gravenhorst (Coleoptera: Staphylinidae) of North and Central America. *Memoirs of the Entomological Society of Canada* **109**: 1-95.

—— 1991. A revision of the genera *Mycetoporus* Mannerheim and *Ischnosoma* Stephens (Coleoptera: Staphylinidae) of North and Central America. *Memoirs of the Entomological Society of Canada* **156**: 1-169.

—— 1993. A revision of the genera *Bryoporus* Kraatz and *Bryophacis* Reitter and two new related genera from America north of Mexico (Coleoptera: Staphylinidae: Tachyporinae). *Memoirs of the Entomological Society of Canada* **166**: 1-85.

CARR, R. 1984a. *Limnebius crinifer* Rey new to Britain, with a revised key to the British *Limnebius* species (Coleoptera: Hydraenidae). *Entomologist's Gazette* **35**: 99-102.

—— 1984b. A *Coelambus* species new to Britain (Coleoptera: Dytiscidae). *Entomologist's Gazette* **35**: 181-184.

—— 1999. *Nebrioporus canaliculatus* (Lacordaire *in* Boisudval & Lacordaire) new to Britain. *Latissimus* **11**: 36.

—— 2000. The occurrence of *Nebrioporus canaliculatus* (Lacordaire) (Coleoptera: Dytiscidae) in Britain, with a note on recent taxonomic changes to related British species. *Entomologist's Gazette* **51**: 125-128.

CARTER, I. & OWEN, J. 1987. *A Numerical List of British Beetles. 2nd Revision.* Cheltenham: Dead Wood Press.

CASALE, A. & BŘEZINA, B. 2003. Checklist. In: Turin, H., Penev, L. & Casale, A. (eds.) *The Genus* Carabus *in Europe: a synthesis.* Sofia: Pensoft Books, 511pp.

CHANDLER, D.S., NARDI, G. & TELNOV, D. 2004. Nomenclatural notes on the Palaearctic Anthicidae (Coleoptera). *Mitteilungen des Internationalen Entomologischen Vereins* **29**: 109-173.

CHUTER, K. 2010. *Polygraphus grandiclava* (Thomson) (Curculionidae) new to Britain and breeding. *The Coleopterist* **19**: 63-64.

CLEMONS, L. 1983. *Gronops inaequalis* Boheman (Col: Curculionidae): a weevil new to Britain. *Entomologist's Record & Journal of Variation* **95**: 213-215.

COIFFAIT, H. 1972. Coléoptères Staphylinidae de la région paléarctique occidentale. I Généralités, Sous familles Xantholininae et Leptotyphlinae. *Nouvelle Revue d'Entomologie* **2** (supplément): 1-651.

—— 1974. Coléoptères Staphylinidae de la région paléarctique occidentale. II Sous famille Staphylininae Tribus Philonthini et Staphylinini. *Nouvelle Revue d'Entomologie* **3** (supplément): 1-593.

—— 1982a. Coléoptères Staphylinidae de la région paléarctique occidentale. IV Sous famille Paederinae Tribu Paederini 1 (Paederi, Lathrobii). *Nouvelle Revue d'Entomologie* **7** (supplément): 1-440.

—— 1982b. Contribution à la connaissance des staphylinides de l'Himalaya (Nepal, Ladakh, Cachemire) (Insecta: Coleoptera: Staphylinidae). *Senckenbergiana Biologica* **62**: 21-179.

—— 1984. Coléoptères Staphylinidae de la région paléarctique occidentale. V Sous famille Paederinae Tribu Paederini 2 Sous famille Euaesthetinae. *Nouvelle Revue d'Entomologie* **8** (supplément): 1-424.

COLLIER, M. 1986. *Choleva elongata* (Paykull) new to Britain. *Coleopterist's Newsletter* **26**: 12.

COLONNELLI, E. 1990. Curculionidae Ceutorrhynchinae from the Canaries and Macaronesia (Coleoptera). *Vieraea* **18**: 317-337.

—— 1993. The Ceutorhynchinae types of I.C. Fabricius and G. von Paykull (Coleoptera: Curculionidae). *Koleopterologische Rundschau* **63**: 299-310.

—— 1998. Systematic and synonymic notes on Ceuthorhynchinae, with lectotype and neotype designations, and descriptions of three new genera. *Fragmenta Entomologica* **30**: 105-175.

—— 2004. *Catalogue of Ceutorhynchinae of the World, with a key to genera (Insecta: Coleoptera: Curculionidae).* Barcelona: Argania Editio.

CONSTANTINE, B. & MAJERUS, M. 1994. *Cryptolaemus montrouzieri* Mulsant (Col., Coccinellidae) in Britain. *Entomologist's Monthly Magazine* **130**: 45-46.

COOTER, J. 1982. *Dendroctonus micans* Kug. *Coleopterist's Newsletter* **10**: 3.

—— 1983. *Dendroctonus micans* Kug. (Col., Scolytidae) in Britain. *Entomologist's Monthly Magazine* **119**: 231.

—— 1991. *Mordellistena pygmeola* Ermisch, 1956 (Coleoptera: Mordellidae) new to Britain. *Entomologist's Gazette* **42**: 97-98.

—— 1992a. *Leiodes strigipenne* Daffner (Leiodidae) in Britain. *The Coleopterist* **1**(2): 4.

—— 1992b. *Anthaxia quadripunctata* (Linnaeus, 1758) (Coleoptera: Buprestidae) in England: an enigma. *Entomologist's Gazette* **43**: 75.

—— 1995. Recent changes in the classification of the British Nitiduloidea. *The Coleopterist* **4**: 37-38.

—— 1996a. *Leiodes pallens* (Sturm, 1807) (Col., Leiodidae) not a British insect. *Entomologist's Monthly Magazine* **132**: 87-90.

—— 1996b. Annotated keys to the British Leiodinae (Col., Leiodidae). *Entomologist's Monthly Magazine* **132**: 205-272.

—— 2004. *Agathidium pisanum* Brisout (Col., Leiodidae) in Britain. *Entomologist's Monthly Magazine* **140**: 59-64.

COQUEMPOT, C. 2006. Asian longhorned beetles (Coleoptera Cerambycidae): original interceptions and introductions in Europe, mainly in France, and notes about recently imported species. *Redia* **89**: 35-50.

COULON, J. 2004. Taxonomy of *Eotachys* of the West Palearctic region: description of eight new species (Coleoptera, Carabidae). *Nouvelle Revue d'Entomologie* **21**: 65-93.

COX, M.L. 1995a. Identification of the *Oulema 'melanopus'* species group (Chrysomelidae). *The Coleopterist* **4**: 33-36.

—— 1995b. *Psylliodes cucullata* (Illiger, 1807) (Coleoptera: Chrysomelidae, Alticinae), a species new to Britain. *Entomologist's Gazette* **46**: 271-276.

—— 2000. Progress report on the Bruchidae/Chrysomelidae Recording Scheme. *The Coleopterist* **9**: 65-74.

—— 2001. Notes on the natural history, distribution and identification of seed beetles (Bruchidae) of Britain and Ireland. *The Coleopterist* **9**: 113-147.

—— 2007. *Atlas of the Seed and Leaf Beetles of Britain and Ireland*. Newbury, Berks.: Pisces Publications. 344 pp.

CROWSON, R.A. 1984 On the systematic position of *Bruchela* Dejean (*Urodon* auctt.) (Coleoptera). *Coleopterists Bulletin* **38**: 91-93.

CSIKI, E. 1915. *Coleopterorum Catalogus* Part 63, Mordellidae. Berlin: W. Junk. pp. 84.

CUCCODORO, G. & LÖBL, I. 1997. Revision of the Palaearctic rove beetles of the genus *Megarthrus* Curtis (Coleoptera: Staphylinidae: Proteininae). *Journal of Natural History* **31**: 1347-1415.

DARBY, M. 2012 in press. Ptiliidae. *In:* DUFF, A.G. *Beetles of Britain and Ireland. Vol. 1: Sphaeriusidae to Silphidae*. West Runton: A.G. Duff.

DELLACASA, G. & DELLACASA, M. 2007. Coleoptera: Aphodiidae, Aphodiinae. *Fauna d'Italia* **41**: 1-484.

DELLACASA, G., BORDAT, P. & DELLACASA, M. 2001. A revisional essay of world genus-group taxa of Aphodiinae. *Memorie della Società Entomologica Italiana* **79**: 1-482.

DELLACASA, M. 1988. Contribution to a world-wide catalogue of Aegialiidae, Aphodiidae, Aulonocnemidae, Termitotrogidae (Coleoptera, Scarabaeoidea). *Memorie della Società Entomologica Italiana* **66** [1987]: 1-455.

DENTON, J. 2004. *Scymnus interruptus* (Goeze) (Coccinellidae) in Hampshire. *The Coleopterist* **13**: 91-92.

—— 2005. *Brachyderes incanus* (Linnaeus) and *Pachyrhinus mustela* (Herbst) (Curculionidae) in Surrey - new to Britain. *The Coleopterist* **14**: 1-5.

—— 2007. *Somotrichus unifasciatus* (Dejean) (Carabidae) in Britain. *British Journal of Entomology and Natural History* **20**: 199-200.

DEUVE, T. 1993. L'abdomen et les genitalia des femelles de Coléoptères Adephaga. *Mémoires du Museum National d'Histoire Naturelle, Zoologie, Paris*. **155**: 1-185.

—— 1994. *Une classification du genre Carabus*. Venette: Sciences Nat.

DIECKMANN, L. 1964. Die mitteleuropäischen Arten aus der Gattung *Bagous* Germ. *Entomologische Blätter für Biologie und Systematik der Käfer* **60**: 88-111.

—— 1976. Revision der *Apion platalea*-Grupp (Coleoptera, Curculionidae). *Entomologische Nachrichten* **20**: 117-128.

—— 1979. *Dorytomus*-Studien (Coleoptera, Curculionidae). *Polskie Pismo Entomologiczne* **49**: 539-546.

—— 1980. Beiträge zur Insektenfauna der DDR: Coleoptera-Curculionidae (Brachycerinae, Otiorhynchinae, Brachyderinae). *Beiträge zur Entomologie* **30**: 145-310.

—— 1986. Beiträge zur Insektenfauna der DDR: Coleoptera-Curculionidae (Erirhinae). *Beiträge zur Entomologie* **36**: 119-181.

—— 1991. Die europäischen *Limnobaris*-Arten (Coleoptera, Curculionidae). *Beiträge zur Entomologie* **41**: 303-311.

DÖBERL, M. 1987. Beitrag zur Kenntnis einiger westpaläarktischer Alticinen (Coleoptera, Chrysomelidae, Alticinae). *Entomologische Blätter für Biologie und Systematik der Käfer* **83**: 115-131.

—— 1995. Der heutige Alticinen-Artenbestand der Schweiz (Coleoptera, Chrysomelidae, Alticinae). *Mitteilungen der Entomologischen Gesellschaft Basel* **45**: 42-96.

DORNING, M., HALSTEAD, D.G.H. & HAMMOND, P.M. 2012. *Typhaea haagi* Reitter and *Typhaea stercorea* (Linnaeus) (Mycetophagidae): their identification and distribution, including the first British records for *T. haagi*. *The Coleopterist* **21**: 1-11.

DRAKE, C.M. 2006. *Hydrovatus cuspidatus* (Kunze, 1818) (Dytiscidae) new to Britain. *The Coleopterist* **15**: 53-57.

DRANE, A.B. 1994. A belated note on *Chloecharis debilicornis* (Wollaston) (Staphylinidae) new to Britain. *The Coleopterist* **3**: 2-3.

DRUGMAND, D. 1994. Le groupe d'espèce proches de *Xantholinus jarrigei* Coiffait, un véritable imbroglio systématique (Coleoplera, Staphylinidae). *Bulletin de la Société Entomologique de France* **99**: 241-252.

DUFF, A.G. 1995. On the genus of *Hypomedon debilicornis* (Wollaston) (Staphylinidae). *The Coleopterist* **4**: 6.

—— (ed.) 2008. *Checklist of Beetles of the British Isles, 2008 edition*. Wells: A.G. Duff.

DUFFY, E.A.J. 1953. Coleoptera (Scolytidae and Platypodidae). *Handbooks for the Identification of British Insects* **5**(15), 20 pp.

DYBAS, H.S. & DYBAS, L. 1990. 36 Insecta: Coleoptera Ptiliidae, pp. 1093-1112. In: Dindal, D.L. (ed.) *Soil Biology Guide* 1300 pp. New York.

ECCLES, T.M. & BOWESTEAD, S. 1987. *Anommatus diecki* Reitter (Coleoptera: Cerylonidae) new to Britain. *Entomologist's Gazette* **38**: 225-227.

ERWIN, T.L. 1985. The taxon pulse: a general pattern of lineage radiation and extinction among carabid beetles. In: Ball, G.E. (ed.) *Taxonomy, phylogeny and zoogeography of beetles and ants*. Dordrecht: Junk, pp. 437-493.

ERWIN, T.L. & SIMS, L.L. 1984. Carabid beetles of the West Indies (Insecta: Coleoptera). A synopsis of the genera and checklist of tribes of Caraboidea and of the West Indian species. *Quaestiones Entomologicae* **20**: 350-466.

EVERSHAM, B. & COLLIER, M. 1997. *Microlestes minutulus* (Goeze) (Carabidae) new to Britain. *The Coleopterist* **5**: 93-94.

FAUVEL, A. 1862. Notice sur quelques aléochariens nouveaux ou peu connus et descriptions de larves de *Phytosus* et *Leptusa*. *Annales de la Société d'Entomologique de France* **2**: 81-94.

FERY, H. 1999. Revision of a part of the *memnonius*-group of *Hydroporus* Clairville, 1806 (Insecta: Coleoptera: Dytiscidae). *Annalen des Naturhistorischen Museums in Wien* **101B**: 217-269.

FIKÁČEK, M., PROKIN, A. & ANGUS, R.B. 2011. A long-living species of the hydrophiloid beetles: *Helophorus sibiricus* from the early Miocene deposits of Karteshevo (Siberia, Russia). *Zookeys* **130**: 239-254.

FOCARILE, A. 1964. Gli *Asaphidion* del gruppo *flavipes* (L.), con particolare riguardo alla fauna Italiana. *Memorie della Società Entomologica Italiana* **43**: 97-120.

FOSTER, A.P., MORRIS, M.G. & WHITEHEAD, P.F. 2001. *Ixapion variegatum* (Wencker, 1864) (Col., Apionidae) new to the British Isles, with observations on its European and conservation status. *Entomologist's Monthly Magazine* **137**: 95-105.

FOSTER, G.N. 1984. Notes on *Enochrus* subgenus *Methydrus* Rey (Coleoptera: Hydrophilidae), including a species new to Britain. *Entomologist's Gazette* **35**: 25-29.

—— 2004. An annotated checklist of British and Irish water beetles, and associated taxa: Myxophaga and Adephaga-Hydradephaga. *The Coleopterist* **13**: 149-160.

—— 2005. An annotated checklist of British and Irish water beetles, and associated taxa: Polyphaga, with an update to Adephaga. *The Coleopterist* **14**: 7-19.

—— 2008. The new Irish water beetle. *Biodiversity Ireland, Bulletin of the National Biodiversity Data Centre* **1**: 4-5.

—— 2009. The *Ochthebius viridis* complex in Britain and Ireland (Hydraenidae). *The Coleopterist* **18**: 85-90.

—— 2012. Notes on Britain & Ireland water beetles (part 1). *Latissimus* **31**: 24-25.

FOSTER, G.N. & FRIDAY, L.E. 2011. Keys to adults of the water beetles of Britain and Ireland (Part 1) (Coleoptera: Hydradephaga: Gyrinidae, Haliplidae, Paelobiidae, Noteridae and Dytiscidae). *Handbooks for the Identification of Brtitish Insects* **4**(5), iv+144 pp.

FOSTER, G.N. & PHILLIPS, E.J. 1983. *Laccobius simulator* D'Orchymont (Coleoptera: Hydrophilidae) confirmed as British. *Entomologist's Gazette* **34**: 25-29.

FOSTER, G.N. & SPIRIT, M. 1986. *Oreodytes alpinus* new to Britain. *Balfour-Browne Club Newsletter* **36**: 1.

FOWLER, W.W. 1891. *The Coleoptera of the British Islands*, Vol. 5. London: Reeve.

FOWLES, A.P. & BOYCE, D.C. 1992. Rare and notable beetles from Cardiganshire (VC46) new to Wales. *The Coleopterist* **1**: 7-15.

FOWLES, A.P. & MORRIS, M.G. 1994. *Apion (Helianthemapion) aciculare* Germar (Col., Apionidae), a weevil new to Britain. *Entomologist's Monthly Magazine* **130**: 177-181.

—— 1999. *Datonychus arquatus* (Herbst, 1795) (Curculionidae) in the British Isles. *The Coleopterist* **8**: 117-119.

FRANCISCOLO, M.E. 1972. Su alcuni generi poco noti di Anaspidinae. *Memorie della Società Entomologica Italiana* **51**:123-155.

FRANK, J.H. 1988. *Paederus*, sensu lato (Coleoptera: Staphylinidae): an index and review of the taxa. *Insecta Mundi* **2**: 97-159.

FRANZ, H. & BESUCHET, C. 1971. Familie: Scydmaenidae. In: Freude, H., Harde, K.W. & Lohse, G.A. (eds) *Die Käfer Mitteleuropas*. Band 3, pp. 271-303. Krefeld: Goecke & Evers.

FREUDE, H, HARDE, K.W. & LOHSE, G.A. 1974. *Die Käfer Mitteleuropas*. Band 5, pp. 7-304. Krefeld: Goecke & Evers.

FÜRSCH, H. 1996. Taxonomy of coccinellids. *Coccinella* **6**: 28-30.

GENTILI, E. & CHIESA, A. 1975. Revisione dei *Laccobius* paleartici. *Memorie della Società Entomologica Italiana* **54**: 5-176.

GENTILI, E. & RIBERA, I. 1998. Description of *Laccobius glorianae* sp. n. from Spain, and notes on *L.*

ytenensis Sharp, 1910 and *L. atrocephalus* Reitter, 1872 (Insecta: Coleoptera: Hydrophilidae). *Annalen des Naturhistorischen Museums in Wien* **100B**: 193-198.

GERSTMEIER, R. 1998. *Checkered Beetles: illustrated key to the Cleridae and Thanerocleridae of the Western Palaearctic.* Weikersheim: Margraf Verlag.

GILDENKOV, M.Yu. 2001. *The Palaearctic* Carpelimus *Fauna (Coleoptera: Staphylinidae). The Problems of Species and the Formation of Species.* (2 vols.) Smolensk: Smolensk State Pedagogical University.

—— 2003. A new system of subfamily Oxytelinae (Coleoptera: Staphylinidae). *Kharkov Entomological Society Gazette* **10**: 32-38.

GØNGET, H. 1997. *The Brentidae (Coleoptera) of Northern Europe.* Fauna Entomologica Scandinavica 34, 289 pp. Leiden: E.J. Brill.

GOOD, J.A. 1998. The identification and habitat of micropterous *Diglotta mersa* (Haliday) and *D. sinuaticollis* (Mulsant & Rey) (Staphylinidae). *The Coleopterist* 7: 73-76.

GREBENNIKOV, V.V. & NEWTON, A.F. 2009. Good-bye Scydmaenidae, or why the ant-like stone beetles should become megadiverse Staphylinidae sensu latissimo (Coleoptera). *European Journal of Entomology* 106: 275-301.

GUSAROV, V.I. 1991. New and little known Palearctic Staphylinidae (Coleoptera). 3rd communication. *Vestnik Leningradskogo Gosudarstvennogo Universiteta Seriia Biologiia* **1991**: 3-12.

—— 1992. New and little known Palearctic Staphylinidae (Coleoptera). *Entomologicheskoe Obozrenie* **71**: 775-788.

—— 1994. New and little known Palaearctic Paederinae (Coleoptera: Staphylinidae). *Annales de la Société Entomologique de France* 30: 431-446.

—— 2003. Revision of some types of North American aleocharines (Coleoptera: Staphylinidae: Aleocharinae), with synonymic notes. *Zootaxa* 353: 1-134.

—— 2004. A revision of the Nearctic species of the genus *Halobrecta* Thomson, 1858 (Coleoptera: Staphylinidae: Aleocharinae), with notes on some Palaearctic species of the genus. *Zootaxa* 746: 1-25.

GUSAROV, V.I. & HERMAN, L.H. 2002. *Leptusa* Kraatz, 1856 (Coleoptera, Staphylinidae, Aleocharinae): designation of the type species. *Entomologische Blätter für Biologie und Systematik der Käfer* 98: 115-119.

HALSTEAD, A.J. 1996. Possible breeding by the Rosemary Beetle, *Chrysolina americana* L. in Britain. *British Journal of Entomology and Natural History* 9: 107-108.

HAMMOND, P.M. 1973. Notes on British Staphylinidae 3. The British species of *Sepedophilus* Gistel (*Conosomus* auct.). *Entomologist's Monthly Magazine* 108: 130-165.

—— 1981. *Aloconota (Aloconota) subgrandis* (Brundin) (Coleoptera: Staphylinidae) new to Britain. *Entomologist's Gazette* 32: 120-122.

—— 1982a. *Cymindis macularis* (Fischer v. Waldheim) (Col., Carabidae) - apparently a British species. *Entomologist's Monthly Magazine* 118: 37-38.

—— 1982b. On the British species of *Phacophallus* Coiffait (Col., Staphylinidae). *Entomologist's Monthly Magazine* 118: 231-232.

—— 1996. *A Taxonomic Review of Possibly Endemic British Non-marine Invertebrates.* Unpublished report to English Nature. London: The Natural History Museum.

—— 2000a. Coastal Staphylinidae (rove beetles) in the British Isles, with special reference to saltmarshes. In: Sherwood, B.R., Gardiner, B.G. & Harris, T. (eds.) *British Saltmarshes* pp. 247-302. Cardigan: Forrest Press.

—— 2000b. Essex Coleoptera in 1999. *Essex Naturalist* (n.s.) **17**: 82-84.

—— 2002. *UK tachyine beetles (Coleoptera: Carabidae) with special reference to the status of* Tachys edmondsi *Moore.* Peterborough: English Nature Research Reports 465, pp. 42.

—— 2007a. *Pentaphyllus testaceus* (Hellwig) (Tenebrionidae): an established and perhaps native British species? *The Coleopterist* 16: 47-52.

—— 2007b. *Paraphloeostiba gayndahensis* (MacLeay) (Staphylinidae) in Britain, with notes on other Omaliinae exhibiting adventive behaviour. *The Coleopterist* 16: 105-109.

—— 2007c. *Cryptophilus integer* (Heer, 1841) (Languriidae) found in the open in Britain. *The Coleopterist* 16: 150-151.

—— 2008. *Cotaster uncipes* (Boheman, 1838) (Curculionidae) apparently established in Britain. *The Coleopterist* 17: 43-46.

HAMMOND, P.M. & BARHAM, C.S. 1982. *Laricobius erichsoni* Rosenhauer (Coleoptera: Derodontidae), a species and superfamily new to Britain. *Entomologist's Gazette* 3: 35-40.

HAMMOND, P.[M.] & HARVEY, P. 2011. The exotic seed beetle *Bruchus brachialis* Fahraeus (Coleoptera: Chrysomelidae: Bruchinae) established in South Essex. *Essex Naturalist* (n.s.) **28**: 29-33.

HANCE, D. 2007. *Astrapaeus ulmi* (Rossi, 1790) (Staphylinidae) in Britain. *The Coleopterist* 16: 1-2.

HANLEY, R.S. 2002. A new species of Mexican *Tinotus* from the refuse piles of *Atta* ants, including an annotated world catalog of *Tinotus* (Coleoptera: Staphylinidae: Aleocharinae: Aleocharini). *Coleopterists Bulletin* 56: 453-471.

HANSEN, M. 1982. Revisional notes on some European *Helochares* Muls. (Coleoptera: Hydrophilidae). *Entomologica Scandinavica* 13: 201-211.

—— 1987. *The Hydrophiloidea (Coleoptera) of Fennoscandia and Denmark.* Fauna Entomologica Scandinavica 18, 254 pp.

—— 1991. *The hydrophiloid beetles: phylogeny, classification and a revision of the genera (Coleoptera, Hydrophiloidea).* Biologiske Skrifter 40, 368 pp.

—— 1996. Katalog over Danmarks biller (Catalogue of the Coleoptera of Denmark). *Entomologiske Meddelelser* **64**(1&2): 1-231.

—— 1997. Phylogeny and classification of the staphyliniform beetle families (Coleoptera). *Royal Danish Academy of Science and Letters, Biologiske Skrifter* **48**: 1-339.

—— 1998. *World Catalogue of Insects. Vol. 1. Hydraenidae (Coleoptera).* Stenstrup: Apollo Books.

—— 1999. *World Catalogue of Insects. Vol. 2. Hydrophiloidea (s. str.) (Coleoptera).* Stenstrup: Apollo Books.

HARRISON, T.D. 1992a. *Tetrops starkii* Chevrolat (Cerambycidae) new to Britain. *The Coleopterist* **1**(1): 3.

—— 1992b. *Tetrops starkii* (Col., Cerambycidae) new to Britain. *Entomologist's Monthly Magazine* **128**: 181-183.

—— 1993. *Trachyphloeus angustisetulus* Hansen and *T. bifoveolatus* (Beck) (Curculionidae) in Britain. *The Coleopterist* **2**: 35-36.

—— 1996. *Eulagius filicornis* (Reitter) (Mycetophagidae) apparently established in Britain. *The Coleopterist* **4**: 65-67.

—— 2008. *Otiorhynchus cribricollis* Gyllenhal (Curculionidae) new to Britain. *The Coleopterist* **17**: 141-143.

—— 2009. Records of notable beetles from Stockbridge Down, Hampshire including *Xyletinus ater* (Creutzer) (Anobiidae) new to Britain. *The Coleopterist* **18**: 75-76.

—— 2010. *Longitarsus symphyti* Heikertinger, 1912 (Chrysomelidae) new to Britain. *The Coleopterist* **19**: 41-43.

HAWKINS, R. 2000. *Rhyzobius chrysomeloides* (Herbst) (Coleoptera: Coccinellidae) new to Britain. *British Journal of Entomology & Natural History* **13**: 193-195.

HEAL, N.F. 1992. The discovery of *Lixus scabricollis* Bohe. (Curculionidae) in Britain. *The Coleopterist* **1**(1): 2.

—— 1993. *Trichiusa immigrata* Lohse (Staphylinidae) - first record for Britain. *The Coleopterist* **2**: 18.

—— 2003. *Scolytus pygmaeus* (Fabricius, 1787) (Scolytidae) - a new arrival to Britain. *The Coleopterist* **12**: 57-60.

—— 2006. *Taphrorychus villifrons* Dufour (Scolytidae) in Britain. *The Coleopterist* **15**: 135-136.

HEBAUER, F. 1985. *Ochthebius (Hymenodes) difficilis* Mulsant - another British species? *Balfour-Browne Club Newsletter* **34**: 1-2.

HERMAN, L.H. 1986. Revision of *Bledius* Part IV. Classification of species groups, phylogeny, natural history, and catalogue (Coleoptera, Staphyliniae, Oxytelinae). *Bulletin of the American Museum of Natural History* **184**, 367 pp.

—— 2001a. Nomenclatural changes in the Staphylinidae (Insecta, Coleoptera). *Bulletin of the American Museum of Natural History* **264**, 83 pp.

—— 2001b. Catalog of the Staphylinidae (Insecta, Coleoptera): 1758 to the End of the Second Millennium.

Bulletin of the American Museum of Natural History **265**, vi+4218 pp.

—— 2002. Case 3231 Staphylinidae Latreille, 1804 (Insecta, Coleoptera): proposed conservation of 17 specific names. *Bulletin of Zoological Nomenclature* **59**: 256-268.

—— 2003. Nomenclatural changes in the Paederinae (Coleoptera: Staphylinidae). *American Museum Novitates* **3416**, 28 pp.

—— 2004. Case 3265 *Lathrobium geminum* Kraatz, 1857 (Insecta, Coleoptera): proposed precedence over *L. volgense* Hochhuth, 1851 and *L. boreale* Hochhuth, 1851; *L. volgense*: proposed precedence over *L. boreale. Bulletin of Zoological Nomenclature* **61**: 25-28.

HINCKS, W.D. 1953. Imported Cassidinae (Col., Chrysomelidae) in Britain. *Entomologist's Monthly Magazine* **89**: 263.

HINTON, H.E. 1945. *A monograph of the beetles associated with stored products.* Vol. 1, 443 pp. London: British Museum (Natural History).

HODGE, P.J. 1997. *Bruchidius varius* (Olivier) (Chrysomelidae) new to the British Isles. *The Coleopterist* **5**: 65-68.

—— 1998. 1996 Annual Exhibition. Imperial College, London SW7 - 2 November 1996. Coleoptera. *British Journal of Entomolgy and Natural History* **10**: 167-176.

—— 1999. 1998 Annual Exhibition. Imperial College, London SW7 - 31 October 1998. Coleoptera. *British Journal of Entomology and Natural History* **12**: 169-179.

—— 2000. 1999 Annual Exhibition. Imperial College, London SW7 - 27 November 1999. Coleoptera. *British Journal of Entomology and Natural History* **13**: 172-182.

—— 2003. 2002 Annual Exhibition. Imperial College, London SW7 - 9 November 2002. Coleoptera. *British Journal of Entomology and Natural History* **16**: 181-186.

—— 2005. 2004 Annual Exhibition. Imperial College, London SW7 - 13 November 2004. Coleoptera. *British Journal of Entomology and Natural History* **18**: 201-208.

—— 2007. 2006 Annual Exhibition. Imperial College, London SW7 - 11 November 2006. Coleoptera. *British Journal of Entomology and Natural History* **20**: 179-191.

—— 2010. *Agrilus cuprescens* (Ménétriés, 1832) and *A. cyanescens* Ratzeburg, 1837 (Buprestidae) established in Britain. *The Coleopterist* **18**: 85-88.

—— 2011. *Pseudoperapion brevirostre* (Herbst, 1797) (Apionidae) new to the British Isles. *The Coleopterist* **20**: 111-115.

—— in prep. 2011 Annual Exhibition. Imperial College, London SW7 - 5 November 2011. Coleoptera. *British Journal of Entomology and Natural History* **25**.

HODGE, P.J. & JONES, R.A. 1995. *New British Beetles. Species not in Joy's practical handbook.* Reading: British Entomological and Natural History Society.

HODGE, P.J. & PARRY, J.A. 1981. *Ptinus dubius* Sturm (Col., Ptinidae) new to Britain. *Entomologist's Monthly Magazine* **117**: 225-226.

HOLMEN, M. 1987. *The aquatic Adephaga (Coleoptera) of Fennoscandia and Denmark.* I. Gyrinidae, Haliplidae,

Hygrobiidae and Noteridae. Leiden: E.J. Brill. Fauna Entomologica Scandinavica, No. 20.

HORÁK, J. 1996. Revision of some little known species of genus *Mordellistena* with description of two new species. Part 2. (Coleoptera: Mordellidae). *Klapalekiana* 32: 171-184.

HORION, A.D. 1963. *Faunistik der Mitteleuropäischen Käfer, IX: Staphylinidae, 1: Micropeplinae bis Euaesthetinae.* Überlingen am Bodensee: A. Feyel.

HŮRKA, K. 1996. *Carabidae of the Czech and Slovak Republics.* Zlín, Czech Republic: V. Kabourek Ltd.

HYMAN, P.S. 1987a. *Otiorhynchus aurifer* Boheman (Col., Curculionidae) new to the British Isles. *Entomologist's Monthly Magazine* 123: 59.

—— 1987b. *Bruchella rufipes* (Olivier) (Col., Anthribidae) rediscovered in Great Britain. *Entomologist's Monthly Magazine* 123: 90.

HYMAN, P.S. (revised PARSONS, M.S.) 1994. *A review of the scarce and threatened Coleoptera of Great Britain.* Part 2. UK Nature Conservation: 12. Peterborough: Joint Nature Conservation Committee.

ICZN 1999. *International Code of Zoological Nomenclature.* Fourth Edition. London. Available online: http://www.iczn.org.

ISRAELSON, G. 1979. The taxonomy of some West European and Macaronesian *Heterothops* Stephens (Coleoptera: Staphylinidae). *Entomologica Scandinavica* 10: 261-268.

JÄCH, M.A. 1989. Revision of the Palearctic species of the genus *Ochthebius* Leach. I. The so-called subgenus "*Bothochius*" (Hydraenidae, Coleoptera). *Koleopterologische Rundschau* 59: 95-126.

JAMES, T.J. 1994. *Agrilus sulcicollis* Lacordaire (Buprestidae): a jewel beetle new to Britain. *The Coleopterist* 3: 33-35.

—— 2011. Coleoptera 2010-2011. *Transactions of the Hertfordshire Natural History Society* 43: 112-121.

JÁSZAY, T. & HLAVAČ, P. 2006. A revision of the Palaearctic species of the genus *Dropephylla* (Coleoptera: Staphylinidae: Omaliinae). *Entomological Problems* 36: 31-62.

JERMIIN, L.S., LOESCHCKE, V., SIMONSEN, V. & MAHLER, V. 1991. Electrophoretic and morphometric analyses of two sibling species pairs in *Trachyphloeus* (Coleoptera: Curculionidae). *Entomologica Scandinavica* 22: 159-170.

JOHNSON, C. 1963. *Chrysolina americana* (L.) (Col., Chrysomelidae) in Britain. *Entomologist's Monthly Magazine* 99: 228-229.

—— 1967. A revised and annotated British list of *Acrotrichis* (Col., Ptiliidae). *Entomologist* 100: 132-136.

—— 1975a. Five species of Ptiliidae (Col.) new to Britain, and corrections to the British list of the family. *Entomologist's Gazette* 26: 211-223.

—— 1975b. Nine species of Coleoptera new to Britain. *Entomologist's Monthly Magazine* 111: 177-183.

—— 1978. Notes on Byrrhidae (Col.); with special reference to, and a species new to, the British fauna.

Entomologist's Record & Journal of Variation 90: 141-147.

—— 1982a. *Phytobius olssoni* Israelson (Coleoptera: Curculionidae) new to Britain. *Entomologist's Gazette* 33: 221-222.

—— 1982b. An introduction to the Ptiliidae (Coleoptera) of New Zealand. *New Zealand Journal of Zoology* 9: 333-376.

—— 1986. Notes on some Palaearctic *Melanophthalma* Motschulsky (Coleoptera: Latridiidae) with special reference to *transversalis* auctt. *Entomologist's Gazette* 37: 117-126.

—— 1987a. Additions and corrections to the British list of Ptiliidae (Coleoptera). *Entomologist's Gazette* 38: 117-122.

—— 1987b. A revised checklist of British *Acrotrichis* Motschulsky (Coleoptera: Ptiliidae). *Entomologist's Gazette* 38: 229-242.

—— 1988. Notes on some British *Cryptophagus* Herbst (Coleoptera: Cryptophagidae), including *confusus* Bruce new to Britain. *Entomologist's Gazette* 39: 329-335.

—— 1991. Coleoptera of Merioneth, North Wales: a second supplement to Skidmore & Johnson's list, 1969. *Entomologist's Gazette* 42: 107-145.

—— 1992a. Further changes in the nomenclature of European *Atomaria* Stephens (Coleoptera: Cryptophagidae). *Entomologist's Gazette* 43: 145-146.

—— 1992b. Additions and corrections to the British list of Coleoptera. *Entomologist's Record & Journal of Variation* 104: 305-310.

—— 1993. *Provisional atlas of the Cryptophagidae-Atomariinae (Coleoptera) of Britain and Ireland* (ed. Harding, P.T. & Dring, J.C.M.). Huntingdon: Biological Records Centre.

—— 1997. *Clambus simsoni* Blackburn (Col., Clambidae) new to Britain with notes on its wider distribution. *Entomologist's Monthly Magazine* 133: 161-164.

—— 2001a. Notes on Palaearctic Ptiliidae (Coleoptera). *Entomologist's Gazette* 52: 129-137.

—— 2001b. *Euryptilium gillmeisteri* Flach (Coleoptera: Ptiliidae) new to Britain. *Entomologist's Gazette* 52: 181-182.

—— 2002. Provisional atlas of the Cryptophagidae: Atomariinae (Coleoptera) of Britain and Ireland: supplement to the 1993 atlas and *Atomaria turgida* Erichson newly recorded for the fauna. *Entomologist's Gazette* 53: 183-189.

—— 2003. Further notes on Palaearctic and other Ptiliidae (Coleoptera). *Entomologist's Gazette* 54: 55-70.

—— 2004. Ptiliidae, pp. 122-131. In: Löbl, I. & Smetana, A. (eds.) *Catalogue of Palaearctic Coleoptera. Vol. 2.* Stenstrup: Apollo Books, 942 pp.

—— 2007. *Stephostethus caucasicus* (Mannerheim, 1844) (= *sinuatocollis* auctt., = *campicola* auctt.) (Latridiidae) new to Britain. *The Coleopterist* 16: 101-103.

—— 2012a in press. A new Himalayan *Sternodea* species together with notes and corrections on Palaearctic Cryptophagidae (Coleoptera). *Entomologist's Gazette.*

—— 2012b in press. *Corticaria saginata* Mannerheim (Col., Latridiidae) erroneously recorded as British. *The Coleopterist.*

—— 2012c in press. *Brachygluta klimschi* Holdhaus, 1902 (Staphylinidae; Pselaphinae) in England. *The Coleopterist.*

JOHNSON, C. & BOOTH, R.G. 2004. *Luperomorpha xanthodera* (Fairmaire): a new British flea beetle (Chrysomelidae) on garden centre roses. *The Coleopterist* **13**: 81-86.

JOHNSON, C. & ECCLES, T.M. 1983. *Plectophloeus erichsoni occidentalis* Besuchet (Coleoptera: Pselaphidae) new to Britain. *Entomologist's Gazette* **34**: 267-269.

JONES, R.A. 1989. 1988 Annual Exhibition. Imperial College, London SW7 - 19 November 1988. Coleoptera. *British Journal of Entomology and Natural History* **2**: 47-53.

—— 2006. *Rhopalapion longirostre* (Olivier 1807) (Apionidae) finally discovered in Britain. *The Coleopterist* **15**: 93-97.

JOY, N.H. 1932. *A Practical Handbook of British Beetles.* London: H.F. & G. Witherby.

JOY, N.H. & TOMLIN, J.R.LeB. 1909. *Micropeplus caelatus*, Erichson: a British insect. *Entomologist's Monthly Magazine* **45**: 149-150.

KASULE, F.K. 1966. The subfamilies of the larvae of Staphylinidae (Coleoptera) with keys to the larvae of British genera of Steninae and Proteininae. *Transactions of the Royal Entomological Society of London* **118**: 261-283.

KAZANTSEV, S.V. 2004. Phylogeny of the tribe Erotini (Coleoptera, Lycidae), with descriptions of new taxa. *Zootaxa* **496**: 1-48.

KENDALL, P. 1981. *Bromius obscurus* (L.) in Britain (Col., Chrysomelidae). *Entomologist's Monthly Magazine* **117**: 233-234.

KEVAN, D.K. 1963. *Atheta (Amidobia) luctuosa* (Mulsant & Rey) (Col., Staphylinidae) new to the British list. *Entomologist's Monthly Magazine* **99**: 137-138.

KEVAN, D.K. & ALLEN, A.A. 1961. Notes on some British species of *Stenus* Latreille (Col., Staphylinidae), with additions and amendments to the British list. *Entomologist's Monthly Magazine* **97**: 211-217.

KEY, R.S. 1983. *Troglops cephalotes* (Olivier) (Col., Melyridae) from Buckinghamshire, possibly a new British species. *Entomologist's Monthly Magazine* **119**: 71-72.

KING, C.J. & FIELDING, N.J. 1989. *Dendroctonus micans* in Britain - its biology and control. *Forestry Commission Bull. 85.* London: HMSO.

KIRK-SPRIGGS, A.H. 1988. *Meligethes gagathinus* Erichson, 1845 (Col., Nitidulidae) a distinct species - a correction to the British checklist. *Entomologist's Monthly Magazine* **124**: 148.

—— 1992. *Meligethes haemorrhoidalis* Förster (Col., Nitidulidae) in Britain. *Entomologist's Monthly Magazine* **128**: 25-29.

—— 1996. Pollen beetles. Coleoptera: Kateretidae and Nitidulidae: Meligethinae. *Handbooks for the Identification of British Insects* **5**(6a), 157 pp.

KISTNER, D.H. 1971. Studies of Japanese myrmecophiles Part 1. The genera *Pella* and *Falagria* (Coleoptera, Staphylinidae). In: Asahima, S., Gressitt, J. L., Hikada, Z., Mishida, T. & Namura, K. (eds.) *Entomological Essays to Commemorate the Retirement of Prof. K. Yasumatsu.* Tokyo: Hokuryukan Publishing Co., Ltd., pp. 141-165.

KLAUSNITZER, B. 1998. Über die *Cyphon*-Arten Henri Tourniers. *Beiträge zur Entomologie, Berlin* **48**: 411-415.

—— 2004. Eine neue Gattung der Familie Scirtidae (Insecta: Coleoptera). *Entomologische Abhandlungen* **62**: 77-82.

KÖHLER, F. & KLAUSNITZER, B. 1998. Verzeichnis der Käfer Deutschlands. *Entomologische Nachrichten und Berichte* Beiheft **4**: 1-185.

KOJIMA, H. & MORIMOTO, K. 1996. Systematics of the flea weevils of the tribe Ramphini (Coleoptera, Curculionidae) from East Asia. II. Phylogenetic analysis and higher classification. *Esakia* **36**: 97-134.

KOLIBÁČ, J. 1992. Revision of Thanerocleridae n.stat. (Coleoptera, Cleroidea). *Mitteilungen der Schweizerischen Entomologischen Gesellschaft* **65**: 303-340.

KORGE, H. 1963. Beiträge zur Koleopterenfauna der Mark Brandenburg (Teil XXVII). *Mitteilungen der Deutschen Entomologischen Gesellschaft* **22**: 76-78.

KRAJČÍK, M. 1998. *Cetoniidae of the World, Catalogue. Part 1: Goliathini, Cetoniini, Gymnetini, Diplognathini, Phaedimini, Taenioderini (Coleoptera: Cetoniidae).* Most, Czech Republic: Typos Studio, 96+36 pp.

—— 1999. *Cetoniidae of the World, Catalogue. Part 2: Trichiinae, Valginae, Cetoniinae (Coleoptera: Cetoniidae).* Most, Czech Republic: Typos Studio, 72+23 pp.

KRÁL, D. 1993. Scarabaeoidea. In: Jelinek, J. (ed.) *Checklist of Czechoslovak Insects IV (Coleoptera)*, Folia Heyrovskyana. Supplementum 1, pp. 66-70.

KRELL, F.-T. 1991. Vorschlag zur Stabilisierung der *Hoplia*-Nomenklatur (Coleoptera: Scaraboidea: Hopliinae). *Entomologische Blätter für Biologie und Systematik der Käfer* **87**: 186-192.

—— 1998. Erganzungen und Berichtigungen zu den Banden 6, 7, 8 und 13 (2. Supplement). Familienreihe Lamellicornia. In: Lucht, W. & Klausnitzer, B. (eds.) *Die Kafer Mitteleuropas.* Bd. 15: 4. Supplementband. Krefeld: Goecke & Evers, pp. 285-295.

—— 2012. On nomenclature and synonymy of *Trichius rosaceus*, *T. gallicus*, and *T. zonatus* (Coleoptera: Scarabaeidae: Cetoniinae: Trichiini). *Zootaxa* **3278**: 61-68.

KRELL, F.-T. & FERY, H. 1992. Trogidae, Geotrupidae, Scarabaeidae, Lucanidae. In: Lohse, G.A. & Lucht, W.H. (eds.) *Die Käfer Mitteleuropas.* Band 13. 2

Supplementband mit Katalogteil. Krefeld: Goecke & Evers, pp. 201-253.

KRELL, F.-T., JOHNSON, C., BOOTH, R. & MENDEL, H. 2005. The British *Dienerella separanda* (Reitter) is *D. clathrata* (Mannerheim): with a compilation of British records of *D. clathrata* and *D. elongata* (Curtis) (Latridiidae). *The Coleopterist* **14**: 117-123.

KRELL, F.-T. & RÖßNER, E. 2009. The British *Amphimallon ochraceum* (Knoch) is *A. fallenii* (Gyllenhal) (Scarabaeidae: Melolonthinae). *The Coleopterist* **18**: 15-16.

KRYZHANOVSKIJ, O.L., BELOUSOV, I.A., KABAK, I.I., KATAEV, B.M., MAKAROV, K.V. & SHILENKOV, V.G. 1995. *A checklist of the ground-beetles of Russia and adjacent lands (Insecta, Coleoptera, Carabidae).* Sofia: Pensoft Books.

KUROCHKIN, A.S. & KIREJTSHUK, A.G. 2006. Notes on the synonymy, variability and bionomy of *Epuraea (Epuraea) biguttata* (Thunberg, 1784) and *E. (E.) unicolor* (Olivier, 1790) (Coleoptera: Nitidulidae). *Russian Entomological Journal* **15**: 393-397.

LACKNER, T. 2005. *Hypocaccus* (s.str.) *crassipes* (Erichson, 1834), a new species for the fauna of Great Britain (Coleoptera: Histeridae). *Entomological Problems* **35**: 46.

LAST, H. 1952. Synonymic notes on species of the genus *Atheta* subgenus *Disopora* (Col., Staphylinidae) with description of a new British species. *Entomologist's Monthly Magazine* **88**: 263-264.

—— 1980. *Aloconota mihoki* Bernh. (Col., Staphylinidae) new to Britain. *Entomologist's Monthly Magazine* **114**: 239-240.

LAWRENCE, J.F. & NEWTON, A.F. 1982. Evolution and classification of beetles. *Annual Review of Ecology and Systematics* **13**: 261-290.

—— 1995. Families and subfamilies of Coleoptera (with selected genera, notes, references and data on family-group names). In: Pakaluk, J. & Slipinski, S.A. (eds.): *Biology, phylogeny and classification of Coleoptera.* Museum of the Zoological Institute PAN, Warsaw: pp 779-1006.

LESCHEN, R.A.B. 2003. Erotylidae (Insecta: Coleoptera: Cucujoidea): phylogeny and review. *Fauna of New Zealand* 47, 108 pp.

LEVEY, B. 1996. *Anaspis septentrionalis* Champion, a senior synonym of *A. schilskyana* Csiki (Scraptiidae). *The Coleopterist* **5**: 58.

—— 1997. *Stephostethus alternans* (Mannerheim) (Lathridiidae), a species new to Britain. *The Coleopterist* **6**: 49-51.

—— 1999. *Mordellistena secreta* Horák (Coleoptera: Mordellidae), a species new to Britain. *British Journal of Entomology and Natural History* **12**: 227-229.

—— 2002. Are *Anaspis septentrionalis* Champion and *A. thoracica* (Linnaeus) a single variable species? *The Coleopterist* **11**:1-5.

—— 2003. Taxonomic notes on European *Anaspis (Nassipa)* (Coleoptera: Scraptiidae), with general notes on the British species. *Entomologist's Gazette* **54**: 197-204.

—— 2005. Some British records of *Chaetarthria simillima* Vorst & Cuppen, 2003 and *C. seminulum* (Herbst) (Hydrophilidae), with notes on their differentiation. *The Coleopterist* **14**: 97-99.

—— 2009. British Scraptiidae. *Handbooks for the Identification of British Insects* 5(18), pp. iv+32.

—— 2012 in press. *Trachys subglaber* Rey, 1891 (Buprestidae) an unrecognised British species. *The Coleopterist.*

LEVEY, B. & PAVETT, P.M. 1999. *Bembidion (Pseudolimnaeum) inustum* Duval, (Coleoptera: Carabidae) an interesting new addition to the British fauna. *British Journal of Entomology and Natural History* **11**: 169-171.

LIKOVSKÝ, Z. 1984. Über die Nomenklatur der Aleocharinen (Coleoptera, Staphylinidae). *Annotationes Zoologicae et Botanicae* **160**: 1-8.

LINDROTH, C.H. 1974. Coleoptera Carabidae. *Handbooks for the Identification of British Insects* **4**(2), 148 pp.

—— 1985. *The Carabidae (Coleoptera) of Fennoscandia and Denmark.* Leiden: E.J. Brill.

LÖBL, I. 1997. *Catalogue of the Scaphidiinae (Coleoptera: Staphylinidae).* Geneva: Muséum d'Histoire Naturelle de Genève, 190 pp.

—— 1998. On new and old replacement names in Palearctic Pselaphinae (Coleoptera: Staphylinidae). *Mitteilungen der Schweizerischen Entomologischen Gesellschaft* **71**: 463-465.

LÖBL, I. & BESUCHET, C. 2004. Pselaphinae. pp. 272-329 In: Löbl, I. & Smetana, A. (eds.) *Catalogue of Palaearctic Coleoptera* Vol. 2. Stenstrup: Apollo Books.

LÖBL, I. & SMETANA, A. (eds.) 2003. *Catalogue of Palaearctic Coleoptera, Vol. 1: Archostemata - Myxophaga - Adephaga.* Stenstrup: Apollo Books, 819 pp.

—— 2004. *Catalogue of Palaearctic Coleoptera. Vol. 2, Hydrophiloidea - Histeroidea- Staphylinoidea.* Stenstrup: Apollo Books, 942 pp.

—— 2006. *Catalogue of Palaearctic Coleoptera. Vol. 3. Scarabaeoidea - Scirtoidea - Dascilloidea - Buprestoidea - Byrrhoidea.* Stenstrup: Apollo Books, 690 pp.

—— 2007. *Catalogue of Palaearctic Coleoptera, Vol. 4: Elateroidea - Derodontoidea - Bostrichoidea - Lymexyloidea - Cleroidea - Cucujoidea.* Stenstrup: Apollo Books, 935 pp.

—— 2011. *Catalogue of Palaearctic Coleoptera, Vol. 7: Curculionoidea I.* Stenstrup: Apollo Books, 373 pp.

LOHSE, G.A.1963. Neue Staphyliniden aus Mitteleuropa und dem Alpengenbiet. *Entomologische Blätter für Biologie und Systematik der Käfer* **59**: 168-178.

—— 1982. 13. Nachtrag zum Verzeichnis der mitteleuropäischen Käfer. *Entomologische Blätter für Biologie und Systematik der Käfer* **78**: 115-126.

—— 1984. *Phloeopora*-Studien (ein nomenklatorischer Horror-Krimi). *Entomologische Blätter für Biologie und Systematik der Käfer* **80**: 153-162.

—— 1985a. Kritische Bemerkungen zur Staphylinidae in *Enumeratio Coleopterorum Fennoscandiae et Daniae* (Helsingfors 1979). *Notulae Entomologicae* **65**: 33-35.

—— 1985b. Betrachtungen über die Gattung *Emplenota* Casey (Coleoptera, Staphylinidae). *Faunistisch-Ökologische Mitteilungen* **5**: 327-330.

—— 1985c. *Diglotta*-Studien. *Entomologische Blätter für Biologie und Systematik der Käfer* **81**: 179-182.

—— 1987. Bemerkungen zur Systematik der Athetae. *Entomologische Blätter für Biologie und Systematik der Käfer* **83**: 17-18.

—— 1988. Stahylinidenstudien II. *Entomologische Blätter für Biologie und Systematik der Käfer* **84**: 179-182.

—— 1989. Staphylinidae. In: Lohse, G.A. & Lucht, W.H. (eds.) *Die Käfer Mitteleuropas*. Band 12, Supplementband mit Katalogteil, pp. 121-240. Krefeld: Goecke & Evers.

LOHSE, G.A., KLIMASZEWSKI, J. & SMETANA, A. 1990. Revision of Arctic Aleocharinae of North America (Coleoptera: Staphylinidae). *Coleopterists Bulletin* **44**: 121-202.

LOHSE, G.A. & LUCHT, W.H. 1989. *Die Käfer Mitteleuropas. 1 Supplementband mit Katalogteil.* Krefeld: Goecke & Evers.

LOHSE, G.A. & SMETANA, A. 1985. Revision of the types of species of Oxypdini and Athetini (*sensu* Seevers) described by Mannerheim and Mäklin from North America (Coleoptera: Staphylinidae). *Coleopterists Bulletin* **39**: 281-300.

LOTT, D.A. 1989. *Hadrognathus longipalpus* (Mulsant & Rey) (Coleoptera: Staphylinidae) new to the British Isles. *Entomologist's Gazette* **40**: 221-222.

—— 1993a. *Ischnoglossa obscura* Wunderle, new to Britain. *The Coleopterist* **2**: 20.

—— 1993b. Two changes to the British list of Staphylinidae. *The Coleopterist* **2**: 20-22.

—— 1993c. The British species of the *Thinobius longipennis* (Heer) group (Coleoptera: Staphylinidae). *Entomologist's Gazette* **44**: 285-287.

—— 1996. Changes to the British List published in 1995. *The Coleopterist* **5**: 1-2.

—— 1998. Changes to the British List published in 1996 and 1997. *The Coleopterist* **7**: 3-5.

—— 2002. A Review of Staphylinidae doubtfully recorded from the British Isles. *The Coleopterist* **11**: 33-39.

—— 2008. The British species of *Ochthephilus* Mulsant & Rey (Staphylinidae). *The Coleopterist* **17**: 17-22.

—— 2009a. The Staphylinidae (rove beetles) of Britain and Ireland. Part 5: Scaphidiinae, Piestinae, Oxytelinae. *Handbooks for the Identification of British Insects* **12**(5), pp. iv+100.

—— 2009b. *Acrotona convergens* Strand (Staphylinidae) new to Ireland. *The Coleopterist* **18**: 131-138.

—— 2010. A new species of *Quedius* Stephens subgenus *Microsaurus* Dejean (Coleoptera:

Staphylinidae) from Scotland and Ireland. *The Coleopterist* **19**: 135-141.

—— 2011. Some recent developments affecting the British list of Oxytelinae (Staphylinidae). *The Coleopterist* **20**: 23-30.

LOTT, D.A. & ANDERSON, R. 2011. The Staphylinidae (rove beetles) of Britain and Ireland. Parts 7 & 8: Oxyporinae, Steninae, Euaesthetinae, Pseudopsinae, Paederinae, Staphylininae. *Handbooks for the Identification of British Insects* **12**(7), pp. iv+340.

LOTT, D.A. & FOSTER, G.N. 1990. Records of terrestrial Coleoptera from wetland sites in 1987, including *Stenus glabellus* Thomson (Staphylinidae) new to the British Isles. *Irish Naturalists' Journal* **23**: 280-282.

LOTT, D.A., HAMMOND, P.M. & WEBB, J.R. 2007. The British species of *Ochthephilum* Stephens (Staphylinidae). *The Coleopterist* **16**: 81-87.

LUFF, M.L. 1990. *Pterostichus rhaeticus* Heer (Col., Carabidae), a British species previously confused with *P. nigrita* (Paykull). *Entomologist's Monthly Magazine* **126**: 245-249.

—— 1993. *The Carabidae (Coleoptera) larvae of Fennoscandia and Denmark*. Leiden: E.J. Brill.

—— 2007. The Carabidae (ground beetles) of Britain and Ireland. *Handbooks for the Identification of British Insects* **4**(2) (2nd edn.), iv+247 pp.

LUNDMARK, M., DROTZ, M.K. & NILSSON, A.N. 2001. Morphometric and genetic analysis shows that *Haliplus wehnckei* is a junior synonym of *H. sibiricus* (Coleoptera: Haliplidae). *Insect Systematics & Evolution* **32**: 241-251.

LYSZKOWSKI, R.M. 1992. *Atheta (s.str.) heymesi* Hubenthal (Staphylinidae) new to Britain. *The Coleopterist* **1**(2): 1-2.

—— 1993. *Pityophthorus lichtensteini* (Ratzeburg), (Col.: Scolytidae) rediscovered in Aberdeenshire, and a problem partially solved. *Entomologist's Record & Journal of Variation* **105**: 229-231.

LYSZKOWSKI, R.M., OWEN, J.A. & TAYLOR, S. 1992. *Corticaria abietorum* Motschulsky (Col.: Lathridiidae) new to Britain. *Entomologist's Record & Journal of Variation* **104**: 67-69.

MAHLER, V. & VAGTHOLM-JENSEN, O. 2002. De danske arter af rovbilleslaegten *Schistoglossa* Kraatz, 1856, med *S. bergvalli* Palm, 1968, som ny for Danmark (Coleoptera, Staphylinidae). *Entomologiske Meddelelser* **70**: 51-55.

MAJERUS, M.E.N., MAJERUS, T.M.O., BERTRAND, D. & WALKER, L.E. 1997. The geographic distribution of ladybirds (Coleoptera: Coccinellidae) in Britain (1984–1994). *Entomologist's Monthly Magazine* **133**: 181-203.

MAJERUS, M.E.N. & ROY, H.E. 2005. Scientific opportunities presented by the arrival of the Harlequin ladybird, *Harmonia axyridis*, in Britain. *Antenna* **29**: 196-208.

MAKRANCZY, G. & SCHÜLKE, M. 2001. Typenstudien an den mitteleuropäischen Vertreten der Artengruppe des *Thinobius linearis* Kraatz, 1857 (Coleoptera, Staphylinidae, Oxytelinae). *Entomologische Blätter für Biologie und Systematik der Käfer* **97**: 185-193.

MANN, D.J. 2000b. Changes to the British List published in 1998 and 1999. *The Coleopterist* 9: 49-53.

—— 2002. Changes to the British Coleoptera List published in 2000 and 2001. *The Coleopterist* 11: 52-63.

—— 2006. *Ptilodactyla exotica* Chapin, 1927 (Coleoptera: Ptilodactylidae: Ptilodactylinae) established breeding under glass in Britain, with a brief discussion on the family Ptilodactylidae. *Entomologist's Monthly Magazine* 142: 67-79.

MANN, D.J. & BOOTH, R.G. 2000. *Brindalus porcicollis* (Illiger) (Coleoptera: Scarabaeidae: Psammodiinae) in Britain. *British Journal of Entomology and Natural History* 13: 137-145

MANN, D.J., HANCOCK, E.G. & MORRIS, M.G. 2005. History of the genus *Lixus* Fabricius (Curculionidae) in Britain with comments on nomenclature. *The Coleopterist* 14: 65-80.

MARTÍN-PIERA, F.Y. & LÓPEZ-COLÓN, J.I. 2000. Coleoptera Scarabaeoidea I. In: Ramos, M.A. *et al.* (eds.) *Fauna Ibérica Vol. 14*. Madrid: Museo Nacional de Ciencias Naturales CSIC, 526pp.

MARUYAMA, M. 2006. Revision of the Palearctic species of the myrmecophilous genus *Pella* : Coleoptera, Staphylinidae, Aleocharinae. *National Science Museum Monographs (Tokyo)* 32: 1-207.

MASON, P.B. 1898. *Cryptohypnus meridionalis*, Lap. an addition to the British list of Elateridae. *Entomologist's Monthly Magazine* 34: 207.

MAUS, C. 2001. On the identity of *Aleochara* (*Coprochara*) *pauxilla* (Mulsant & Rey, 1874) (Coleoptera: Staphylinidae). *Beiträge zur Entomologie* 51: 223-229.

MAUS, C., PESCHKE, K. & ASHE, J.S. 2000. *Aleochara* In: The Tree of Life web site. Available online: http://phylogeny.arizona.edu/tree/phylogeny.html.

MAZUR, S. 1984. A world catalogue of Histeridae. *Polskie Pismo Entomologiczne* 54(3-4): 1-379.

MENDEL, H. 1982. *Hemicoelus nitidus* (Hbst.) (Col., Anobiidae) new to Britain. *Entomologist's Monthly Magazine* 118: 253-254.

—— 1990. *Zorochros meridionalis* (Lap.) (Coleoptera: Elateridae) a British species? *Coleopterist's Newsletter* 39: 2-3.

—— 1994. *Rhynchaenus calceatus* (Germar) (Curculionidae) - new to Ireland. *The Coleopterist* 3: 38-39.

—— 2002. Notes on British Elateridae: *Dicronychus equisetioides* Lohse, 1976 and *Negastrius arenicola* (Boheman, 1853) recorded from Britain. *The Coleopterist* 11: 77-80.

—— 2004. *Melanotus villosus* (Geoffroy in Fourcroy, 1785) and *Melanotus castanipes* (Paykull, 1800) (Elateridae) in Britain. *The Coleopterist* 13: 121-124.

MENDEL, H. & BARCLAY, M.V.L. 2008. *Semanotus russicus* (Fabricius, 1776) (Cerambycidae) breeding in Britain. *The Coleopterist* 17: 1-4.

MENDEL, H. & HATTON, J. 2012 in press. *Gastrallus laevigatus* (Olivier) (Col., Anobiidae) a British species. *The Coleopterist*.

MENDEL, H., JEFFERY, P. & PLEDGER, M.J. 2011. *Isorhipis melasoides* (Laporte, 1835) (Eucnemidae) breeding and probably established in the British Isles. *The Coleopterist* 20: 41-43.

MENDEL, H. & OWEN, J.A. 1987. *Cicones undata* Guérin (Coleoptera: Colydiidae) new to Britain. *Entomologist's Record & Journal of Variation* 99: 93-95.

—— 1991. *Dorcatoma ambjoerni* Baranowski (Col., Anobiidae), another Windsor specialty? *Coleopterist's Newsletter* 43: 12-13.

MENZIES, I. & SPOONER, B. 2000. *Henosepilachna argus* (Geoffroy) (Coccinellidae, Epilachninae), a phytophagous ladybird new to the U.K., breeding at Molesey, Surrey. *The Coleopterist* 9: 1-4.

MONRÓS, F. 1956. Revisión genérica de Lamprosominae con descripción de algunos généros y especies nuevas (Col., Chrysomelidae). *Revista Agronómica del Noroeste Argentino* 2: 25-77.

MORRIS, M.G. 1977. The British species of *Anthonomus* Germar (Col., Curculionidae). *Entomologist's Monthly Magazine* 112: 19-40.

—— 1987. *Rhynchaenus pseudostigma* Tempère in Britain - a preliminary note. *Coleopterist's Newsletter* 29: 5.

—— 1989. The identity of four unique British specimens of *Gymnetron* (Col., Curculionidae). *Entomologist's Monthly Magazine* 125: 25-30.

—— 1990. Orthocerous weevils. Coleoptera: Curculionoidea (Nemonychidae, Anthribidae, Urodontidae, Attelabidae, and Apionidae). *Handbooks for the Identification of British Insects* 5(16), 108 pp.

—— 1991a. A taxonomic check list of the British Ceutorhynchinae, with notes, particularly on host plant relationships (Coleoptera: Curculionidae). *Entomologist's Gazette* 42: 255-265.

—— 1991b. *Phytobius zumpti* Wagner, 1939 (Curculionidae) in Britain - a preliminary note and appeal for records. *Coleopterist's Newsletter* 42: 1-2.

—— 1992. Five weevil species new to Ireland. *Irish Naturalists' Journal* 24: 65-69.

—— 1993a. 'British Orthocerous Weevils': corrections and new information (Coleoptera, Curculionoidea). *Entomologist's Monthly Magazine* 129: 23-29.

—— 1993b. A review of the British species of Rhynchaeninae (Col., Curculionidae). *Entomologist's Monthly Magazine* 129: 177-197.

—— 1995. Recent advances in the higher systematics of Curculionoidea, as they affect the British fauna. *The Coleopterist* 4: 21-30.

—— 1996. *Polydrusus chrysomela* (Olivier) not a British species (Curculionidae). *The Coleopterist* 5: 26-27.

—— 1997. Broad-nosed weevils. Coleoptera: Curculionidae (Entiminae). *Handbooks for the Identification of British Insects* 5(17a), 106 pp.

—— 2002. True weevils (Part I). Coleoptera: Curculionidae (subfamilies Raymondionyminae to Smicronychinae). *Handbooks for the Identification of British Insects* 5(17b), 149 pp.

—— 2003. An annotated check list of British Curculionoidea (Col.). *Entomologist's Monthly Magazine* **139**: 193-225.

—— 2004a. *Tychius crassirostris* Kirsch (Curculionidae) rediscovered in Dorset. *The Coleopterist* **13**: 27-28.

—— 2004b. The weevils (Insecta: Coleoptera, Curculionoidea) of the Dorset coast and their conservation. *Dorset Proceedings* **126**: 85-109.

—— 2011. Taxonomic and nomenclatural changes in the British Trachyphloeini (Curculionidae, Entiminae). *The Coleopterist* **20**: 77-81.

MORRIS, M.G. & BOOTH, R.G. 1997. Notes on the nomenclature of some British weevils (Curculionoidea). *The Coleopterist* **6**: 91-99.

MORRIS, M.G. & JOHNSON, C. 2005. Sidebotham's weevils (Curculionoidea). *The Coleopterist* **14**: 101-113.

MORRIS, M.G. & OWEN, J.A. 1997. Notes on *Rhynchaenus calceatus* (Germar) (Curculionidae) in Ireland. *The Coleopterist* **6**: 72-73.

MORRIS, M.G. & PÉRICART, J. 1988. A propos d'*Apion (Aspidapion) soror* Rey, 1895, espèce jumelle d'*Apion radiolus* (Marsham, 1802) méconnue en France (Col. Apionidae). *Bulletin de la Société Entomologique de France* **92**: 221-224.

MOUND, L. (ed.) 1989. *Common Insect Pests of Stored Food Products.* 7th edn. London: British Museum (Natural History), 68 pp.

MUGGLETON, J. 1999. *Coccinula 14-pustulata* (L.) and *Exochomus nigromaculatus* (Goeze) (Col., Coccinellidae) in Britain. *Entomologist's Monthly Magazine* **135**: 169-172.

MÜLLER-MOTZFELD, G. 2004. Adephaga 1: Carabidae (Laufkäfer). In: Freude, H., Harde, K.W., Lohse, G.A. & Klausnitzer, B. (eds.): *Die Käfer Mittteleuropas.* Bd. 2. Heidelberg/Berlin: Spektrum-Verlag, 2nd edition, 521 pp.

MUONA, J. 1979a. Staphylinidae In: Silfverberg, H. *Enumeratio Coleopterorum Fennoscandiae, Daniae et Baltiae.* Helsinki, pp. 79.

—— 1979b. The aleocharine types of Mulsant and Rey. *Annales Entomologici Fennici* **45**: 47-58.

—— 1987. Some aspects of Aleocharinae systematics – a response to Dr. G. A. Lohse. *Entomologische Blätter für Biologie und Systematik der Käfer* **83**: 19-24.

—— 1990. The Fennoscandian and Danish species of the genus *Amischa* Thomson (Coleoptera, Staphylinidae). *Entomologisk Tidskrift* **111**: 17-24.

—— 1991. The North European and British species of the genus *Meotica* Mulsant & Rey (Coleoptera, Staphylinidae). *Deutsche Entomologische Zeitschrift* **38**: 225-246.

—— 1995. Taxonomic notes on the genus *Philhygra* Mulsant & Rey (Coleoptera, Staphylinidae). *Entomologiske Meddelelser* **63**: 11-16.

NASH, D.R. 1982. *Epierus comptus* (Erichson) (Col., Histeridae) new to Britain. *Entomologist's Record & Journal of Variation* **94**: 165-167.

NASH, R., ANDERSON, R. & O'CONNOR, J.P. 1997. Recent additions to the list of Irish Coleoptera. *Irish Naturalists' Journal* **25**: 319-325.

NELSON, H.G. 1990. *Pomatinus*, a name reclaimed for *Helichus susbtriatus* (Ph. Müller) (Coleoptera: Dryopidae). *Coleopterists Bulletin* **44**: 233-234.

NEWBERY, E.A. 1914. *Philhydrus halophilus*, Bedel: an addition to the British list of Coleoptera. *Entomologist's Monthly Magazine* **50**: 79.

NEWTON, A.F. & CHANDLER, D.S. 1989. World catalog of the genera of Pselaphidae (Coleoptera). *Fieldiana, Zoology* **53**: 1-93.

NEWTON, A.F. & FRANZ, H. 1998. World catalog of the genera of Scydmaenidae. *Koleopterologische Rundschau* **68**: 137-165.

NEWTON, A.F. & THAYER, M.K. 1992. Current classification and family-group names in Staphyliniformia (Coleoptera). *Fieldiana, Zoology* **67**: 1-92.

—— 1995. Protopselaphinae new subfamily for *Protopselaphus* new genus from Malaysia, with a phylogenetic analysis and review of the Omaliinae group of Staphylinidae including Pselaphidae (Coleoptera). In: Pakaluk, J. & Slipinski, S.A. (eds.) *Biology, Phylogeny and Classification of Coleoptera.* Museum of the Zoological Institute PAN, Warsaw: pp 219-320.

NIKITSKY, N.B. 1998. *Generic classification of the beetle family Tetratomidae (Coleoptera, Tenebrionoidea) of the world, with description of new taxa.* Faunistica 9, 80 pp. Sofia: Pensoft Books.

NIKITSKY, N.B., SEMENOV, V.B. & DOLGIN, N.M. 1998. The beetles of the Prioksko-Terrasny Biosphere Reserve – xylobiontes, mycetobiontes, and Scarabaeidae (with the review of the Moscow region fauna of these groups). Supplement 1 (with remarks on nomenclature and systematics of some Melandryidae of the world fauna). *Archives of the Zoological Museum of Moscow State University* **36** (Supplement 1): 1-55.

NILSSON, A.N. 2000. A new view of the generic classification of the *Agabus*-group of genera of the Agabini, aimed at solving the problem with a paraphyletic *Agabus* (Coleoptera: Dytiscidae). *Koleopterologische Rundschau* **70**: 17-36.

—— 2003. Noteridae. In: Löbl, I. & Smetana, A. (eds.) *Catalogue of Palaearctic Coleoptera. Vol. 1.* Stenstrup: Apollo Books, pp 33-35. Corrected and updated version available online: http://www.emg.umu.se/biginst/andersn/Not_idae.htm.

—— 2003[2006]. Dytiscidae. In: Löbl, I. & Smetana, A. (eds.) *Catalogue of Palaearctic Coleoptera. Vol. 1.* Stenstrup: Apollo Books, pp 35-78. Corrected and updated version available online: http://www.emg.umu.se/biginst/andersn/ CPD_061112.pdf [2006].

NILSSON, A.N. & HOLMEN, M. 1995. *The aquatic Adephaga (Coleoptera) of Fennoscandia and Denmark.* II. Dytiscidae. Leiden: E.J. Brill. Fauna Entomologica Scandinavica, No. 32.

NILSSON, A.N. & VONDEL, B.J., van 2005. *World Catalogue of Insects. Vol. 7 Amphizoidae, Haliplidae,*

Noteridae and Paelobiidae (Coleoptera, Adephaga). Stenstrup: Apollo Books.

O'BRIEN, C.W. 1970. A taxonomic revision of the weevil genus *Dorytomus* in North America (Coleoptera: Curculionidae). *University of California Publications in Entomology* **60**: 1-80.

O'CALLAGHAN, E., FOSTER, G.N., BILTON, D.T. & REYNOLDS, J.D. 2009. *Ochthebius nilssoni* Hebauer new for Ireland (Hydraenidae, Coleoptera), including a key to Irish *Ochthebius* and *Enicocerus*. *Irish Naturalist's Journal* **30**: 19-23.

ODEGAARD, F. 2001. Taxonomic status and geographical range of some recently revised complex-species of Coleoptera in Norway. *Norwegian Journal of Entomology* **48**: 237-249.

ORLEDGE, G.M. 2009. Recent changes to the names of some British Ciidae. *The Coleopterist* **18**: 139-140.

ORLEDGE, G.M. & BOOTH, R.G. 2006. An annotated checklist of British and Irish Ciidae with revisionary notes. *The Coleopterist* **15**: 1-16.

OSELLA, G. 1977. Revisione della sottofamiglia Raymondionyminae (Coleoptera, Curculionidae). *Memorie del Museo Civico di Storia Naturale di Verona (IIa serie) Sezione Scienze della Vita* **1**: 1-162.

OSTOJÁ-STARZEWSKI, J.C. 2005. The western corn rootworm *Diabrotica virgifera virgifera* LeConte (Col., Chrysomelidae) in Britain: distribution, description and biology. *Entomologist's Monthly Magazine* **141**: 175-182.

OSTOJÁ-STARZEWSKI, J.C. & HAMMON, R.P. 2008. *Chlorophorus* spp. (Cerambycidae) in the Central Science Laboratory collection, with particular reference to *Chlorophorus annularis* (Fabricius). *The Coleopterist* **17**: 125-133.

OWEN, J.A. 1979. *Carpelimus lindrothi* Palm (Col., Staphylinidae) new to Britain. *Entomologist's Monthly Magazine* **114**: 102.

—— 1982. *Meotica lohsei* Benick (Col., Staphylinidae) new to Britain. *Entomologist's Monthly Magazine* **118**: 44.

—— 1983. *Atheta hansseni* Strand. (Col., Staphylinidae) new to Britain. *Entomologist's Monthly Magazine* **119**: 192.

—— 1988. Indoor meetings. 14 January 1988. *British Journal of Entomology and Natural History* **1**: 117.

—— 1990a. BENHS indoor meetings. 25 April 1990. *British Journal of Entomology and Natural History* **3**: 180-181.

—— 1990b. *Schistoglossa benicki* Lohse in Britain. *Coleopterist's Newsletter* **38**: 11.

—— 1990c. Notes on three species of *Aleochara* (s.g. *Coprochara* Mulsant & Rey) (Col.: Staphylinidae), including two new to Britain. *Entomologist's Record & Journal of Variation* **102**: 227-232.

—— 1991. *Pseudomicrodota jelineki* Krasa (Staphylinidae) new to Britain. *Coleopterist's Newsletter* **44/45**: 7-8.

—— 1993a. An annotated list of recent additions and deletions affecting the recorded beetle fauna of the British Isles. *The Coleopterist* **2**: 1-18.

—— 1993b. Use of a flight interception trap in studying the beetle fauna of a Surrey wood over a three year period. *Entomologist* **112**: 141-160.

—— 1994a. *Pityophthorus lichtensteini* (Ratz.) (Col., Scolytidae) in S. Aberdeenshire and Elgin, with notes on its ecology and reported occurrences in Britain. *Entomologist's Monthly Magazine* **130**: 139-140.

—— 1994b. *Ischnoglossa turcica* Wunderle (Staphylinidae) new to Britain. *The Coleopterist* **2**: 65.

—— 1994c. Corrections to: an annotated list of recent additions and deletions affecting the recorded beetle fauna of the British Isles. *The Coleopterist* **2**: 67.

—— 1994d. *Ischnoglossa turcica* Wunderle (Col.: Staphylinidae) in Britain. *Entomologist's Record & Journal of Variation* **106**: 241-244.

—— 1995. A pitfall trap for repetitive sampling of hypogean arthropod faunas. *Entomologist's Record & Journal of Variation* **107**: 225-228.

—— 1996. *Harpalus (Pseudophonus) griseus* Panzer (Col.: Carabidae) at Wimbledon, Surrey – the first definitely British record? *Entomologist's Record & Journal of Variation* **108**: 69-72.

—— 1997a. Observations on *Raymondionymus marqueti* (Aubé) (Col: Curculionidae) in north Surrey. *Entomologist* **116**: 122-129.

—— 1997b. Some uncommon beetles from Headley Warren, Surrey. *Entomologist's Record & Journal of Variation* **109**: 301-307.

—— 1999. *Mordellistena pseudoparvula* Ermisch (Col.: Mordellidae) new to Britain. *Entomologist's Record & Journal of Variation* **111**: 101-102.

OWEN, J.A., ALLEN, A.A., CARTER, I.S. & HAYEK, C.M.F. von 1985. *Panspoeus guttatus* Sharp (Col., Elateridae) new to Britain. *Entomologist's Monthly Magazine* **121**: 91-95.

OWEN, J.A., LYSZKOWSKI, R.M., PROCTOR, R. & TAYLOR, S. 1992a. *Agabus wasastjernae* (Dytiscidae) Sahlberg new to Scotland. *The Coleopterist* **1**(2): 2-3.

—— 1992b. *Agabus wasastjernae* Sahlberg (Col.: Dytiscidae) new to Scotland. *Entomologist's Record & Journal of Variation* **104**: 225-230.

OWEN, J.A. & MENDEL, H. 1990. *Cercyon alpinus* Vogt at Braemar. *Coleopterist's Newsletter* **41**: 1-2.

PACE, R. 2006. Note sulle Aleocharinae della Spagna (Coleoptera, Staphylinidae). *Elytron* **15-16**: 15-22.

PACHECO, F. 1964. *Sistemàtica, Filogenia y Distribución de los Heteroceridos de América (Coleoptera: Heteroceridae)*. Monografias del Colegio de Post-Graduados Escuela Nacional de Agricultura 1, 209 pp. Chapingo, México: Escuela Nacional de Agricultura.

PARRY, J.A. 1980. *Oulimnius major* (Rey) (= *O. falcifer* Berthélemy) (Col.: Elmidae) new to Britain. *Entomologist's Record & Journal of Variation* **92**: 248.

—— 1981. *Rhynchaenus populi* (F.) (Col., Curculionidae) new to Britain. *Entomologist's Monthly Magazine* **117**: 253.

—— 1982a. The new broom. *Balfour-Browne Club Newsletter* **23**: 1-2.

—— 1982b. A weevil new to Britain: *Apion intermedium* Eppelsheim (Col., Curculionidae) in Kent. *Entomologist's Monthly Magazine* **118**: 227-229.

—— 1983. *Haliplus varius* Nicolai (Col., Haliplidae) new to Britain. *Entomologist's Monthly Magazine* **119**: 13-16.

—— 1990. *Meligethes haemorrhoidalis* Förster (Col., Nitidulidae) new to Britain. *Entomologist's Monthly Magazine* **126**: 237.

PAŚNIK, G. 2006a. A revision of the World species of the genus *Tachyusa* Erichson, 1837 (Coleoptera, Staphylinidae: Aleocharinae). *Zootaxa* **1146**: 1-152.

—— 2006b. Taxonomy and phylogeny of the world species of the genus *Ischnopoda* Stephens, 1837 (Coleoptera, Staphylinidae: Aleocharinae). *Zootaxa* **1179**: 1-96.

PEACOCK, E.R. 1977. Coleoptera Rhizophagidae. *Handbooks for the Identification of British Insects* **5**(5a), 19 pp.

—— 1979. *Attagenus smirnovi* Zhantiev (Coleoptera: Dermestidae) a species new to Britain, with keys to adults and larvae of British *Attagenus*. *Entomologist's Gazette* **30**: 131-136.

—— 1993. Adults and larvae of hide, larder and carpet beetles and their relatives (Coleoptera: Dermestidae) and of derodontid beetles (Coleoptera: Derodontidae). *Handbooks for the Identification of British Insects* **5**(3), 144 pp.

PEARCE, E.J. 1957. Coleoptera (Pselaphidae). *Handbooks for the Identification of British Insects* **4**(9), 32 pp.

PESARINI, C. 1981. Le species paleartiche occidentali della tribu Phyllobiini (Coleoptera Curculionidae). *Bollettino di Zoologia Agraria e di Bachicoltura* (ser. 2) **15**[1979-80]: 49-230.

PETITPIERRE, E. 2000. In: Ramos, M.A. *et al.* (eds.) *Coleoptera, Chrysomelidae I*. Fauna Iberica, Vol. 13. Madrid: Museo Nacional de Ciencias Naturales CSIC, pp. 461-462.

PIPER, R.W., COMPTON, S.G., RASPLUS, J.-Y. & PIRY, S. 2001. The species status of *Cathormiocerus britannicus*, an endemic, endangered British weevil. *Biological Conservation* **101**: 9-13.

PITTINO, R. 1996. *Psammoporus latipunctus* (Gredler), specie misconosciuta della fauna italiana (Coleoptera Scarabaeoidea: Aegialiidae). *Giornale Italiano di Entomologia* **8**(42): 55-62.

—— 2006. A revision of the genus *Psammoporus* Thompson, 1859 in Europe, with description of two species (Coleoptera, Scarabaeoidea: Aegialiidae). *Giornale Italiana d'Entomologia* **11**: 325-342.

PLANT, C.W., MORRIS, M.G. & HEAL, N.F. 2006. *Pachyrhinus lethierryi* (Desbrochers, 1875) (Curculionidae) new to Britain and evidently established in south-east England. *The Coleopterist* **15**: 59-65.

POOLE, R.W. & GENTILI, P. (eds.) 1996. *Nomina Insecta Nearctica*. Rockville: Entomological Information Services.

POPE, R.D. 1977. Kloet and Hincks. A Check List of British Insects. Second edition (completely revised). Part 3: Coleoptera and Strepsiptera. *Handbooks for the Identification of British Insects* **11**(3), xiv+105 pp.

PRANCE, D.A. 2001. *Cybocephalus fodori* Endrödy-Younga, 1965 (Coleoptera: Cybocephalidae) new to Britain. *Entomologist's Gazette* **52**: 125-127.

PUTHZ, V. 1967. Über *Stenus (Parastenus) alpicola* Fauvel und andere abweichend gebaute paläarktische Arten. *Annales Entomologici Fennici* **33**: 226-256.

—— 1968. Die *Stenus-* und *Megalopinus*-Arten Motschulskys und Bemerkungen über das Subgenus *Tesnus* Rey, mit einer Tabelle der paläarktischen Vertreter (Coleoptera, Staphylinidae). *Notulae Entomologicae* **48**: 197-219.

—— 1972. Das Subgenus "*Hemistenus*" (Col., Staphylinidae). *Annales Entomologici Fennici* **38**: 75-92.

—— 1974. Was ist *Stenus rogeri* Kraatz, 1857? *Beiträge zur Entomologie* **24**: 311-314.

—— 1993. Zur Synonymie und Stellung einiger Steninen VI (Coleoptera, Staphylinidae). *Entomologische Blätter für Biologie und Systematik der Käfer* **89**: 139-153.

—— 1999. Zwei neue afrikanische Arten der Gattung *Stenus* Latreille, 1976 und eine taxonomische Bemerkung (Col., Staphylinidae) 259. Beitrag zur Kenntnis der Steninen. *Zeitschrift der Arbeitsgemeinschaft Oesterreichischer Entomologen* **89**: 139-153.

—— 2001. Beiträge zur Kenntniss der Steninen. – CCLXIX. Zur Ordnung in der Gattung *Stenus* Latreille, 1796 (Staphylinidae, Coleoptera). *Philippia* **10**: 33-42.

READ, R.W.J. 1981. *Furcipus rectirostris* (L.) (Coleoptera: Curculionidae) new to Britain. *Entomologist's Gazette* **32**: 51-58.

REJZEK, M. 2004. Check-list of Cerambycidae (Col.) of the British Isles. *Entomologist's Monthly Magazine* **140**: 51-57.

ROIG-JUÑENT, S. 1998. Cladistic relationships of the tribe Broscini (Coleoptera: Carabidae). In: Ball, G.E., Casale, A. & Vigna Taglianti, A. (eds.): *Phylogeny and Classification of Caraboidea (Coleoptera: Adephaga)*. Torino: Museo Regionale di Scienze Naturali, pp. 343-358.

RÖSSNER, E., SCHÖNFELD, J. & AHRENS, D. 2010. *Onthophagus (Palaeonthophagus) medius* (Kugelann, 1792)—a good western palaearctic species in the *Onthophagus vacca* complex (Coleoptera: Scarabaeidae: Scarabaeinae: Onthophagini). *Zootaxa* **2629**: 1-28.

ROY, H.E., BROWN, P.M.J., FROST, R. & POLAND, R.L. 2011. *The Ladybirds (Coccinellidae) of Britain and Ireland: an atlas of the ladybirds of Britain, Ireland, the Isle of Man and the Channel Islands*. FSC Publications, pp. x+198.

RŮŽIČKA, J & VÁVRA, J. 2003. A revision of the *Choleva agilis* species group (Coleoptera, Leiodidae, Cholevinae). In: Cuccodoro, G. & Leschen, R.A.B. (eds.) *Systematics of Coleoptera: papers celebrating the retirement of Ivan Löbl*. Memoirs on Entomology, International 17, pp. 143-256. Associated Publishers.

RYE, E.C. 1869. Notes upon Gemminger and von Harold's "*Catalogus Coleopterorum*" Tom ii. *Entomologist's Monthly Magazine* **5**: 247-250.

SABELLA, G., BÜCKLE, D., BRACHAT, V. & BESUCHET, C. 2004. *Revision der paläarktischen Arten der Gattung* Brachygluta *Thomson 1859. 1 Teil. (Coleoptera, Staphylinidae)*. Genève: Muséum d'Histoire Naturelle, 283 pp.

SALISBURY, A., MALUMPHY. C., & HALSTEAD, A.J. 2012. First record of blue mint beetle *Chrysolina coerulans* (Scriba, 1791) (Chrysomelidae) breeding in Britain. *Coleopterist* **21**: 35-37.

SCHEDL, K.E. 1981. Familie: Scolytidae (Borken- und Ambrosiakäfer) (Ipidae). In: Freude, H., Harde, K.W. & Lohse, G.A. (eds.) *Die Käfer Mitteleuropas*. Band 10, pp. 34-101. Krefeld: Goecke & Evers.

SCHEERPELTZ, O. 1968. *Catalogus Faunae Austriae. Teil XV fa: Coleoptera-Staphylinidae*. Wien: Österreichische Akademie der Wissenschaften, 279 pp.

SCHILLHAMMER, H. 1997. Taxonomic revision of the Oriental species of *Gabrius* Stephens (Coleoptera: Staphylinidae). *Monographs on Coleoptera* **1**: 1-139.

SCHILTHUIZEN, M. 1990. A revision of *Choleva agilis* (Illiger, 1798) and related species (Coleoptera: Staphylinoidea: Cholevidae). *Zoologische Mededelingen Leiden* **64**(10): 121-153.

—— 2010. *Catops borealis* Krogerus, 1931, new to the British fauna, and some notes on other British Cholevinae (Leiodidae). *The Coleopterist* **19**: 1-9.

SCHÖDL, S. 1997. Taxonomic studies on the genus *Enochrus* (Coleoptera: Hydrophilidae). *Entomological Problems* **28**: 61-66.

SCHOLTZ, C.H. & BROWNE, D.J. 1996. Polyphyly in the Geotrupidae (Coleoptera: Scarabaeidae): a case for a new family. *Journal of Natural History* **30**: 597-614.

SCHÜLKE, M. 1998. Zur Identität einiger westpaläarktischer Arten der Gattung *Thinobius* Kiesenwetter, 1844 (Col., Staphylinidae, Oxytelinae). *Entomologische Nachrichten und Berichte* **42**: 127-138.

—— 1999. Zur Taxonomie der Gattung *Bolitobius* Leach in Samouelle, 1819 (Coleoptera, Staphylinidae, Tachyporinae). *Linzer Biologischen Beiträge* **31**: 975-985.

—— 2005. Über *Staphylinus rufipes* Linnaeus, 1758 (Coleoptera, Staphylinidae, Tachyporinae). *Entomologische Blätter für Biologie und Systematik der Käfer* **100**: 201-204.

—— 2009. Zwei neue Arten der Gattung *Anotylus* Thomson aus der Verwandtschaft von *A. complanatus* (Erichson) (Coleoptera, Staphylinidae, Oxytelinae). *Linzer Biologische Beiträge* **41**: 2009-2024.

—— 2010. Zur Taxonomie und Systematik einiger Arten der Untergattung *Bledius* LEACH 1819 (Coleoptera, Staphylinidae, Oxytelinae). *Linzer Biologische Beiträge* **42**: 1495-1509.

—— 2011a. Zur Kenntnis der Verwandtschaft von *Bledius* (*Hesperophilus*) *atricapillus* (Germar) (Coleoptera, Staphylinidae: Oxytelinae). *Linzer Biologische Beiträge* **43**: 1595-1608.

—— 2011b. *Mycetoporus ambiguus* Luze, 1901 and *M. reichei* (Pandellé 1869) (Staphylinidae) - First records for the British Isles. 77th Contribution to the knowledge of Tachyporinae. *The Coleopterist* **20**: 105-109.

—— 2012 in press. Bryophacis maklini (J. Sahlberg, 1871), not *B. rugipennis* (Pandellé, 1869) in Great Britain (Staphylinidae). *The Coleopterist*.

SEEVERS, C.H. 1978. A generic and tribal revision of the North American Aleocharinae (Coleoptera: Staphylinidae). *Fieldiana, Zoology* **71**: 1-275.

SHARP, D. 1914. Notes on British *Philhydrus*. *Entomologist's Monthly Magazine* **50**: 80-83.

SHAW, M.R. 1999. *Trogoderma angustum* (Solier, 1849) (Coleoptera: Dermestidae), a museum and herbarium pest new to Britain. *Entomologist's Gazette* **50**: 99-102.

SHIRT, D.B. (ed.) 1987. *British Red Data Books: 2. Insects*. Peterborough: Nature Conservancy Council.

SILFVERBERG, H. 1977. Nomenclatural notes on Coleoptera Polyphaga. *Notulae Entomologicae* **57**: 91-94.

—— 1979. *Enumeratio Coleopterorum Fennoscandaie et Daniae*. Helsinki: Helsingin Hyönteisvaintoyhdistys Helsingfors Entomologiska Bytesförening.

—— 1991. Nomenclatural corrections in north European Coleoptera. *Entomologia Fennica* **2**: 21-22.

—— 1992. *Enumeratio Coleopterorum Fennoscandiae, Daniae et Baltiae*. Helsinki: Helsingin Hyönteisvaihtoyhdistys, 94 pp.

—— 2004. Enumeratio Coleopterorum Fennoscandiae, Daniae et Baltiae. *Sahlbergia* **9**: 1-111.

SINCLAIR, M. & HUTCHINS, D. 2009. *Aphthona pallida* (Bach, 1856) (Chrysomelidae) is a British species. *The Coleopterist* **18**: 155-157.

SKIDMORE, P. 1985a. *Exochomus nigromaculatus* (Goeze) (Col., Coccinellidae) in Britain. *Entomologist's Monthly Magazine* **121**: 239-240.

—— 1985b. *Cyphon kongsbergensis* Munster (Col., Scirtidae) in Scotland. *Entomologist's Monthly Magazine* **121**: 249-252.

—— 1988. *Othius lapidicola* Kiesenwetter (Col., Staphylinidae) in Scotland, a first British record. *Entomologist's Monthly Magazine* **124**: 92.

SKIDMORE, P. & HUNTER, F.A. 1980. *Ischnomera cinerascens* Pand. (Col. Oedemeridae) new to Britain. *Entomologist's Monthly Magazine* **116**: 129-132.

SMETANA, A. 1966. *Philonthus varipes* Mulsant and Rey, a staphylinid (Col.) new to the British list. *Entomologist's Monthly Magazine* **102**: 47-48.

—— 1982. Revision of the subfamily Xantholininae of America north of Mexico (Coleoptera: Staphylinidae). *Memoirs of the Entomological Society of Canada* **120**: 1-389.

—— 1995. Rove beetles of the subtribe Philonthina of America north of Mexico (Coleoptera: Staphylinidae) classification, phylogeny and taxonomic revision. *Memoirs on Entomology, International* **3**:1-946.

—— 2004a. Omaliinae. pp. 237-268 In: Löbl, I. & Smetana, A. (eds.) *Catalogue of Palaearctic Coleoptera. Vol. 2.* Stenstrup: Apollo Books.

—— 2004b. Micropeplinae. pp. 271-272 In: Löbl, I. & Smetana, A. (eds.) *Catalogue of Palaearctic Coleoptera. Vol. 2.* Stenstrup: Apollo Books.

—— 2004c. Tachyporinae. pp. 330-352 In: Löbl, I. & Smetana, A. (eds.) *Catalogue of Palaearctic Coleoptera. Vol. 2.* Stenstrup: Apollo Books.

—— 2004d. Aleocharinae. pp. 353-494 In: Löbl, I. & Smetana, A. (eds.) *Catalogue of Palaearctic Coleoptera. Vol. 2.* Stenstrup: Apollo Books.

—— 2004e. Oxytelinae. pp. 511-535 In: Löbl, I. & Smetana, A. (eds.) *Catalogue of Palaearctic Coleoptera. Vol. 2.* Stenstrup: Apollo Books.

—— 2004f. Steninae. pp. 537-564 In: Löbl, I. & Smetana, A. (eds.) *Catalogue of Palaearctic Coleoptera. Vol. 2.* Stenstrup: Apollo Books.

—— 2004g. Paederinae. pp. 579-624 In: Löbl, I. & Smetana, A. (eds.) *Catalogue of Palaearctic Coleoptera. Vol. 2.* Stenstrup: Apollo Books.

—— 2004h. Staphylininae. pp. 624-698 In: Löbl, I. & Smetana, A. (eds.) *Catalogue of Palaearctic Coleoptera. Vol. 2.* Stenstrup: Apollo Books.

SMETANA, A. & DAVIES, A. 2000. Reclassification of the north temperate taxa associated with *Staphylinus* sensu lato, including comments on relevant subtribes of Staphylinini (Coleoptera: Staphylinidae). *American Museum Novitates* **3287**, 88 pp.

SMETANA, A. & HERMAN, L.H. 1999. *Philonthus immundus* of Gyllenhal, a misidentification (Coleoptera, Staphylinidae, Staphylinini, Philonthina). *Coleopterists Bulletin* **53**: 297-298.

SMITH, K.G.V. 1990. *Pyrrhalta luteola* (Mull.) (Col., Chrysomelidae) accidentally introduced into Britain. *Entomologist's Monthly Magazine* **126**: 190.

SÖRENSSON, M. 2003. New records of featherwing beetles in North America. *Coleopterists Bulletin* **57**: 369-381.

SPEIGHT, M.C.D. 1986. *Asaphidion curtum, Dorylomorpha maculata, Selatosomus melancholicus* and *Syntormon miki*; insects new to Ireland. *Irish Naturalists' Journal* **22**: 20-23.

SPEIGHT, M.C.D., MARTINEZ, M. & LUFF, M.L. 1986. The *Asaphidion* (Col.: Carabidae) species occurring in Great Britain and Ireland. *Proceedings & Transactions of the British Entomological & Natural History Society* **19**: 17-21.

STEBNICKA, Z. 1977. A revision of the world species of the tribe Aegialiini (Coleoptera, Scarabaeidae, Aphodiinae). *Acta Zoologica Cracoviensia* **22**: 397-506.

STEEL, W.O. 1948. The British species of *Staphylinus* subgenus *Ocypus* Steph. (Col., Staphylinidae). *Entomologist's Monthly Magazine* **84**: 271-275.

—— 1957. *Acrolocha minuta* (Ol.) (= *striata* Grav.) in Kent and Sussex. *Entomologist's Monthly Magazine* **93**: 99.

—— 1970. The larvae of the genera of the Omaliinae (Coleoptera: Staphylinidae) with particular reference to the British fauna. *Transactions of the Royal Entomological Society of London* **122**: 1-47.

STRAND, A. & VIK, A. 1964. Die Genitalorgane der nordischen Arten der Gattung *Atheta* Thoms. (Col., Staphylinidae). *Norsk Entomologisk Tidsskrift* **12**: 327-335 (Taf. I-XXI).

STÜBEN, P.E., BEHNE, L. & BAHR, F. 2003. Analytischer Katalog der westpaläarktischen Cryptorhynchinae / Analytical Catalogue of Westpalaearctic Cryptorhynchinae, Teil 2/Part 2: *Acalles, Allocrates* (Col.: Curculionidae: Cryptorhynchinae). *Snudebiller* **4** [CD-ROM].

TELFER, M.G. 2001a. *Bembidion coeruleum* Serville (Carabidae) new to Britain and other notable carabid records from Dungeness, Kent. *The Coleopterist* **10**: 1-4.

—— 2001b. *Ophonus subsinuatus* Rey (Carabidae) new to Britain, with a discussion of its status. *The Coleopterist* **10**: 39-43.

—— 2003. *Acupalpus maculatus* Schaum, 1860: another carabid new to Britain from Dungeness. *The Coleopterist* **12**: 1-6.

—— 2007a. *Xyleborus monographus* (Fabricius) (Curculionidae: Scolytinae) new to Britain. *The Coleopterist* **16**: 41-45.

—— 2007b. A second British locality for *Tachyusa objecta* (Mulsant & Rey) (Staphylinidae). *The Coleopterist* **16**: 120.

TELNOV, D. 2010. Ant-like flower beetles (Coleoptera: Anthicidae) of the UK, Ireland and Channel Isles. *British Journal of Entomology and Natural History* **23**: 99-117.

TEMPÈRE, G. 1979. Sur divers *Leiosoma* de la faune française, notamment des Pyrénées (Col. Curculionidae). *Nouvelle Revue d'Entomologie* **9**: 271-286.

—— 1982. *Rhynchaenus stigma* (Germar) et *R. pseudostigma* nov. sp. (Col., Curculionidae). *Nouvelle Revue d'Entomologie* **12**: 245-254.

THAYER, M.K. 2003. Omaliinae of Mexico: new species, combinations and records (Coleoptera, Staphylinidae). *Memoirs on Entomology, International* **17**: 311-358.

THOMPSON, J., 1978. First record of *Trogoderma variabile* Ballion (Col., Dermestidae) in Britain. *Entomologist's Monthly Magazine* **113**[1977]: 29.

THOMPSON, R.T. 1992. Observations on the morphology and classification of weevils (Coleoptera, Curculionoidea) with a key to major groups. *Journal of Natural History* **26**: 835-891.

—— 1994. *Rhynchaenus erythropus* (Germar) (Curculionidae) not (yet) British. *The Coleopterist* **2**: 68-69.

—— 1995. Raymondionymidae (Col., Curculionoidea) confirmed as British. *Entomologist's Monthly Magazine* **131**: 61-64.

—— 2006. A revision of the weevil genus *Procas* Stephens (Coleoptera: Curculionoidea: Erirhinidae). *Zootaxa* **1234**: 1-63.

THOMPSON, R.T. & ALONSO-ZARAZAGA, M.A. 1988. On some weevil species described by Linnaeus (Coleoptera, Curculionoidea). *Entomologica Scandinavica* **19**: 81-86.

TOMASZEWSKA, K.W. 2000. Morphology, phylogeny and classification of adult Endomychidae (Coleoptera: Cucujoidea). *Annales Zoologici (Warszawa)* **50**(4): 449-558.

TOTTENHAM, C.E. 1939. Some notes on the nomenclature of the Staphylinidae (Coleoptera). *Proceedings of the Royal Entomological Society of London (B)* **8**: 521-564.

―― 1954. Coleoptera: Staphylinidae (Piestinae to Euaesthetinae). *Handbooks for the Identification of British Insects* **4**(8a), 79 pp.

TURIN, H. 1990. Naamlijst voor de Nederlandse loopkevers (Coleoptera: Carabidae). *Entomologische Berichten, Amsterdam* **50**(6): 61-72.

―― 2000. De Nederlandse Loopkevers. Verspreiding en Oecologie [The Netherlands Ground Beetles. Distribution and Ecology]. KNNV Uitgeverij.

TURIN, H., CASALE, A., KRYZHANOVSKIJ, O.L., MAKAROV, K.V. & PENEV, L.D. 1993. *Checklist and Atlas of the Genus* Carabus *Linnaeus in Europe (Coleoptera, Carabidae).* Leiden: Backhuys.

TURNER, C.R. 2011. *Pentarthrum elumbe* (Boheman) (Curculionidae, Cossoninae) in Britain. *The Coleopterist* **20**: 20-22.

UHTHOFF-KAUFMANN, R.R. 1989. The occurrence of *Grammoptera* Serville and *Alosterna* Mulsant (Col.: Cerambycidae) in the British Isles. *Entomologist's Record & Journal of Variation* **101**: 87-103.

ULLRICH, W.G. 1975. Monographie der Gattung Tachinus *Gravenhorst (Coleoptera Staphylinidae) mit Bermerkungen zur Phylogenie und Verbreitung der Arten.* Keil: W.G. Ullrich.

VELÁZQUEZ, A.J.deC., ALONSO-ZARAZAGA, M.A. & OUTERELO, R. 2007. Systematics of Sitonini (Coleoptera: Curculionidae: Entiminae), with a hypothesis on the evolution of feeding habits. *Systematic Entomology* **32**: 312-331.

VOGEL, J. 2004. Bemerkungen zur Systematik und Taxonomie der Athetini (Coleoptera: Staphylinidae, Aleocharinae). *Linzer Biologische Beiträge* **36**: 1115-1123.

VORST, O. 2009. *Cercyon castaneipennis* sp. n., an overlooked species from Europe. *Zootaxa* **2054**: 59-68.

WALLIN, H., NYLANDER, U. & KVAMME, T. 2009. Two sibling species of *Leiopus* Audinet-Serville, 1835 (Coleoptera: Cerambycidae) from Europe: *L. nebulosus* (Linnaeus, 1758) and *L. linnei* sp. nov. *Zootaxa* **2010**: 31-45.

WANAT, M. 1995. Systematics and phylogeny of the tribe Ceratapiini (Coleoptera: Curculionoidea: Apionidae). *Genus* supplement 3, 406 pp.

WARCHAŁOWSKI, A. 1995. Bemerkungen zur Systematik und Nomenklatur der Erdflöhe (Coleoptera: Chrysomelidae: Halticinae). *Genus* **6**: 463-468.

WELCH, R.C. 1969. *Aleochara verna* Say (Col., Staphylinidae) new to Britain. *Entomologist* **102**: 207-209.

―― 1990a. *Macrorhyncolus littoralis* (Broun) (Col., Curculionidae), a littoral weevil new to the Palaearctic Region, from two sites in Kent. *Entomologist's Monthly Magazine* **126**: 97-101.

―― 1990b. *Aleochara binotata* Kr., not *A. verna* Say (Col.: Staphylinidae), a British insect. *Entomologist's Record & Journal of Variation* **102**: 225-226.

―― 1995. *Cypha tarsalis* Luze (Col: Staphylinidae) new to Britain. *Entomologist's Record & Journal of Variation* **107**: 185-187.

―― 1997. The British species of the genus *Aleochara* Gravenhorst (Staphylinidae). *The Coleopterist* **6**: 1-45.

―― 1998. *Cyphea curtula* (Erichon) (Col., Staphylinidae) new to the British Isles. *Entomologist's Monthly Magazine* **134**: 339-343.

―― 2000. *Gyrophaena rousi* Dvorák 1966 (Col., Staphylinidae) new to Britain. *Entomologist's Monthly Magazine* **136**: 247-251.

―― 2008. *Colotes punctatus* (Erichson, 1840) (Melyridae, Malachiinae) new to Britain from the Isles of Scilly. *The Coleopterist* **17**: 77-81.

WHITEHEAD, P.F. 1992. *Melöe (Eurymelöe) mediterraneus* Müller, 1925 (Coleoptera: Meloidae) new to the British fauna. *Entomologist's Gazette* **43**: 65.

―― 1994. *Phloeopora bernhaueri* Lohse 1984, not *P. teres* (Gravenhorst) (Col., Staphylinidae), a British species. *Entomologist's Monthly Magazine* **130**: 173-174.

―― 2002. The taxonomy of *Heterothops praevius* Erichson, 1939 (Col., Staphylinidae) in Britain with refrence to a distinctive male from Kemerton, Worcestershire. *Entomologist's Monthly Magazine* **138**: 71-76.

WILLIAMS, B.S. 1929. A new species of *Bledius* Mannh. *Entomologist's Monthly Magazine* **65**: 28-29.

WILLIAMS, S.A. 1968. *Raymondionymus marqueti* (Aubé) typical form (Col., Curculionidae) in Surrey. *Entomologist's Monthly Magazine* **104**: 112.

―― 1979a. *Amischa simillima* (Sharp) a synonym of *A. soror* (Kraatz) (Col., Staphylinidae). *Entomologist's Monthly Magazine* **113**[1977]: 250.

―― 1979b. *Ocyusa nitidiventris* (Fagel) new to Britain. *Proceedings & Transactions of the British Entomological & Natural History Society* **12**: 46-48.

―― 1979c. The genus *Oligota* Mannerheim (Col., Staphylinidae) in the Ethiopian Region. *Entomologist's Monthly Magazine* **114**[1978]: 177-190.

―― 1980. Further notes on British *Oligota* (Col., Staphylinidae). *Entomologist's Monthly Magazine* **116**: 57-58.

―― 1990. *Oxypoda praecox* Erichson (Coleoptera: Staphylinidae) an addition to the British fauna. *British Journal of Entomology and Natural History* **3**: 114.

WILSON, C.R. 2001. *Aphodius pedellus* (DeGeer), a species distinct from *A. fimetarius* (Linnaeus)

(Coleoptera: Aphodiidae) *Tijdschrift voor Entomologie* **144**: 137-143.

WINTER, T. 1990. *Crypturgus subcribrosus* Eggers (Col., Scolytidae) a bark beetle new to Britain. *Entomologist's Monthly Magazine* **126**: 209-211.

—— 1991. Interceptions of exotic bark beetles (Col., Scolytidae) on timber imports into Great Britain, 1980-1988. *Entomologist's Monthly Magazine* **127**: 13-17.

—— 1998. *Phloeosinus aubei* (Perris) (Scolytidae) in Surrey, the first record of this bark beetle breeding in Britain. *The Coleopterist* **7**: 1-2.

WOOD, S.L. 1982. The bark and ambrosia beetles of North and Central America (Coleoptera: Scolytidae), a taxonomic monograph. *Great Basin Naturalist Memoirs* **6**, 1359 pp.

WOOD, S.L. & BRIGHT, D.E. 1992. A catalog of Scolytidae and Platypodidae (Coleoptera), Part 2: Taxonomic Index Volume A & B. *Great Basin Naturalist Memoirs* **13**, 1553 pp.

ZANETTI, A. 1987. Coleoptera Staphylinidae Omaliinae. *Fauna d'Italia* 25. Bologna: Edizioni Calderini, 490 pp.

ZERCHE, L. 1991. Was ist *Oxypoda lividipennis* Mannerheim, 1831 (Coleoptera, Staphylinidae)? *Entomologische Blätter für Biologie und Systematik der Käfer* **87**: 79-82.

—— 1994. Die Revision der *Oxypoda*-Typen aus der Sammlung Claudius Rey im Musée Guimet d'Histoire naturelle de Lyon und einiger anderer Typen der Gattung sowie die Beschreibung von vier neuen *Oxypoda*-arten (Coleoptera, Staphylinidae, Aleocharinae). *Schwanfelder Coleopterologische Mitteilungen* **6**: 1-36.

—— 1998. Phylogenetisch-systematische Revision der westpalaarktischen Gattung *Metopsia* Wollaston 1854. *Beiträge zur Entomologie* **48**: 3-101.

ZUNINO, M. 1979. Gruppi artificiali e gruppi naturali negli *Onthophagus* (Coleoptera, Scarabaeoidea). *Bollettino del Museo di Zoologia dell'Universita di Torino* **1**: 1-18.

—— 1984. Sistematica generica dei Geotrupinae (Coleoptera, Scarabaeoidea: Geotrupidae), filogenesi della sottofamiglia e considerazioni biogeografiche. *Bollettino del Museo Regionale di Scienze Naturali, Torino* **2**: 9-162.

Index to family-group and genus-group names

A

SYNCHITINI, 83
SYNDESINAE, 55
Synechostictus, 18
Syntomium, 47
Syntomus, 23, 128, 131
Synuchus, 19

T

Tachinus, 37, 129
Tachyerges, 102, 130
TACHYPORINAE, 37
TACHYPORINI, 37
Tachyporus, 38
Tachypus, 15, 17
Tachys, 17
Tachyta, 131
Tachyusa, 46
Tachyusida, 43
TACHYUSINA, 46
Taeniapion, 98
TANYMECINI, 108
Tanymecus, 108
TANYSPHYRINI, 100
Tanysphyrus, 100
Tapeinotus, 105
Taphria, 19
Taphrorychus, 111
Tapinotus, 105
TARSOSTENINAE, 69
Tarsostenus, 69
Tarsostinus, 22
Tarulus, 22
Tasgius, 54
Tatianaerhynchites, 97
Telephorus, 64
Telmatophilus, 74
Temnocerus, 97
Temnochila, 129
Tenebrio, 84, 133
Tenebrioides, 69
TENEBRIONIDAE, 84, 129, 133
TENEBRIONINAE, 84
TENEBRIONINI, 84
TENEBRIONOIDEA, 81
Tenebroides, 69, 129, 132
Tentyria, 133
TEREDINAE, 77
Teredus, 77
TERETRIINI, 26
Teretrius, 26, 131
Teropalpus, 48
Tesarius, 57
Tesnus, 50
Testedium, 17
Tetartopeus, 51
Tetralaucopora, 46
Tetrapriocera, 132
Tetratoma, 82
TETRATOMIDAE, 82
TETRATOMINAE, 82
TETROPINI, 90
Tetropium, 89, 134
Tetropla, 39
Tetrops, 90
Teuchestes, 57, 129
Thalassophilus, 16
Thalycra, 72
Thamiaraea, 42

THAMIARAEINA, 42
Thamiocolus, 105
Thanasimus, 69
Thanatophilus, 33, 128
THANEROCLERINAE, 69
Thaneroclerus, 69
Thaumaphrastus, 65
Thea, 79
Thectura, 44
Thecturota, 43
Thes, 80
Thiasophila, 46
Thinobaena, 42
THINOBIINI, 47
Thinobius, 48
Thinodromus, 48
Thinonoma, 46
Thomsoneonymus, 108
THORICTINAE, 65
Thorictodes, 65, 131
Thorictus, 131
THROSCIDAE, 61
Throscus, 61
Thryogenes, 100
Thyasophila, 46
Thylodrias, 65
THYLODRIINAE, 65
Thymalus, 68
Thymapion, 98
Tigrinellus, 109
TILLINAE, 69
Tilloidea, 69
Tillus, 69, 132
Timarcha, 92
Tinocyba, 99
Tinotus, 38
Tipnus, 67
Tiresias, 66
Titan, 29
Tomicus, 111, 112
Tomocarabus, 128, 131
Tomoxia, 83
Tournotaris, 100
Toxotus, 88
TRACHEINI, 59
Trachodes, 110
TRACHODINI, 110
Trachyphlaeus, 109
TRACHYPHLOEINI, 108
Trachyphloeus, 109
Trachys, 59
Traumoecia, 42
TRECHINI, 16
Trechoblemus, 16
Trechus, 16, 128
Trepanedoris, 18
Trepanes, 18
Triaena, 20
Triarthron, 31
TRIBALINAE, 27
TRIBOLIINI, 84
Tribolium, 84, 133
TRICHAPIINA, 99
Trichapion, 99
Trichelaphrus, 16
Trichelophorus, 23
TRICHIINI, 58
Trichius, 58
Trichiusa, 42

Trichocellus, 21
Trichoderma, 54
Trichodes, 69
Trichoferus, 134
Trichohydnobius, 31
Trichonotulus, 131
TRICHONYCHINA, 36
TRICHONYCHINI, 36
Trichonyx, 36
Trichophya, 38
TRICHOPHYINAE, 38
Trichoplataphus, 17
Trichopteryx, 30
Trichosirocalus, 105
Trichosphaerula, 31
Tricorynus, 132
Trigonogenius, 67
trigonorhinus, 134
TRIMIINA, 36
Trimium, 36
Trimorphus, 22
Trinodes, 66
TRINODINAE, 66
Trinophyllum, 89
Trinophylum, 89
Triodonta, 129
Triphyllus, 81
Triplax, 76
Trissemus, 36
Tristaria, 132
Tritoma, 76, 81
TRITOMINI, 76
Trixagus, 61
Trochalus, 131
Trochoideus, 132
Trochorhopalus, 134
TROGIDAE, 55, 131
Troginus, 48
Troglops, 70
Trogoderma, 66, 131
Trogophloeus, 48
TROGOSSITIDAE, 68, 129, 132
TROGOSSITINAE, 69
Trogoxylon, 132
Tropideres, 96
TROPIPHORINI, 109
Tropiphorus, 109
Trox, 55, 131
Trypocopris, 55
Trypodendron, 111, 135
Trypophloeus, 111, 135
Trypopitys, 132
TYCHIINI, 102
TYCHINI, 36
Tychius, 102, 135
Tychobythinus, 36
Tychus, 36
TYLODINA, 106
Tylodrusus, 108
Typhaea, 81
Typhaeus, 55
Typhlolinus, 55
Typhoeus, 55
TYPODERINA, 110
Tyrtaeus, 133
Tytthaspis, 78

Notes

Notes